VALUES OF NON-ATOMIC GAMES

A Rand Corporation Research Study
A list of other Rand publications appears
at the back of this book.

Values of Non-Atomic Games

BY R. J. AUMANN AND L. S. SHAPLEY

Princeton University Press, Princeton, New Jersey

L.C. Card No.: 72-4038

ISBN: 0-691-08103-4

This book is composed in Monotype Bodoni

Printed in the United States of America

To Esther and Marian

Contents

Preface

This work had its inception in the summer of 1963. By that time, multiperson game theory had outgrown its early preoccupation with small numbers of players and with purely abstract models. The basic connection between the "core" concept and Edgeworthian economics had been made, and the study of the limiting behavior of the core, in economic games with many players, was well under way, as well as the study of related economic models with a continuum of economic agents. In a parallel development, the "value" concept had been applied to several types of games with many players—mainly voting games—including some special infinite-person situations. Thus, the time was ripe for a systematic attack on the "value" problem for economic and other games with a continuum of players.

But many difficulties presented themselves. A rough working note, saved from that summer, is revealing: it is headed *Eight ways to define values of general oceanic games*. In fact, this note was a testament of frustration, as each "way" was beset by thorny counter-examples or crippling impossibility theorems or conceptual holes or tough, unsolved mathematical problems. It took a decade to resolve these difficulties and build a coherent mathematical theory, which we present here; except for Sections 17 and 18, none of the material has been published before.

In this endeavor, we are indebted to the assistance of many people, whose specific contributions can only rarely be pin-pointed in the final text. From the outset our approach has been influenced by close personal contacts with those people involved with us in the preceding theoretical developments, as sketched above; we mention especially Gerard Debreu, John Milnor, Herbert Scarf, Norman Shapiro, and Martin Shubik. As the work progressed, many others became interested and "chipped in" with helpful ideas, references, corrections and provocative questions; these include S. Agmon, S. Amitsur, Y. Kannai, Y. Lindenstrauss, B. Peleg, B. Shitovitz, R. Weber, B. Weiss, and, surely, many more. Special thanks are due to Harry Furstenberg, who resolved a major difficulty (thus, Lemma 8.6 and its proof are essentially his); to Uriel Rothblum, who went carefully over much of the text and suggested several important improvements (for example, the present definition of "mixing value" appeared first in his master's thesis); to Louis Billera, who conducted a graduate seminar at Cornell based on the manuscript, which generated a number of corrections and useful comments; and, finally, to Zvi Artstein, whose painstaking reading of the final text has done a great deal to improve the quality and consistency of the exposition.

We would like to thank all the institutions and agencies that supported and funded this research. First and foremost, of course, we have been supported by our home institutions, The Hebrew University of Jerusalem and The Rand Corporation. Much of Aumann's research at the Hebrew University was funded by the Logistics and Mathematical Statistics Branch of the (United States) Office of Naval Research, and much of Shapley's at The Rand Corporation was funded by the United States Air Force Project RAND. Portions of Aumann's research were done also at The Rand Corporation, as consultant to the Mathematics Department; at Yale University, with the Department of Statistics and The Cowles Foundation for Research in Economics; at the University of California at Berkeley, as Ford Visiting Research Professor of Economics; at Stanford University, with the Institute for Mathematical Studies in the Social Sciences; and at the Catholic University of Louvain, with the Center for Operations Research and Econometrics. The National Science Foundation funded a considerable part of the research, at many of the institutions listed. We are most grateful for all this support.

In addition, our collaborative effort has benefitted appreciably from participation in several "workshops," namely: the First International Workshop in Game Theory, held in Jerusalem in 1965 and sponsored by the Israel Academy of Sciences and Humanities and the Hebrew University; the Workshop in Mathematical Economics and Game Theory, held in Santa Monica in 1969 and sponsored by the Mathematical Social Sciences Board and The Rand Corporation; the Second International Workshop in Game Theory, held in Berkeley in 1970 and sponsored by the National Science Foundation and the University of

California; and the Workshop in Mathematical Economics, held in Luminy, France, in 1972 and sponsored by the University of Marseilles. Such gatherings of specialists and advanced students perform an important function in the growth and maturation of "young" fields of inquiry.

Finally, we would be remiss if we did not salute Oskar Morgenstern, Albert W. Tucker, and John D. Williams for their indispensable service to the whole area of mathematical research exemplified by this book. They had steady confidence in the ultimate value of the "pure" mathematical approach to game theory, and this confidence took tangible form on many occasions during the early years of theoretical development, both through the institutional support that they repeatedly mobilized on behalf of mathematicians in the field and through their personal advice and encouragement.

Jerusalem and Santa Monica
December 1973

VALUES OF NON-ATOMIC GAMES

Introduction

Interaction between people—as in economic or political activity—usually involves a subtle mixture of competition and cooperation. Thus bargaining for a purchase is cooperative, in that both sides want to consummate the transaction, but also competitive, in that each side wants terms that are more favorable to itself, and so less favorable to the other side. People cooperate to organize corporations, then compete with other corporations for business and with each other for positions of power within the corporation. Political parties compete for the voter's favor, but cooperate in forming ruling coalitions and in "log-rolling." Often it is impossible to draw a clear border between "co-operation" and "competition."

When analyzed from a rational (as distinguished from psychological or sociological) viewpoint, such interactive situations have come to be known as *games*. The participants are called *players*, and the "winnings" of the players are called *payoffs*. The mathematical theory of games, founded by John von Neumann (1928), gained impetus with the publication of the classic book of von Neumann and Morgenstern (1944) and was subsequently developed by the contributions of many others. In its early years the theory was concerned largely with n-player games for small values of n (e.g. $n = 2$, 3, or 4). Since about 1960, attention has focussed more and more on games with large masses of players, in which no individual player can affect the overall outcome. Such games arise naturally in the social sciences, as models for situations in which there are large numbers of very "small" individuals, like consumers in an economy or voters in an election. Mathematically, it is often convenient to represent these games with the aid of a "continuum" of players—like the continuum of points on a line or the continuum of drops in a liquid. Represented thus, such games are called *non-atomic*.

One of the objects of the theory is to provide an a priori evaluation of games, i.e., to define an operator that assigns to each player of a game a number that purports to represent what he would be willing to pay in order to participate. Such an operator is called a *value*. Value theory for finite games—i.e., n-player games with n finite—was first studied by Shapley (1953a), and is by now a well established branch of game theory. It is the purpose of this book to develop a corresponding theory for non-atomic games.

A game can be mathematically represented in many forms, corresponding to various different aspects of the game that one might wish to study. In this book we are interested in the processes of coalition-forming and payoff distribution. For this aspect the most appropriate is the "coalitional" form. When there are finitely many players, this consists of a function[1] v that associates a real number $v(S)$ with each set S of players; intuitively, $v(S)$ expresses the potential worth of S, should its members decide to act in concert. The set function v is in general *not* additive; an example is the three-person "majority game," in which $v(S) = 1$ if S has at least two members, and 0 otherwise. Payoff distributions are represented by vectors with one component for each player, representing the amount he obtains under the particular distribution in question. A value for n-player games can then be defined as a mapping from set functions (games) to n-vectors (payoff distributions) that satisfies certain plausible or desirable conditions, like symmetry and joint optimality. For example, to the three-player majority game described above, the value assigns the payoff vector $(\frac{1}{3}, \frac{1}{3}, \frac{1}{3})$, as might have been expected.

To generalize this coalitional form so that it will apply to non-atomic (and other infinite) games, we start with a measurable space $(I, \mathcal{C})$, where I is interpreted as the space of players, and the σ-field $\mathcal{C}$ as the family of possible coalitions. A game is then a function from $\mathcal{C}$ to the reals. As in the finite case, these set functions are typically non-additive; thus though a measure on $(I, \mathcal{C})$ is an example of a game, a more interesting example is the square of a measure or the product of distinct measures. Since individual "winnings" are normally infinitesimal in a non-atomic situation, we represent payoff distributions by measures

[1] Usually called "characteristic function" in the literature; see Note 1 to Appendix A. A more general coalitional form in which $v(S)$ is not a real number but a set of payoff distributions (intuitively, those "achievable" by S) has also been widely studied; we shall not be concerned with this in the present volume.

on I, or, more precisely, by (finitely) additive functions on $\mathcal{C}$. A *value* is then a mapping, φ, from set functions to additive set functions; our "plausible conditions" are that φ be linear, monotonic (or continuous in an appropriate topology), symmetric in a natural sense, and, finally, "efficient" in the sense that always $\varphi v(I) = v(I)$. It is remarkable that these simple postulates suffice to determine a unique value, in many cases of interest.

Complementing this "axiomatic" approach to the value, there is an alternative, "constructive" approach, which has run like a counter-motif through the development of value theory from its beginnings. Rather than deducing what the value of a game must be, from supposedly plausible or desirable conditions, one instead describes some specific, "fair" way of playing the game, or at least of calculating the outcome, and then asks whether the resulting payoff distribution has the properties desired in a value. Note that this approach considers individual games, while the axiomatic approach requires some kind of linear space of games. Each method sheds light on the other, and we shall exploit both in this book. After first studying the axiomatic approach at some length, we consider several constructive definitions of the value and explore their relationships to each other and to the axiomatic definition. In the latter half of the book we make a connection between the value and another game-theoretic solution concept, called the *core*, and then follow with an extended application to a non-atomic model of economic equilibrium.

It must be stressed that this is first and foremost a book of mathematics—a study of non-additive set functions together with associated linear operators. While the ultimate objective is to build a theory that will be useful in economics and other fields, a major share of the present motivation is supplied by simple intellectual curiosity about the mathematical objects under study. To get the most out of the book, the reader should be both mathematically inclined and mathematically knowledgeable. In particular, a familiarity with the elements of measure theory (as found, say, in Halmos, 1950) and functional analysis (e.g. Royden, 1966) is almost indispensable. On the other hand, we assume no previous experience with either game theory or mathematical economics; indeed, this volume may serve some readers as an effective, if specialized, introduction to both of these fields at an advanced mathematical level.

In the remainder of this introduction we shall indicate the contents of the eight chapters that follow, and then map out some "itineraries" for readers with special interests.

Chapter I (Sections 1–10) sets forth the axiomatic approach. The stage is set by defining BV, the Banach space of set functions of bounded variation, endowed with the variation norm. The center of attention is a linear space of set functions called pNA, which can be characterized as the closed subspace of BV that is spanned by powers of non-atomic measures, but is actually much larger than this characterization would at first seem to indicate. For example, it contains all set functions of the form $v(S) = f(\mu(S))$, where μ is a non-atomic measure and f is absolutely continuous. There is a unique value on pNA, and for an important subclass of games in pNA called "vector measure games" there is a remarkably simple formula for it. The unique value on pNA can be uniquely extended to a considerably larger space of games, containing set functions $f(\mu(S))$ where μ is non-atomic but f may be singular and may even have jump discontinuities. This chapter contains many other results; a more detailed guide will be found in Section 3.

Chapter II (Sections 11–16) takes up the first of the constructive approaches. The idea is to "play the game" by shuffling the players into a random order, like a deck of cards, and paying each player the amount that he adds to the worth of the coalition of those players that come before him in the order. In the finite case this method is known to yield the same value as that deduced from the axioms, but serious difficulties intervene in the infinite, non-atomic case. In fact, a direct analogue of the finite approach is impossible: we prove in this chapter that random orders having the requisite measurability and symmetry properties cannot be defined on $(I, \mathcal{C})$. Fortunately, there is another way to capture the idea of "shuffling the players", namely, through the use of *mixing transformations*. The value that can then be defined, which we call the "mixing value," exists for a wide class of games, including pNA, where it coincides with the value introduced in Chapter I.

Chapter III (Sections 17–19) treats another direct way of defining the value, due to Y. Kannai. The infinite-person game is approximated by finite games with the aid of sequences of increasingly fine partitions of the player space, and if the values of these finite games converge to the same limiting payoff distribution, regardless of the sequence of partitions used, then this limit is called the "asymptotic value" of the original infinite game. The asymptotic value proves to be closely related to the mixing value, but an example is provided of a non-atomic game for which the former but not the latter exists.

Chapter IV (Sections 20–25) develops the principle that a player's value should be based on his incremental worth to "homogeneous" coalitions, i.e. coalitions whose "composition" is the same as that of the all-player set, though their size may vary. To formalize this idea, we extend the set functions v to a space that includes not only the ordinary coalitions (measurable sets) but also certain "ideal" coalitions—intuitively, sets to which a player can "belong" at any intensity between 0 and 1. The above-mentioned homogeneous coalitions are then represented by ideal coalitions to which all players belong with the same intensity t, where $0 \leqq t \leqq 1$. For games in pNA, we can now characterize the value in terms of derivatives of the extended set function, evaluated on the homogeneous coalitions.

In Chapter V (Sections 26 and 27) we investigate the relationship between the value and another basic game theoretic concept, called the "core." Intuitively speaking, the core is the set of outcomes (payoff distributions) that no coalition of players can improve upon; it has been studied widely, by both game theorists and economists. In this chapter, we prove under suitable assumptions (superadditivity, homogeneity of degree 1), that the core of a non-atomic game in pNA consists of a single measure and that this measure is the value.

In Chapter VI (Sections 28–41), the theory developed in the previous Chapters is applied to a class of economic models, which may be interpreted either as exchange economies or as productive economies. The chief result is that under fairly wide conditions, the game derived from such a model is in pNA and its value coincides with the unique point in the core, and hence with the competitive equilibrium of the economic model. A number of examples are discussed in detail.

In Chapter VII (Sections 42–46) we show how the results obtained in

Chapters IV and V can be generalized to apply not only to pNA, with its unique value, but to other spaces of games on which the mixing value (Chapter II) or the asymptotic value (Chapter III) are defined. It turns out that there is a basic common property of all these values, called the "diagonal property," that is closely related to the kind of homogeneity investigated in Chapter IV; this property is the keystone of the proofs in this chapter. The results are then used to extend and strengthen the economic applications studied in Chapter VI.

Up to this point it has been assumed throughout that the underlying measurable space $(I, \mathcal{C})$ is isomorphic to the unit interval with its Borel subsets. In Chapter VIII (Sections 47 and 48) we examine the consequences of removing this assumption. It turns out that much of the theory stays intact, though results relating to the uniqueness of values are considerably weakened. In the course of constructing counter-examples, one encounters some pathologies that may have general measure-theoretic interest.

Three appendices, on values of finite games, on ϵ-monotonic set functions, and on an alternative characterization of the mixing value, complete the text.

It will have been realized that the book contains a complex of inter-related approaches and applications. Readers with different interests who do not wish to read the whole book can find different "itineraries" that will take them through parts of the book, each of which forms a more or less coherent and connected whole. Here are some possibilities:

1. *The Axiomatic Theory*. This would include at least Sections 2 through 8. Optional continuations would be: Sections 20 through 23, perhaps followed by Chapter V; and/or Chapter VIII; and/or Section 9 (and maybe 10).

2. *The Differentiable Theory*. This might also be called the "pNA itinerary," pNA being the single most important space of set functions defined in the book. It would include at least Sections 2, 3, 4, 6, and 7. Recommended additions are Section 5; and/or Sections 20 through 23, followed by Chapter V. Optional continuations would be Section 24; and/or Chapter VIII; and/or Chapter VI (perhaps without Sections 32 and 33); and/or Section 9 (and maybe 10); and/or the Random Order

itinerary described immediately below (Number 3). In case the reader takes the last option, he will probably wish to read Section 25 as well.

3. *The Random Order Theory*. This itinerary starts with Sections 2, 3, 4 through Proposition 4.4, 5 through Corollary 5.3, and 7 through Lemma 7.4. Then come Chapters II and III, with Sections 16 and 19 optional; the reader not interested in the proofs of the impossibility results enunciated in Chapter II may also skip Section 12 after Corollary 12.16 and Section 13 after Lemma 13.5. Optional additions are Sections 42 and 43, and possibly Section 44.

4. *The Economist's Itinerary*. This is meant to lead as fast as possible to the economic results in Chapters VI and VII. If one wants to do this conscientiously, then it isn't very fast, as these results depend on much of what has gone before. A minimal itinerary that is logically complete consists of Sections 2, 3, 4, 6, 7, 10, 20, 21, 22 through Formula (22.11), 23, 26, 27, 28, 30, 31 (except for Proposition 31.7), 35, 36, 37, 38, 39, and 40. Highly recommended are the conceptual discussion in Section 29, the discussion of competitive equilibrium in Section 32, and the examples in Section 33 (perhaps without subsection D, much of which concerns the asymptotic value).

The economist who wishes to see how this ties in with random order notions should add Sections 5 through Corollary 5.3, 11, 12 through Corollary 12.16, 13 through Lemma 13.5, 14, 15, 17, 18, 25, Proposition 31.7, perhaps subsection D of Section 33, and Chapter VII.

5. *For the Browser*. Some readers may be interested in a description of the theory, with statements of the major results, but without proofs. A good panorama can be had by reading Sections 2, 3, 11, 12 through the statement of Theorem D, 13 through Proposition 13.3, 14, 17, 20, 21, 26, 28, 29, 30, 31, 32 and 42. The browser primarily interested in the economic results may limit himself to a more restricted view, consisting of Sections 2, 3, 28, 29, 30, 31 (without Proposition 31.7), 32, 33 (without Subsection D), and 34.

6. *Set-Function Theory*. Part of this book develops the rudiments of a theory of non-additive set-functions, without referring to the symmetric operators that are called values. Readers interested in this should read Sections 2, 3, 4 through Proposition 4.5, 7 through Lemma 7.4,

and Sections 5, 9, and 10. The extension operator discussed in Sections 20, 21 (Theorem G), and 22 is also of interest independent of value theory. Of course much of Sections 2 and 3 refer to the value, but an itinerary through these sections that provides the necessary basics without mentioning the value would be a rather tortuous affair. We trust that the reader interested in this itinerary only will not suffer too greatly from his brief forced aquaintance with the value concept.

This completes our description of the itineraries. We also mention some items touched on in the book that may be of mathematical interest outside of value or set-function theory. The first of these is the treatment of certain variational problems with inequality constraints, and in particular the differentiability properties of their maxima (Sections 36 through 38). The second area involves the investigation of various aspects of the theory of measurable spaces. Thus in Section 12 and Appendix C, we discuss "measurable" orders on measurable spaces; in Section 13 we demonstrate the impossibility of imposing an appropriately invariant measure on the set of all measurable orders on a measurable space; and in Section 48 we construct a non-denumerable subspace of the unit interval, which when considered as a measurable space, has no non-trivial measurable automorphisms. The third and last area involves the notion of the "variation" of a real function of several real variables; most of the discussion of this is concentrated in Section 7 (see especially Lemma 7.14 and Note 3) and in Section 10 (see especially the example at the beginning of the section).

We close this introduction with a list of some open problems. The problem that is perhaps most important for the theory is whether the diagonal property described in Section 42 is a consequence of the axioms (see Section 42). Another problem that has proved very stubborn is whether $bv'NA \subset ASYMP$, or even whether the simplest single-jump functions are in $ASYMP$ (see Section 18). Open problems connected with the economic model of Chapter VI are listed in Section 41. Other open problems are mentioned in Footnote 3 of Section 7 and immediately after the proof of Example 19.2.

Chapter I. The Axiomatic Approach

1. Preliminaries

The symbol $\| \quad \|$ for norm is used in many different senses throughout the book; but it is never used in two different senses on the same space, so no confusion can result. In particular, when x is in a euclidean space of finite dimension (i.e. it is a finite-dimensional vector), then $\|x\|$ will always mean the maximum norm, i.e.

$$\|x\| = \max_i |x_i|.$$

It is important to distinguish notationally between functions and their values. For example, if μ is a measure, then $\|\mu\|$ is its total variation, whereas $\|\mu(S)\|$ is simply the absolute value of the real number $\mu(S)$.

Occasionally it will be necessary to use more than one norm on a given space; in that case, the norms will be distinguished in various ways, for example, by subscript, as in $\| \quad \|_1$.

Closure will be denoted by a bar; thus $\bar{A}$ is the closure of A.

Composition will usually be denoted by the symbol $\circ$; thus if f is defined on the range of μ, then the function whose value at S is $f(\mu(S))$ will be denoted $f \circ \mu$. When no confusion can result, especially in the case of compositions of linear operators, the symbol $\circ$ will occasionally be omitted.

The origin of any linear space (including, of course, the real line) will be denoted 0; no confusion can result. In euclidean n-space E^n, $x \cdot y$ will denote the scalar product of two vectors x and y, e^i will denote the i-th unit vector $(0, \ldots, 0, 1, 0, \ldots, 0)$, and e will denote the vector $(1, \ldots, 1)$.

The symbol $\subset$ will be used for inclusion that is not necessarily strict. Set-theoretic subtraction will be denoted by $\setminus$, whereas $-$ will be reserved for algebraic subtraction. $f|A$ will mean "f restricted to A." The cardinality of a set A is denoted $|A|$.

Closed and open intervals are denoted $[a, b]$ and (a, b) respectively; $[a, b)$ and $(a, b]$ denote half-open intervals.

W.l.o.g. means "without loss of generality." W.r.t. means "with respect to."

A *measurable space* is a pair $(I, \mathcal{C})$ where I is a set and $\mathcal{C}$ is a σ-field[1] of subsets of I; the members of $\mathcal{C}$ are called *measurable sets*. When no confusion can result, we shall sometimes denote the measurable space $(I, \mathcal{C})$ simply by I. A function f from one measurable space $(I, \mathcal{C})$ into another one $(J, \mathcal{D})$ is called *measurable* if $T \in \mathcal{D}$ implies $f^{-1}(T) \in \mathcal{C}$. Two measurable spaces are called *isomorphic* if there is a one-one function from one onto the other that is measurable in both directions;[2] the mapping is called an *isomorphism*. The measurable space consisting of the closed unit interval with its Borel subsets will be denoted $([0, 1], \mathcal{B})$; Lebesgue measure on this space will be denoted λ.

Proposition 1.1. *Any uncountable Borel subset of any euclidean space, and indeed of any complete separable metric space, when considered as a measurable space,*[3] *is isomorphic to* $([0, 1], \mathcal{B})$.

For a proof, see Mackey (1957), or Parthasarathy (1967), Theorems 2.8 and 2.12, pp. 12 and 14.

Proposition 1.2. *Let f be a one-one measurable function from a Borel subset of* $[0, 1]$ *into the real line, both considered as measurable spaces.*[3] *Then the range of f is a Borel set.*

For a proof, see Mackey (1957), p. 139, Theorem 3.2, or Parthasarathy (1967), Theorem 3a, p. 21.

Unless otherwise specified, the word "measure" in this book refers to countably additive totally finite signed scalar measures. Recall that a measure ξ on a measurable space $(I, \mathcal{C})$ is *non-atomic* if for all S in $\mathcal{C}$ with $\xi(S) \neq 0$, there is a $T \subset S$ with $\xi(S) \neq \xi(T) \neq 0$. A *vector measure* is an n-tuple $\mu = (\mu_1, \ldots, \mu_n)$ of measures μ_i, with n finite; if the μ_i are non-atomic then also μ is called *non-atomic*. The *range* of μ is the set $\{\mu(S): S \in \mathcal{C}\}$; it is a subset of E^n. The following proposition will be used repeatedly throughout the book:

[1] Note that we insist on $I \in \mathcal{C}$, a condition that is sometimes not demanded in defining "measurable space."

[2] I.e., both it and its inverse are measurable.

[3] The measurable subsets being the Borel subsets.

PROPOSITION 1.3 (*Lyapunov's Theorem*). *The range of a non-atomic vector measure is convex and compact.*

Since the original proof of Lyapunov (1940), this theorem has been reproved many times. For an elementary proof, the reader is referred to Halmos (1948), and for a quick, though deeper proof, to Lindenstrauss (1966).

NOTES

1. If $(I, \mathcal{C})$ is isomorphic to $([0, 1], \mathcal{B})$, then it may be verified that a measure ξ is non-atomic if and only if $\xi(\{s\}) = 0$ for all s in I.

2. Definitions of Game and Value

At the beginning of the section we shall concentrate on the formal basic definitions of "game" and "value," interpolating only a minimum of discussion and illustration. These basic definitions—which form a self-contained unit—will be slightly indented, in order to set them off from the discussion. In the second part of the section we will interpret and motivate the definitions from the game-theoretic point of view.

Let $(I, \mathcal{C})$ be a measurable space; it will be fixed throughout and will be referred to as the *underlying space*. The term *set function* will always mean a real-valued function v on $\mathcal{C}$ such that $v(\emptyset) = 0$.

In most of this book (until Chapter VIII) we will make the following

(2.1) *Standardness Assumption:* The underlying space $(I, \mathcal{C})$ is isomorphic to $([0, 1], \mathcal{B})$.

Because of Proposition 1.1, this assumption is not as drastic as it seems. We add that for much of the material of this book, (2.1) is not needed; see Section 47.

In the interpretation, a set function is a *game*, I is the *player space*, and the members of $\mathcal{C}$ are *coalitions*. The number $v(S)$, for $S \in \mathcal{C}$, is interpreted as the total payoff that the coalition S, if it forms, can obtain for its members; it will be called the *worth* of S. This way of representing a game is an obvious generalization of the standard representation in "characteristic function" (or "coalitional") form of a game with finitely many players (cf. von Neumann and Morgenstern, 1953, or Appendix A, below). Sometimes the phrase "game with a continuum of players" is used to distinguish these games from the finite ones. It should be stressed, however, that our treatment is measure-theoretic; the player space is a measurable, not a topological space.

By a *carrier* of a game (or set function) v, we mean a coalition I' such that $v(S) = v(S \cap I')$ for all $S \in \mathcal{C}$; the complement of a carrier is said to be *null*. An *atom* of v is a non-null coalition S such that for every coalition $T \subset S$, either T or $S \backslash T$ is null. If v has no atoms, v is called *non-atomic;* this agrees with the above definition of "non-atomic" if v happens to be a measure. Though we will have little further technical use in this book for the notion of a general non-atomic game, most of the games to be considered here will in fact be non-atomic, and their non-atomicity is more or less the crux of the matter.

A set function v is said to be *monotonic* if $S \supset T$ implies $v(S) \geqq v(T)$. The difference between two monotonic set functions is said to be *of bounded variation.* The set of all set functions of bounded variation forms a linear space over the field of real numbers, which will be called BV.

An example of an element of BV is any set function of the form

$$v(S) = f(\mu(S))$$

where μ is a finite non-negative measure and f is a function of bounded variation on the real interval $[0, \mu(I)]$ with $f(0) = 0$. Our investigations will focus on the space BV and certain of its subspaces.

The subspace of BV consisting of all bounded, finitely additive set functions, i.e. the bounded, finitely additive, signed measures on $(I, \mathcal{C})$, will be denoted FA. Note that an element μ of FA is monotonic if and only if $\mu(S) \geqq 0$ for all $S \in \mathcal{C}$.

Let Q be any subspace of BV. The set of monotonic set functions in Q will be denoted Q^+. A mapping of Q into BV is called *positive* if it maps Q^+ into BV^+, i.e. if it transforms monotonic set functions into monotonic set functions. It may of course happen that Q has no monotonic elements other than 0; in that case all linear mappings of Q into BV are positive.

Let $\mathcal{G}$ denote the group of *automorphisms* of the underlying space $(I, \mathcal{C})$, that is, isomorphisms of that space onto itself. Each Θ in $\mathcal{G}$ induces a linear mapping Θ_* of BV onto itself, defined by

$$(\Theta_* v)(S) = v(\Theta S).$$

A subspace Q of BV is called *symmetric* if $\Theta_* Q = Q$ for all Θ in $\mathcal{G}$.

We come at last to the definition of "value." Let Q be a symmetric subspace of BV. A *value* on Q is a positive linear mapping φ from Q into FA, such that for all Θ in $\mathcal{G}$ and v in Q we have

$$\varphi\Theta_* = \Theta_*\varphi \tag{2.2}$$

and

$$(\varphi v)(I) = v(I). \tag{2.3}$$

We shall refer to φv as the *value* of the game v, and to $(\varphi v)(S)$ as the *value* of the coalition S.

Essentially, the "symmetry" condition (2.2) says that the value does not depend on how the players are named, and the "efficiency" condition (2.3) says that it distributes to the players the entire amount available to the all-player set. The positivity condition says that in monotonic games all coalitions have non-negative values; this is plausible because no coalition has negative worth itself, nor can it decrease the worth of another coalition when joining it. There is a close correspondence between this definition of value and the axiomatization (cf. Shapley, 1953a, or Appendix A, below) of the value for finite games; this will be explored further below.

Another reasonable condition that we might wish to impose on φ is continuity; but before that can be done, we must define a topology on BV. This will be done in Section 3. It will turn out that there is a close relationship between positivity and continuity and that little would be changed if we replaced the positivity condition in the definition of value by an appropriate continuity condition.

Still another likely condition, plausible but unneeded at present, is that of "invariance under strategic equivalence" (cf. von Neumann and Morgenstern, 1953, p. 245), i.e. if $v \in Q$ and $a \in FA \cap Q$, then $\varphi(v + a) = \varphi v + a$. With linearity, this amounts to saying that φ is a *projection* on FA, i.e. for any $v \in FA \cap Q$ we have $\varphi v = v$. (In the usual terminology, a game $v \in FA$ would be called "inessential.") Some discussion of the use of a projection axiom, in connection with the relaxation of assumption (2.1), will be found near the end of Section 48.

A slightly different approach to the axiomatization of value is as follows: One speaks only of monotonic games v; the value φv is always a nonnegative measure (finitely additive); and the linearity condition is replaced by the condition $\varphi(\alpha v + \beta w) = \alpha\varphi v + \beta\varphi w$, where α and β are non-negative real numbers. Conditions (2.2) and (2.3) remain unchanged. This approach is entirely equivalent to the one we have adopted.

A non-atomic game is meant to represent a game with many players, each of whom is individually insignificant. Examples are economies with many economic agents (see Chapter VI), elections with many voters, and so on. In such a game, it is sometimes useful to think of an individual player not as a single point s in I, but as an "infinitesimal subset" ds of I; see Section 29 for a discussion of this viewpoint.

In a finite game with player space N, a *payoff vector* is simply a member x of E^N, i.e. a function from N to the reals (see Appendix A). When we are thinking in terms of coalitions rather than individuals, it is convenient to think of the payoff vector x as a measure on N, defined for all $S \subset N$ by

$$x(S) = \sum_{i \in S} x(i);$$

here $x(S)$ signifies the total payoff to S under the outcome x. The measures on N are of course in natural one-one correspondence with the points in E^N. Thinking of a payoff vector as a measure is especially useful in connection with non-atomic games, or more generally with arbitrary games on $(I, \mathcal{C})$. In such games a payoff vector may often be represented by a non-atomic measure μ; this means that the individual player ds gets the infinitesimal payoff $\mu(ds)$, whereas the total payoff to a coalition is often a positive number. For the sake of generality, we define a *payoff vector* of a game on $(I, \mathcal{C})$ to be any member of FA.

Just as the games v defined here generalize the games with finite player sets, so the value defined here generalizes the value on finite games (cf. Shapley, 1953a, or Appendix A, below). The parallelism is indeed very close. In both cases the value assigns to each game a payoff vector, is a linear operator, and satisfies symmetry ((2.2) or (A.2)) and efficiency ((2.3) or (A.3)) conditions. The positivity condition does not appear explicitly in the finite treatment, but only because, in that context, it follows from the other axioms.[1] Similarly, the dummy axiom (A.4) does not appear explicitly in this section, but only because, for non-atomic games obeying (2.1), it follows from the other conditions (see Note 4).

Two basic assumptions about the nature of a game are implicit in the use of a real-valued set function to describe it, namely the assumptions of *unrestricted side payments* and *fixed threats*. These are well known in the literature[2] for the case of finite player sets, and they are not essentially different when there is a continuum of players. It is perhaps worthwhile, however, simply to restate the side payment condition in the continuum case. This says that not only can each coalition S obtain for its members a total of $v(S)$, but it can also distribute this total among its members in any way it pleases. Thus, if ν is any member of FA with $\nu(S) = v(S)$, then S can act so that each $T \subset S$ will obtain $\nu(T)$, or in other words, so that each "member" ds of S will obtain $\nu(ds)$.

NOTES

1. By the Standardness Assumption (2.1), all underlying spaces are isomorphic to ([0, 1], $\mathcal{B}$). Another possible candidate for a "canonical" underlying space would have been ([0, 1], $\mathcal{L}$), where $\mathcal{L}$ is the σ-field of Lebesgue-measurable subsets of [0, 1]. We chose ([0, 1], $\mathcal{B}$) instead because the set functions that are non-atomic measures play a decisive role in our work, and the latter admits a much wider class of such measures. Indeed, if S is any set of Lebesgue measure 0, then all subsets of S are in $\mathcal{L}$. Hence from Proposition C_{53} in Sierpinski (1956), it follows that, under the continuum hypothesis, any non-atomic measure on $\mathcal{L}$ must vanish on S and on its subsets. Thus ([0, 1], $\mathcal{L}$) admits only measures that have no non-singular component, whereas ([0, 1], $\mathcal{B}$) admits non-atomic measures that are singular with respect to Lebesgue measure.

2. The class of games of the form $f \circ \mu$, where μ is a finite non-negative measure and f is of bounded variation on $[0, \mu(I)]$, includes the generalized weighted majority voting games, with denumerably many players or a continuum of players; value theories for these games have been investigated in

[1] See Corollary A.10.

[2] See e.g. Harsanyi (1959) or Aumann (1967a).

papers by Milnor and Shapley (1961), Shapley (1961, 1962b), Artstein (1971), Hart (1973), and elsewhere. It also includes all games with finite carriers, which are essentially the finite games in coalitional form. On the space of games with finite carrier there is a unique value (in the sense of this section), and it coincides with the finite value (Appendix A); this follows easily from Proposition A.5 and Note 4 below.

3. Contrary to the view taken above, in which an individual player is identified with an "infinitesimal set," it is sometimes convenient to think of a player as a single point in I. In this spirit, let us define a *null player* to be a point $s \in I$ such that $\{s\}$ is a null coalition. In a non-atomic game, all points in I are null players. Readers familiar with the game theoretic notion of "dummy" should note that null players are dummies, but dummies are not always null players; see Note 4 to Appendix A.

4. In any game with infinitely many null players, and in particular in a non-atomic game, every null coalition must have value zero. To prove this, assume that $(I, \mathcal{C}) = ([0, 1], \mathcal{B})$, which we may do w.l.o.g. by (2.1). First deduce from (2.2) that all null players must get the same value, which must therefore be 0, since the value is bounded. Hence all finite null coalitions get value 0. Now let S be an infinite null coalition, and let S_1 and S_2 be disjoint subcoalitions of S having the same cardinality as S. Whether they are denumerable or non-denumerable (and hence Borel sets with the power of the continuum), there is a symmetry of the game that takes S_1 onto S_2; hence they have the same value. It then follows easily that the value of S must be zero, as claimed.

If there are only finitely many null players, however, the argument fails. An example is the "unanimity" game (see Section 3), with a single null player added. This game can be imbedded in a symmetric subspace of BV on which there is a unique value, which assigns value 1 to the null player in question.

From the opening remark of this note it follows that in a game with infinitely many null players, $(\varphi v)(I') = v(I')$ for all carriers I' of v. This is the form in which the efficiency and dummy axioms (A.3) and (A.4) are combined into a single axiom in Shapley (1953a) (see Note 2 to Appendix A).

5. In the text, we defined a payoff vector of a game on $(I, \mathcal{C})$ to be a member of FA; we have not found it necessary to allow more generality, e.g. to allow unbounded measures. Neither is it convenient, on the other hand, to restrict the generality, e.g. to consider only countably additive measures. This is because FA is a subspace of BV, and, if we look at a member μ of FA as a game, then the payoff vectors naturally associated with this game cannot be expected to be completely additive if μ is not; for example, the core of μ (see Chapter V) consists of the unique point μ itself.

3. Statement of Chief Results

In this section we shall state and briefly motivate and discuss some of the main results of this chapter. No proofs will be given in this section.

The first remark to be made about our definition of value is that there is no value on BV itself. Indeed, consider the "unanimity game" v defined by

$$v(S) = \begin{cases} 1 & \text{if } S = I \\ 0 & \text{otherwise;} \end{cases}$$

it is invariant under all automorphisms of the player space. Therefore if there is a value φ on BV, then φv must also be invariant under all automorphisms of the underlying space, by (2.2). But no member of FA can have this property unless $(\varphi v)(I) = 0$, in which case (2.3) is violated.[1]

Although there is no value on BV, we will find that there are important subspaces of BV on which there *is* a value, which is moreover unique. In this chapter we shall characterize certain such subspaces and investigate their properties and the properties of the values on them.

To state our results in their fullest generality, it will be useful to define a norm on BV. For v in BV define

$$\|v\| = \inf(u(I) + w(I)),$$

where the inf ranges over all monotonic set functions u and w such that $v = u - w$. The quantity $\|v\|$ will be called the *variation*[2] of v; it is easily seen that it is a norm. Unless otherwise specified, all references in the sequel to topological notions (such as closure) on BV will be in the sense of the topology induced by the variation norm; in particular, if A is a subset of BV, then $\bar{A}$ will always denote the closure of A in the variation norm. Similarly, the word "spanned" will always be used in the topological linear sense; i.e. the space spanned by a subset of BV is the closure of the set of all linear combinations of elements of that subset.

Denote by NA the space of non-atomic measures on the underlying space $(I, \mathcal{C})$; clearly $NA \subset BV$. In conformance with our notation, NA^+ will denote the cone of all non-negative measures in NA, and

[1] Note that the unanimity game is not non-atomic. However, we could define $v(S)$ to be 1 when S differs from I by at most a finite set, and 0 otherwise. This, too, is invariant under all automorphisms of the underlying space.

[2] A characterization of this norm that is more directly related to the intuitive notion of "variation" will be given in Section 4 (Proposition 4.1).

NA^1 will denote the set of all μ in NA^+ with $\mu(I) = 1$ (i.e. the non-atomic "probability measures"). The space of all real-valued functions f of bounded variation on the unit interval [0, 1] that obey $f(0) = 0$ will be denoted *bv*.

Set functions defined with the aid of non-atomic measures form the central subject of interest in this chapter and in much of the rest of the book. Of particular importance are set functions of the form $f \circ \mu$, where $f \in bv$ and $\mu \in NA^1$; they are called *scalar measure games*.[3] The subspace of BV spanned by all scalar measure games will be denoted[4] $bvNA$.

The question arises as to whether there is a value on $bvNA$. Unfortunately, the answer is no. Indeed, let $\mu \in NA^1$, define f in *bv* by

$$f(x) = \begin{cases} 1 & \text{if } x = 1 \\ 0 & \text{if } 0 \leqq x < 1, \end{cases}$$

and let $v = f \circ \mu$. This may be called the "μ-almost unanimity game"; it is invariant under all automorphisms that preserve the sets of μ-measure 0. Therefore if there is a value φ on $bvNA$, then φv must also be invariant under all such automorphisms; and again no member of FA satisfying (2.3) has this property.

The difficulty in this example is caused by the discontinuity of f at 1; a similar difficulty occurs when there is a discontinuity at 0. Let us therefore define bv' to be the set of all *bv* functions that are continuous at 0 and at 1, and $bv'NA$ to be the subspace of BV spanned by all games of the form $f \circ \mu$ where $f \in bv'$ and $\mu \in NA^1$.

THEOREM A. *There is a unique value φ on $bv'NA$; the range of φ is NA, and*[5] $\|\varphi\| = 1$. *Furthermore, if $f \in bv'$ is such that $f(1) = 1$ and $\mu \in NA^1$, then $\varphi(f \circ \mu) = \mu$.*

Theorem A is proved in Section 8.

[3] I.e., games defined with the aid of a (non-negative) scalar measure, as distinguished from games defined with the aid of a vector measure.

[4] More precise, but clumsier, would be "$bvNA^1$" or "$bv(NA^+)$." If we allowed $\mu \in NA \backslash NA^+$ we would introduce additional set functions, including some that are not even in BV; see Examples 9.3 and 9.5. The same remarks apply to $bv'NA$, defined next.

[5] $\|\varphi\|$ is the norm of the operator φ, i.e. the supremum of $\|\varphi v\|/\|v\|$ over all non-zero v in the domain of φ.

The fact that $\|\varphi\| = 1$ implies in particular that $\|\varphi\|$ is finite, i.e. that φ is continuous. The relation between the positivity of φ and its continuity in the variation norm turns out to be basic to our investigation. This relation will be more closely examined in Section 4. In particular, if in the definition of value we replace the positivity condition by the condition that φ be continuous, then Theorem A remains true exactly as stated above.

The second sentence of Theorem A (the assertion that $\varphi(f \circ \mu) = \mu$) can be intuitively understood as follows: In the game $f \circ \mu$, the worth of a coalition depends only on its μ-measure. There would therefore seem to be no reason to "discriminate" between coalitions having equal μ-measure, given assumption (2.1), and it seems natural to conjecture that they should get the same value. Because of the non-atomicity and the normalization condition (2.3), this implies that the value actually equals the μ-measure. This is proved relatively easily in Section 6 on the assumption that a value exists, whereas the proof of existence itself is much more difficult and will be given only in Section 8.

There is an analogy between the situation described in Theorem A and that in finite games that is worth pursuing. Recall that the "symmetry" axiom for finite games, on which (2.2) is modeled, may be replaced by an axiom that says that in games in which the worth of a coalition depends only on the number of non-null players in the coalition, the total value is divided equally among the non-null players, i.e. the value is proportional to the number of non-null players. Now in a finite game, the number of players in a coalition is a measure; moreover it is a measure that is in a certain sense "natural" or "distinguished." In games with a continuum of players, however, there is nothing in the underlying player space to distinguish one non-atomic probability measure from another.[6] Therefore the continuous analogue of the finite situation would be that the value is proportional to the μ-measure whenever the payoff is a function of the μ-measure, for all μ-measures. This is precisely the content of the second sentence of Theorem A.

The space $bv'NA$ is maximal among those subspaces of BV that are spanned by monotonic scalar measure games and on which there is a

[6] Because of (2.1); cf. Lemma 6.2 and Chapter VIII.

value (see Note 2). We turn next to a space spanned by a far smaller class of scalar measure games: the space pNA, defined as the subspace of BV spanned by all powers of NA^+ measures. Powers of measures are in a sense the "simplest" set functions that are not themselves measures, and that is what motivates the definition of pNA. This space will play a very important role in the theory: On the one hand, all the approaches to the value concept that will be developed throughout the book apply to it;[7] on the other hand, though it is defined in terms of powers of measures only, pNA contains many other set functions of interest.[8] For example, we shall now see that pNA contains all "vector measure games" obeying appropriate differentiability conditions, i.e. all set functions of the form $f \circ \mu$, where $\mu = (\mu_1, \ldots, \mu_n)$ is now a non-atomic finite-dimensional vector measure and f is an appropriately differentiable real-valued function defined on the range of μ, with $f(0) = 0$.

To describe the situation we need a number of definitions. Let X be a convex subset of a euclidean space E^n. A vector z is said to be *X-admissible* if $z = x - y$ for some $x, y \in X$. Let f be a continuous real function on X and let z be X-admissible. We shall say that f is *continuously differentiable on X in the direction z* if there is a real function on X which equals the derivative[9] $df(x + \theta z)/d\theta$ at each point x in the relative interior of X, and which is continuous at each point in X. Such a function, if it exists, is unique; it will be denoted f_z, and will be called the *derivative of f in the direction z.*[10] We shall say that f is *continuously differentiable on X* if, for all X-admissible z, it is continuously differentiable on X in the direction z. Essentially, what the definition demands is that the directional derivatives of f exist and are continuous in the relative interior of X, and can be continuously extended to the boundary points in X, for all those directions that do not "lead out" of the smallest linear manifold in which X lies. Of course, if X has full dimension, then these are simply all the directions. For example, if X is the closed unit cube in E^n, the definition is equivalent to saying that each of the partial

[7] See, for example, Theorems B, E, F, and H (Sections 3, 14, 18 and 21).

[8] See, for example, Theorems B, C, and J (Sections 3 and 31).

[9] Of course this involves the assumption that the derivative exists.

[10] Note that f_z depends on the magnitude as well as the direction of z; thus we have $f_{\alpha z} = \alpha f_z$ if z and αz are both X-admissible.

derivatives of f exists and is continuous on all of X, where $\partial/\partial x_j$ is defined in the one-sided sense when $x_j = 0$ or 1.

THEOREM B. *There is a unique value* φ *on* pNA, *and* $\|\varphi\| = 1$. *Furthermore, let* μ *be a vector of measures in* NA, *and let* f *be continuously differentiable on the range*[11] *of* μ, *with* $f(0) = 0$. *Then* $f \circ \mu \in pNA$, *and*

$$\varphi(f \circ \mu)(S) = \int_0^1 f_{\mu(S)}(t\mu(I))\, dt, \tag{3.1}$$

where $f_{\mu(S)}$ *is the derivative of* f *in the direction* $\mu(S)$.

The unique value on pNA provided by Theorem B coincides on pNA with the unique value on $bv'NA$ provided by Theorem A (see Note 1). Thus no confusion can result from denoting both these values by φ. Theorem B is proved in Section 7.

A restatement of formula (3.1) is of interest. Let R denote the range of μ. When R has full dimension, we have

$$\int_0^1 f_{\mu(S)}(t\mu(I))\, dt = \sum_{i=1}^{n} \mu_i(S) \int_0^1 f_i(t\mu(I))\, dt, \tag{3.2}$$

where f_i denotes $\partial f/\partial x_i$. More generally, let R have a dimension $r \leqq n$, and let $z_1, \ldots, z_r$ be a basis for the smallest linear subspace containing R. Then there are measures $\nu_1, \ldots, \nu_r$ such that

$$\mu(S) = \sum_{j=1}^{r} \nu_j(S) z_j,$$

so we obtain the following restatement of (3.1):

$$\varphi(f \circ \mu)(S) = \sum_{j=1}^{r} \nu_j(S) \int_0^1 f_{z_j}(t\mu(I))\, dt.$$

Since each ν_j is a linear combination of the μ_i's, we see that *the value of a continuously differentiable vector measure game is always a linear combination of the component measures.*

These formulas have an additional startling aspect: they show that the value is completely determined by the behavior of f near the *diagonal*, $[0, \mu(I)]$ (i.e. the set $\{t\mu(I): 0 \leqq t \leqq 1\}$—see Figure 1); the

[11] Which is convex by Lyapunov's theorem (Proposition 1.3).

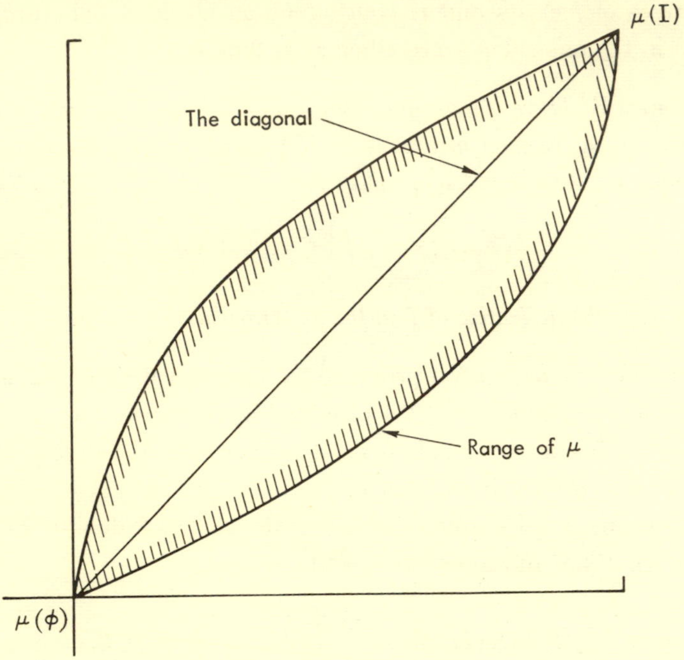

Figure 1. The range of a vector measure

behavior of f away from the diagonal is totally irrelevant.[12] Intuitively, the reason is that a coalition S chosen "at random" from $\mathcal{C}$ will probably be a good "sample" of I, since I represents a large population and S constitutes a considerable proportion of I. For example, $\mu_1(S)$ might denote the total income of the members of S; then the average income over S will be about the same as the average income over I, making $\mu_1(S)/\mu_1(I)$ about the same as the ratio of populations. But now if we let $\mu_2(S)$ be, say, the total weight of S, then we will reach the same conclusion about μ_2, and similarly for all non-atomic μ_i. Hence the ratios $\mu_i(S)/\mu_i(I)$, $i = 1, \ldots, n$, tend to be equal, so that "almost all" of the points $\mu(S)$ fall on the diagonal.

In fact, we do not even have to assume that f is differentiable far from the diagonal; interpreting φ to be the value on $bv'NA$, we shall[13]

[12] See Chapter VII for an extended discussion of this phenomenon.

[13] See Section 8.

actually prove $f \circ \mu \in pNA$ and (3.1) under the following condition, which is weaker than that stated in Theorem B:

(3.3) $f \circ \mu$ in $bv'NA$ is such that μ is a vector of measures in NA and f is continuously differentiable on some convex[14] neighborhood (in R) of the diagonal $[0, \mu(I)]$.

It is, however, impossible to dispense entirely with the differentiability condition. For example, consider the vector measure game defined by

$$v(S) = \max(\mu_1(S), \mu_2(S)), \tag{3.4}$$

where the underlying space is $[-1, 1]$ with its Borel subsets, λ is Lebesgue measure, and

$$\mu_1(S) = \lambda(S \cap [0, 1])$$
$$\mu_2(S) = \lambda(S \cap [-1, 0]);$$

then[15] $v \notin bv'NA$.

Theorem B can be generalized in a number of other directions. First, the condition that f be continuously differentiable on R can be slightly relaxed on the boundary[16] of R; for details, see Section 10 (Proposition 10.17). Second, there is an analogue of formula (3.1) for v in pNA that are not necessarily of the form $f \circ \mu$; for details see Chapter IV (Theorem H, Section 21).

Returning to scalar measure games, it follows from Theorem B that if $\mu \in NA^+$ then a sufficient condition for $f \circ \mu \in pNA$ is that f be continuously differentiable. This condition is, however, not necessary; a necessary and sufficient condition is given in the following theorem, with which we close this section.

THEOREM C. *Let $v = f \circ \mu$, where $\mu \in NA^+$ and f with $f(0) = 0$ is a real-valued function (not necessarily of bounded variation) on the range R of μ. Then $v \in pNA$ if and only if f is absolutely continuous on R.*

[14] We have defined "continuously differentiable" only for convex domains. Of course, any neighborhood of the diagonal contains a convex neighborhood of the diagonal.

[15] See Example 9.4, which is essentially the same game as (3.4).

[16] (3.3) enables a far greater relaxation of the differentiability condition, but assumes a priori that $f \circ \mu \in bv'NA$.

Theorem C is proved in Section 5. In Section 9 we explore the effects of weakening the hypothesis from $\mu \in NA^+$ to $\mu \in NA$. Additional results concerning the structure of and values on BV, $bv'NA$, pNA, and other spaces yet to be defined will be proved in the body of the book.

NOTES

1. Let Q and R be symmetric subspaces of BV with $Q \supset R$. If there is a value on Q, then its restriction to R is also a value. Hence if there is a value on Q and a unique value on R, then they must coincide on R.

2. We wish to prove our assertion that $bv'NA$ is maximal among those subspaces of BV that are spanned by monotonic scalar measure games and on which there is a value. Let Q be a symmetric space containing $bv'NA$ and spanned by monotonic scalar measure games. Unless $Q = bv'NA$, there must be a μ in NA^1 and a monotonic f in bv that has a discontinuity at 0 or at 1, or both, such that $f \circ \mu \in Q$. The function $f|(0, 1)$ has an extension f' to $[0, 1]$ that is continuous at 0 and at 1 and is monotonic; set $f'' = f' - f'(0)$. Then $f'' \in bv'NA$, and so $f - f'' \in Q$. Moreover $f - f''$ is monotonic, has a discontinuity at 0 or at 1, or both, and is constant on $(0, 1)$. Hence reasoning as in the discussion of the "μ-almost unanimity game" above, we deduce that there can be no value on Q.

The situation is different if we do not insist that the spanning set functions be monotonic. This will be taken up again in Note 2 to Section 8.

4. Basic Properties of the Variation Norm and the Space BV

A non-decreasing sequence of sets of the form

$$\emptyset = S_0 \subset S_1 \subset \cdots \subset S_m = I$$

will be called a *chain*. A *link* of this chain is a pair of successive elements $\{S_{i-1}, S_i\}$. A *subchain* is a set of links; the chain itself may and will be identified with the subchain consisting of all the links. If v is a set function and Λ is a subchain of a chain Ω, then the *variation of v over* Λ is defined by

$$\|v\|_\Lambda = \Sigma |v(S_i) - v(S_{i-1})|,$$

where the sum ranges over all indexes i such that $\{S_{i-1}, S_i\}$ is a link in the subchain Λ. For fixed Λ, the functional $\|v\|_\Lambda$ is a pseudonorm on BV, i.e. it enjoys all the properties of a norm except $\|v\| = 0 \Rightarrow v = 0$.

PROPOSITION 4.1. *Let v be a set function. A necessary and sufficient condition that $v \in BV$ is that $\|v\|_\Omega$ be bounded over all chains Ω. If $v \in BV$, then*

$$\|v\| = \sup\|v\|_\Omega,$$

where the sup is taken over all chains Ω.

Remark. This characterization of BV and of the variation norm is basic throughout the entire book. It will be used at least as often as the original definition; because of the frequent use, we will usually not refer explicitly to Proposition 4.1 when using it.

Proof. Necessity: Let $v = u - w$, where u and w are monotonic. Then for any chain Ω

$$\|v\|_\Omega \leqq \|u\|_\Omega + \|w\|_\Omega = u(I) + w(I).$$

This proves the necessity, and also that

$$\sup\|v\|_\Omega \leqq \|v\|.$$

Sufficiency: Assume that $\|v\|_\Omega$ is bounded. Define u to be the "upper variation" of v, i.e.

$$u(S) = \sup \sum_i \max\{(v(S_i) - v(S_{i-1})), 0\},$$

where the sup is taken over all non-decreasing sequences $\emptyset = S_0 \subset S_1 \subset \cdots \subset S_k = S$. Let $w = u - v$. Then both u and w are monotonic. Indeed, if $T \supset S$, then

$$u(T) \geqq \max\{(v(T) - v(S)), 0\} + u(S).$$

The right side of this inequality is $\geqq u(S)$, proving that u is monotonic. But it is also $\geqq v(T) - v(S) + u(S)$, whence by transposing we obtain

$$w(T) = u(T) - v(T) \geqq u(S) - v(S) = w(S),$$

proving that w is monotonic as well. This proves the sufficiency.

Finally, for a given $v \in BV$ let u and w be as in the sufficiency proof. For a given $\epsilon > 0$ let Ω be a chain $\emptyset = S_0 \subset S_1 \subset \cdots \subset S_m = I$ such that

$$\sum_i \max\{(v(S_i) - v(S_{i-1})), 0\} \geqq u(I) - \epsilon.$$

Let Λ be the subchain consisting of all links such that $v(S_i) \geqq v(S_{i-1})$, and Γ the subchain consisting of all the remaining links; then

$$\|v\|_\Lambda \geqq u(I) - \epsilon.$$

Furthermore, we have

$$\|v\|_\Lambda - \|v\|_\Gamma = \sum_i (v(S_i) - v(S_{i-1})) = v(I),$$

and hence

$$\|v\|_\Gamma = \|v\|_\Lambda - v(I) \geqq u(I) - v(I) - \epsilon = w(I) - \epsilon.$$

Hence from the definition of $\|v\|$ it follows that

$$\|v\|_\Omega = \|v\|_\Lambda + \|v\|_\Gamma \geqq u(I) + w(I) - 2\epsilon \geqq \|v\| - 2\epsilon.$$

Since ϵ is arbitrary, it follows that $\sup\|v\|_\Omega \geqq \|v\|$. As the opposite inequality has already been proved, the proof of the proposition is complete.

Corollary 4.2. *The inf in the definition of norm is achieved, i.e. there exist monotonic u and w with $v = u - w$ and $\|v\| = \|u\| + \|w\|$.*

Proof. This follows immediately from the foregoing proof.

Proposition 4.3. *BV is complete, hence a Banach space.*

Proof. Let $\{v_n\}$ be a Cauchy sequence. Then $\{v_n(S)\}$ is a Cauchy sequence for each $S \in \mathcal{C}$; denote its limit by $v(S)$. We must first show that v is of bounded variation. Let N be such that $\|v_n - v_N\| \leqq 1$ whenever $n \geqq N$. Then for each chain Ω and each $n \geqq N$ we have

$$\|v_n\|_\Omega - \|v_N\| \leqq \|v_n\|_\Omega - \|v_N\|_\Omega \leqq \|v_n - v_N\|_\Omega \leqq \|v_n - v_N\| \leqq 1.$$

Letting $n \to \infty$, we deduce

$$\|v\|_\Omega \leqq 1 + \|v_N\|;$$

hence v is of bounded variation. That $\|v_n - v\| \to 0$ is now easily verified, so the proposition is proved.

Proposition 4.4. *NA, FA, and the space of all countably additive members of FA are all closed subspaces of BV.*

The straightforward proof is omitted.

PROPOSITION 4.5. *For* $v_1, v_2 \in BV$, *we have*

$$\|v_1v_2\| \leqq \|v_1\|\,\|v_2\|.$$

Proof. For monotonic u and w we have $\|uw\| = \|u\|\,\|w\|$. From this and Corollary 4.2, it follows that for any v_1 and v_2 in BV,

$$\begin{aligned}\|v_1v_2\| &= \|(u_1 - w_1)(u_2 - w_2)\| \leqq \|u_1u_2\| + \|u_1w_2\| + \|w_1u_2\| + \|w_1w_2\| \\ &= (\|u_1\| + \|w_1\|)(\|u_2\| + \|w_2\|) = \|v_1\|\,\|v_2\|.\end{aligned}$$

This completes the proof.

Proposition 4.5 shows that BV is a Banach algebra. In particular, it shows that multiplication is continuous, so that if X is a subalgebra of BV (linear subspace closed under multiplication), then the closure of X is also a subalgebra.

PROPOSITION 4.6. *Let* Q *be a linear subspace of* BV, *and let* φ *be a linear operator from* Q *into* FA *obeying the normalization condition* (2.3) *and having* $\|\varphi\| \leqq 1$. *Then* φ *is positive.*

Proof. Let v be monotonic and suppose that, contrary to the proposition, there is a coalition S with $(\varphi v)(S) < 0$. Then because of the monotonicity of v, $\|\varphi\| \leqq 1$, and the normalization condition (2.3) we have

$$\begin{aligned}v(I) = \|v\| \geqq \|\varphi v\| &\geqq |(\varphi v)(S)| + |(\varphi v)(I) - (\varphi v)(S)| \\ &> |v(I) - (\varphi v)(S)| = v(I) - (\varphi v)(S) > v(I);\end{aligned}$$

this contradiction establishes the proposition.

We would now like to prove a converse to Proposition 4.6, i.e. a theorem that asserts that, under appropriate conditions, the positivity of φ implies $\|\varphi\| \leqq 1$. This, however, is not as straightforward as Proposition 4.6, and we must first introduce the concept of an "internal" subspace of BV.

A linear subspace Q of BV is said to be *reproducing*[1] if $Q = Q^+ - Q^+$. It is said to be *internal* if for all $v \in Q$ the u and w appearing in the definition of $\|v\|$ can be chosen to be members of Q; explicitly, this means that $\|v\| = \inf(u(I) + w(I))$, where u and w are members of Q^+

[1] The term "reproducing" has been used by Krasnoselskii (1964) in a sense closely related to but not quite identical to the current one.

(and not just of BV^+) such that $v = u - w$. Clearly every internal space is reproducing.

PROPOSITION 4.7. *Let Q be an internal subspace of BV, and let φ be a positive linear operator from Q into BV obeying the normalization condition* (2.3). *Then* $\|\varphi\| \leqq 1$.

Proof. Let $v \in Q$, and for given $\epsilon > 0$ let u and w in Q^+ be such that $v = u - w$ and $\|v\| + \epsilon \geqq u(I) + w(I)$. Then

$$\begin{aligned}\|\varphi v\| &= \|\varphi u - \varphi w\| \leqq \|\varphi u\| + \|\varphi w\| \\ &= (\varphi u)(I) + (\varphi w)(I) = u(I) + w(I) \leqq \|v\| + \epsilon,\end{aligned}$$

and $\|\varphi\| \leqq 1$ follows easily. This completes the proof.

Internal subspaces of BV play a central role in this book; in particular, we shall prove that both pNA and $bv'NA$ are internal. Unfortunately, it is not always easy to prove directly that a given space is internal; sometimes it is easier to establish the internality of a dense subspace. We therefore now prove (Proposition 4.12) that the closure of an internal space is internal.

If Q is a reproducing subspace of BV, define a norm $\| \quad \|_Q$ on Q by

$$\|v\|_Q = \inf(u(I) + w(I)),$$

where the inf ranges over all u and w in Q^+ such that $v = u - w$. Clearly

$$\|v\| \leqq \|v\|_Q, \tag{4.8}$$

and equality holds for all v in Q if and only if Q is internal. The norm $\| \quad \|_Q$ will be called the *internal norm* on Q or the *Q-norm*. Unless otherwise specified, topological terms (such as "closed") will continue to refer to the variation norm.

PROPOSITION 4.9. *Let Q be a closed subspace of BV, and let $B = Q^+ - Q^+$. Then B is reproducing, and is complete in the B-norm.*

Proof. We have

$$Q^+ \subset B \cap BV^+ = B^+ = (Q^+ - Q^+)^+ \subset Q^+, \tag{4.10}$$

whence $Q^+ = B^+$ and $B = Q^+ - Q^+ = B^+ - B^+$; hence B is reproducing.

To prove the completeness, recall[2] that a normed linear space is complete if and only if the absolute convergence of a series implies its convergence (i.e. $\Sigma\|x_i\| < \infty$ implies that the partial sums of Σx_i converge in norm). Let $v_i \in B$ be such that $\Sigma\|v_i\|_B < \infty$; we must show that Σv_i converges in the B-norm. Let u_i and w_i in Q^+ ($= B^+$) be such that $u_i - w_i = v$ and

$$u_i(I) + w_i(I) \leqq \|v_i\|_B + (\tfrac{1}{2})^i. \tag{4.11}$$

Then $\Sigma u_i(I)$ and $\Sigma w_i(I)$ converge; hence by the monotonicity of the u_i and w_i, $\Sigma u_i(S)$ and $\Sigma w_i(S)$ converge for all S; denote the sums by $u(S)$ and $w(S)$, respectively. Then

$$\left\|u - \sum_{i=1}^{n} u_i\right\| = \left\|\sum_{i=n+1}^{\infty} u_i\right\| = \sum_{i=n+1}^{\infty} u_i(I) \to 0,$$

since $\Sigma u_i(I)$ converges; since $\Sigma_{i=1}^{n} u_i \in Q$ and Q is closed, it follows that $u \in Q$. Similarly $w \in Q$. Since u and w are clearly monotonic, it follows that they are in Q^+. Setting $v = u - w$, we deduce that $v \in B$, and that

$$\begin{aligned}\left\|v - \sum_{i=1}^{n} v_i\right\|_B \leqq \left\|u - \sum_{i=1}^{n} u_i\right\|_B + \left\|w - \sum_{i=1}^{n} w_i\right\|_B \\ = \sum_{i=n+1}^{\infty} u_i(I) + \sum_{i=n+1}^{\infty} w_i(I) \to 0;\end{aligned}$$

i.e. Σv_i converges to v. This completes the proof of Proposition 4.9.

PROPOSITION 4.12. *The closure of an internal space is internal.*

Proof. Let A be internal, $Q = \bar{A}$, $B = Q^+ - Q^+$; by Proposition 4.9, B is reproducing and is complete in the B-norm, and $A \subset B$. Now let A^* be the closure of A, as a subspace of B, in the B-norm; we have

$$A \subset A^* \subset B \subset \bar{B} = Q \subset BV.$$

Then we claim that

$$\|v\| = \|v\|_B \text{ for all } v \in A^*. \tag{4.13}$$

Indeed, when $v \in A$ then (4.13) follows from

$$\|v\| \leqq \|v\|_B \leqq \|v\|_A = \|v\|;$$

[2] See, for example, Royden (1968), p. 116, Proposition 4.

and the case $v \in A^*\backslash A$ follows from the case $v \in A$ by a simple limiting argument.

Now A^* is a closed subspace of B in the B-norm, and, since B is complete in this norm, it follows that A^* also is. But then from (4.13) it follows that A^* is complete in the variation norm as well, and hence it is a closed subspace of BV. Thus

$$A^* = \bar{A}^* \supset \bar{A} = Q \supset B \supset A^*;$$

hence equality holds throughout, and in particular $Q = B = Q^+ - Q^+$. This proves that Q is reproducing. Hence the Q-norm is defined on Q, and from (4.13) and $A^* = B = Q$ we obtain

$$\|v\| = \|v\|_Q \tag{4.14}$$

for all $v \in Q$. This completes the proof of Proposition 4.12.

The remainder of this section will not be heavily used in the sequel; but it is of some interest in its own right.

Proposition 4.15. *Let Q be a closed reproducing subspace of BV; then the variation norm and the internal norm on Q are equivalent, i.e. for some γ,*

$$\|v\|_Q \leqq \gamma\|v\|$$

for all $v \in Q$. Furthermore, if φ is a positive linear operator from Q into BV, then φ is continuous.

Remark. The first sentence of this proposition is a kind of converse to the remark that every internal space is reproducing.

Proof. Since Q is closed, it is complete in the variation norm. Since it is reproducing, $Q = Q^+ - Q^+$; hence by Proposition 4.9, Q is complete in the Q-norm. Then since $\|v\| \leqq \|v\|_B$, it follows[3] from the interior mapping principle that the variation norm and the Q-norm are equivalent.

To prove the second sentence of the proposition, we first show that

(4.16) $(\varphi v)(I)/v(I)$ is bounded for $v \in Q^+$.

If not, then for each n, we can find v_n in Q^+ such that

$$(\varphi v_n)(I)/v_n(I) > n.$$

[3] See, for example, Royden (1968), p. 195, Proposition 11 or, slightly less explicitly, Dunford and Schwartz (1958), Section II.2, p. 55 ff.

Furthermore, we may choose v_n so that $v_n(I) = 1/n^2$. Then $v = \sum_{n=1}^{\infty} v_n \in Q^+$; but from the positivity of φ it then follows that

$$(\varphi v)(I) \geqq \varphi\left(\sum_{n=1}^{k} v_n\right)(I) > \sum_{n=1}^{k} 1/n$$

for all k. Hence $(\varphi v)(I)$ is greater than any number—a contradiction. This proves (4.16); denote the bound by K.

To complete the proof of the proposition, we proceed as in Proposition 4.7. Specifically, for $v \in Q$, let u and w in Q^+ be such that $v = u - w$ and $2\|v\|_Q \geqq u(I) + w(I)$. Then

$$\begin{aligned} \|\varphi v\| &= \|\varphi u - \varphi w\| \leqq \|\varphi u\| + \|\varphi w\| \\ &= (\varphi u)(I) + (\varphi w)(I) \leqq K(u(I) + w(I)) \\ &\leqq 2K\|v\|_Q \leqq 2K\gamma\|v\|, \end{aligned}$$

by (4.16) and the first sentence of the proposition. This completes the proof of Proposition 4.15.

In connection with values, the chief consequence of Proposition 4.15 is that, on reproducing spaces, every value is continuous. This is interesting because the positivity condition in the definition of value has content only to the extent that Q is reproducing. Thus the positivity condition has no direct bearing on set functions v in Q that are not the difference of monotonic functions in Q; and if Q has no non-trivial reproducing subspaces—i.e. $Q^+ = \{0\}$—then the positivity condition is vacuous. Thus Proposition 4.15 can be interpreted to mean that whenever the positivity condition is fully effective, then it implies continuity.

Continuity of the value is the chief tool in proving a number of uniqueness theorems, for example the uniqueness of the value on *pNA* (or *bv′NA*); the importance of knowing that a space is reproducing is therefore evident. But the condition is an elusive one because often we can establish it only via Proposition 4.12—i.e. by first proving internality; and, once we know the latter, we no longer need the reproducingness and can proceed directly (cf. the end of Section 7). Under certain conditions one can show directly (i.e. without considering internal norms) that a space is in a sense "almost reproducing"; this will be discussed in Appendix B. Unfortunately, we have not succeeded in deducing from this notion of "almost reproducing" that the value must be continuous.

A good deal of this section has been devoted to finding conditions under which the value is continuous—in the variation norm. In Section 3, we saw that on the spaces that are of particular interest in this paper, such as $bv'NA$, the value is indeed continuous—again in the variation norm. It is therefore of some interest to note that the variation norm is crucial here; that, for example, the value is not continuous in the supremum norm: $\|v\|' = \sup_S |v(S)|$. Indeed, let $\{g_n(x)\}$ be a sequence of polynomials with $(dg_n/dx)(0) = 1$ for all n, and such that $g_n(x) \to 0$ as $n \to \infty$, uniformly for $|x| \leqq 1$. For example, take $g_n(x) = x(1 - x^2)^n$. Suppose that $I = [-1, 1]$, that

$$\mu_1(S) = \lambda(S \cap [-1, 0])$$
$$\mu_2(S) = \lambda(S \cap [0, 1]),$$

and that

$$v_n(S) = g_n(\mu_1(S) - \mu_2(S)).$$

Then $\|v_n\|' = \max_{|x| \leqq 1} |g_n(x)| \to 0$ as $n \to \infty$; but from formula (3.1) applied to $f(x_1, x_2) = g_n(x_1 - x_2)$ we deduce $\varphi(v_n) = \mu_1 - \mu_2$, and hence

$$\|\varphi(v_n)\|' = \sup|\mu_1(S) - \mu_2(S)| = 1$$

for all n, so $\|\varphi(v_n)\|' \not\to 0$.

It might be thought that the above phenomenon is a consequence of our indirectly building the variation into our value notion, via the positivity condition. But this is not the case. Indeed, on the space of all linear combinations of powers of measures in NA^+—of which the above v_n are members (see Lemma 7.2)—the value is uniquely determined *even without the positivity condition.*[4] Thus, at least for this rather restricted space, considerations of positivity and the variation norm arise naturally out of the other conditions for the value.

NOTES

1. Together, Propositions 4.7 and 4.15 demonstrate a close relationship between continuity and positivity. Such relationships have been known for

[4] The value on this space is determined by Proposition 6.1, whose proof makes no use of the positivity condition.

some time; an example is the following theorem of Bakhtin, Krasnoselskii, and Stetzenko (1962):[5]

> Let E_1 and E_2 be two Banach spaces, and let K_1 and K_2 be cones in E_1 and E_2 respectively (i.e. if x, $y \geqq K_i$ and α, β are non-negative real numbers, then $\alpha x + \beta y \in K_i$). Assume
>
> (i) K_1 and K_2 are closed;
> (ii) $E_1 = K_1 - K_1$;
> (iii) $x, y \in K_2$ implies $\|x\| \leqq \|x + y\|$.
>
> Let ψ be an operator from E_1 to E_2 such that $\psi K_1 \subset K_2$. Then ψ is continuous.

Note that Proposition 4.15 is a consequence of this theorem. Indeed, for the first sentence of the proposition, we may take E_1 to be Q with the variation norm, E_2 to be Q with the internal norm, ψ to be the identity, and $K_1 = K_2 = Q^+$. For the second sentence, take $E_1 = Q$, $E_2 = FA$ (both with the variation norm), $K_i = E_i^+$, and $\psi = \varphi$.

5. The Space AC

If v and w are set functions, then v is said to be *absolutely continuous with respect to* w (written $v \ll w$) if for every $\epsilon > 0$ there is a $\delta > 0$ such that for every chain Ω and every subchain Λ of Ω,

$$\|w\|_\Lambda \leqq \delta \Rightarrow \|v\|_\Lambda \leqq \epsilon. \tag{5.1}$$

Note that the relation $\ll$ is transitive and that, if v and w are measures, it coincides with the usual notion of absolute continuity.

A set function v is said to be *absolutely continuous* if there is a measure $\mu \in NA^+$ such that $v \ll \mu$. The set of all absolutely continuous set functions in BV is denoted AC.

PROPOSITION 5.2. *AC is a closed subspace of BV.*

Proof. AC is easily seen to be a linear space. To prove $AC \subset BV$, let $v \ll \mu$, $\mu \in NA^+$, and let δ correspond to $\epsilon = 1$ in accordance with (5.1). Given a chain Ω, we claim that $\|v\|_\Omega \leqq (2\mu(I)/\delta) + 1$. Indeed, because of the non-atomicity of μ we may assume w.l.o.g. that

$$|\mu(S_i) - \mu(S_{i-1})| < \tfrac{1}{2}\delta$$

for all i. We may then partition Ω into subchains Λ, for example by taking sets of consecutive links, such that for all but the last subchain,

[5] See also Krasnoselskii (1964), the footnote on p. 64.

we have

$$\tfrac{1}{2}\delta \leqq \|\mu\|_\Lambda < \delta,$$

and for the last subchain we have

$$0 < \|\mu\|_\Lambda < \delta.$$

Since the variation of v over each such subchain is at most 1, and the number of such subchains is at most one more than $2\mu(I)/\delta$, our claim is proved. Hence $v \in BV$.

It remains to show that AC is closed (i.e. as a subspace of BV). Let $\|v_i - v\| \to 0$, where $v_i \ll \mu_i$ and $\mu_i \in NA^+$. W.l.o.g. assume $\mu_i(I) = 1$, all i, and set

$$\mu = \sum_{i=1}^{\infty} (\tfrac{1}{2})^i \mu_i.$$

Then $\mu_i \ll \mu$ for all i, and hence $v_i \ll \mu$ for all i. Now for given ϵ, let v_i be such that $\|v_i - v\| < \frac{1}{2}\epsilon$, and let δ be such that for any subchain Λ,

$$\|\mu\|_\Lambda < \delta \Rightarrow \|v_i\|_\Lambda < \tfrac{1}{2}\epsilon.$$

Then $\|\mu\|_\Lambda < \delta$ implies

$$\|v\|_\Lambda \leqq \|v_i\|_\Lambda + \|v - v_i\|_\Lambda < \tfrac{1}{2}\epsilon + \tfrac{1}{2}\epsilon = \epsilon.$$

Thus $v \ll \mu$, and so $v \in AC$. This completes the proof of the proposition.

Corollary 5.3. *$pNA \subset AC$.*

Proof. Clearly AC contains all powers of non-atomic measures; the corollary then follows from Proposition 5.2.

The next lemma is a simple consequence of Lyapunov's theorem (Proposition 1.3).

Lemma 5.4. *Let $\mu = (\mu_1, \ldots, \mu_n)$ be a vector of measures in NA, and let S^1 and S^0 in $\mathcal{C}$ be such that $S^1 \supset S^0$. Then we may construct a family of sets $\{S^\alpha: 0 \leqq \alpha \leqq 1\} \subset \mathcal{C}$, in such a way that*

$$\mu(S^\alpha) = \alpha\mu(S^1) + (1 - \alpha)\mu(S^0),$$

and that $\alpha > \beta$ implies $S^\alpha \supset S^\beta$.

Proof. Apply Lyapunov's theorem to the measure space consisting of $S^1 \backslash S^0$ and its measurable subsets to obtain a set $S^{1/2}$ such that $S^1 \supset$

$S^{1/2} \supset S^0$ and

$$\mu(S^{1/2}) = \tfrac{1}{2}\mu(S^1) + \tfrac{1}{2}\mu(S^0).$$

Then apply the theorem again to $S^1 \backslash S^{1/2}$ and $S^{1/2} \backslash S^0$. Continuing in this way, we may define S^α for each dyadic rational α, in a way that satisfies the conditions of the lemma. Now if β is an arbitrary member of $[0, 1]$, define

$$S^\beta = \bigcup_{\alpha \leqq \beta} S^\alpha,$$

the union being extended over all α that are dyadic rationals. This completes the proof of the lemma.

We are now ready for the

Proof of Theorem C (see Section 3). First suppose f is absolutely continuous. Then $f(x) = \int_0^x g(t)\, dt$ for all x in the range R of μ, where $g \in L^1 = L^1(R)$. It is well known that g can be approximated in L^1 by polynomials; this follows from the fact that it can be approximated by step functions, which can be approximated by continuous functions, which can be approximated by polynomials. If q is a polynomial with $\int_R |q(t) - g(t)|\, dt < \epsilon$, and if $p(x) = \int_0^x q(t)\, dt$, then the variation of $p - f$ is $< \epsilon$. Hence $f(x)$ can be approximated in variation by polynomials in x, and so from the non-negativity of μ it follows that v can be approximated in variation by the corresponding polynomials in μ. Hence $v \in pNA$.

To complete the proof, let $v = f \circ \mu \in pNA$. By Corollary 5.3, $f \circ \mu \in AC$, so there is a $\nu \in NA^+$ such that $f \circ \mu \ll \nu$. Without loss of generality, we may assume that $\mu(I) = \nu(I) = 1$. Applying Lemma 5.4 to the vector measure (μ, ν), we may assign to each $\alpha \in [0, 1]$ a set S^α in such a way that $S^0 = \emptyset$, $S^1 = I$, $\alpha > \beta$ implies $S^\alpha \supset S^\beta$, and $\mu(S^\alpha) = \nu(S^\alpha) = \alpha$. If we now apply the definition of $f \circ \mu \ll \nu$ to chains of sets of the form S^α, we deduce that for any finite union of disjoint intervals in $[0, 1]$, the sum of the variations of f over these intervals tends to 0 as their total length tends to 0; this is precisely one of the definitions of absolute continuity of f. This completes the proof of Theorem C.

If $\mu \in NA$ is permitted to take negative as well as positive values, the forward implication in Theorem C remains true, i.e. $f \circ \mu \in pNA$ implies the absolute continuity of f. Indeed, let $I = I^+ \cup I^-$ be a Hahn decomposition of I, that is, μ is non-negative on subsets of I^+ and non-positive on subsets of I^- (Halmos, 1950, p. 121). Clearly the range R of μ is $[\mu(I^-), \mu(I^+)]$. Now apply the theorem to the underlying spaces I^+ and I^- separately; it follows that f is absolutely continuous on $[\mu(I^-), 0]$ and on $[0, \mu(I^+)]$, and hence on all of R.

However, the converse is no longer true if μ is permitted to be in $NA \backslash NA^+$. In fact, absolutely continuous functions f exist for which $f \circ \mu \notin BV$, and, a fortiori, $f \circ \mu \notin pNA$; they also exist for which $f \circ \mu$ *is* in BV but is still not in pNA. This situation will be thoroughly explored in Section 9.

6. The Value of Scalar Measure Games

It is convenient to introduce the term *normalized bv function* for a function f in bv such that $f(1) = 1$.

PROPOSITION 6.1. *Let Q be a symmetric subspace of BV, and let φ be a value on Q. If $\mu \in NA^1$ and f is a normalized bv function such that $f \circ \mu \in Q$, then*

$$\varphi(f \circ \mu) = \mu.$$

For the proof we need the following

LEMMA 6.2. *If μ is an NA^1 measure on the measurable space* $([0, 1], \mathcal{B})$, *then there is an automorphism Φ of* $([0, 1], \mathcal{B})$ *such that*

$$\Phi_* \mu = \lambda,$$

where λ is Lebesgue measure.

Proof. This lemma is surely well known, but we did not find an explicit reference. It follows immediately from the much stronger Lemma C.6 (Appendix C), to which the reader is referred for a complete—if slightly roundabout—proof. A more direct proof can be based on the function from $[0, 1]$ onto itself that takes x to $\mu([0, x])$; after a little doctoring, the inverse of this function fulfills the requirements for Φ. This completes what we have to say about the proof of the lemma.

Proof of Proposition 6.1. We assume that $(I, \mathcal{C}) = ([0, 1], \mathcal{B})$, as we may do without loss of generality by the Standardness Assumption

(2.1). Consider first the case in which μ is Lebesgue measure λ. If S_1 and S_2 are any two sets of equal λ-measure, let Θ be a λ-measure preserving automorphism of $(I, \mathcal{C})$ such that the symmetric difference $(\Theta S_1 \backslash S_2) \cup (S_2 \backslash \Theta S_1)$ is of λ-measure 0 (see Halmos, 1956, p. 74). Let $v = f \circ \mu = f \circ \lambda$. Then any coalition of λ-measure 0 is a null coalition, and hence, by Note 4 to Section 2, has value 0. Hence $(\varphi v)(\Theta S_1) = (\varphi v)(S_2)$. But since Θ is measure preserving, we have $\Theta_* v = v$. Hence by the symmetry condition (2.2), we obtain

$$(\varphi v)(S_2) = (\varphi v)(\Theta S_1) = (\Theta_* \varphi v)(S_1) = (\varphi \Theta_* v)(S_1) = (\varphi v)(S_1).$$

Hence, φv coincides on any two sets of equal λ-measure, i.e. $(\varphi v)(S)$ is a function of $\lambda(S)$ only. If we write $(\varphi v)(S) = g(\lambda(S))$, then from $\varphi v \in FA$ it follows that g is bounded and is additive on $[0, 1]$, i.e. $g(x_1 + x_2) = g(x_1) + g(x_2)$ whenever x_1, x_2, and $x_1 + x_2$ are in $[0, 1]$. Hence, a well-known result yields that $g(x) = g(1)x$. Since $g(1) = (\varphi v)(I) = v(I) = f(\lambda(I)) = f(1) = 1$, it follows that $(\varphi v)(S) = g(\lambda(S)) = \lambda(S)$; i.e. $\varphi v = \lambda$.

In the general case, when $\mu \neq \lambda$, let Φ be the automorphism of Lemma 6.2. Then $\Phi_*(f \circ \mu) = f \circ (\Phi_* \mu) = f \circ \lambda$, and so, by what we have just proved, $\varphi \Phi_*(f \circ \mu) = \varphi(f \circ \lambda) = \lambda$. Hence by (2.2) and Lemma 6.2,

$$(6.3) \qquad \varphi(f \circ \mu) = \Phi_*^{-1} \Phi_* \varphi(f \circ \mu) = \Phi_*^{-1} \varphi \Phi_*(f \circ \mu) = \Phi_*^{-1} \lambda = \mu,$$

and the proof of the proposition is complete.

The following is a converse to Proposition 6.1:

PROPOSITION 6.4. *Let Q be a closed symmetric subspace of BV. Let F be a set of normalized bv functions, such that Q is spanned by the set of all games $f \circ \mu$, where $f \in F$ and $\mu \in NA^1$. Let φ be a linear mapping from Q into FA with $\|\varphi\| \leq 1$, such that $\varphi(f \circ \mu) = \mu$ whenever f and μ are as in the previous sentence. Then φ is a value on Q.*

Proof. Let $v = f \circ \mu$, where f and μ are as stated. Then for each Θ in the group $\mathcal{G}$ of automorphisms, $\Theta_* \mu \in NA^1$; so from the hypothesis it follows that

$$\varphi \Theta_* v = \varphi(f \circ \Theta_* \mu) = \Theta_* \mu = \Theta_* \varphi v.$$

Since both φ and Θ_* are continuous, it follows that $\varphi \Theta_* - \Theta_* \varphi$ is a continuous linear operator on Q that vanishes on a spanning subset. Therefore it vanishes on all of Q, establishing condition (2.2). Similarly, the mapping that takes v to $(\varphi v)(I) - v(I)$ is a continuous linear functional on Q that vanishes on a spanning subset, and so on all of Q, thus establishing condition (2.3). Finally, from Proposition 4.6 it follows that φ is positive. This completes the proof.

NOTES

1. B. Weiss has suggested a different version of the proof of Proposition 6.1, in which one does not start out by assuming that $\mu = \lambda$. Rather, when $0 < \mu(S_1) = \mu(S_2) < 1$, one uses Lemma 6.2 and Proposition 1.1 to construct directly a μ-measure preserving isomorphism from S_1 onto S_2 and from $I\backslash S_1$ onto $I\backslash S_2$. The automorphism Θ is then constructed by combining these two isomorphisms. (When $\mu(S_1) = \mu(S_2) = 0$ or $\mu(S_1) = \mu(S_2) = 1$, one uses Note 4 to Section 2 to show that $(\varphi v)(S_2) = (\varphi v)(S_1)$.) The rest of the proof is as above, except that there is no need for considering the "general case," in which $\mu \neq \lambda$.

2. The definition of value may be extended to non-symmetric subspaces of BV as follows: Let Q be a subspace of BV. A *value* on Q is a positive linear mapping φ from Q into FA satisfying the following conditions:

(6.5) If $v \in Q$ and $\Theta \in \mathcal{G}$ are such that $\Theta_* v \in Q$, then $\varphi\Theta_* v = \Theta_* \varphi v$;

(6.6) $$(\varphi v)(I) = v(I) \text{ for all } v \in Q.$$

We then assert that *Proposition 6.1 holds as it stands if the word "symmetric" is omitted.*

To prove this, assume, as in the proof of Proposition 6.1, that $(I, \mathcal{C}) = ([0, 1], \mathcal{B})$. If μ is Lebesgue measure λ, the proof proceeds as in that of Proposition 6.1. Otherwise, use Lemma 6.2 to obtain an automorphism Φ of the player space such that $\Phi_*\mu = \lambda$, which induces a positive linear one-one operator Φ_* of BV onto itself. Define a positive linear operator φ' from $\Phi_* Q$ into FA by

$$\varphi' = \Phi_* \varphi \Phi_*^{-1}.$$

We claim that φ' is a value on $\Phi_* Q$. Indeed, let Θ in $\mathcal{G}$ and w in $\Phi_* Q$ be such that $\Theta_* w \in \Phi_* Q$. Then if $v = \Phi_*^{-1} w$, then $v \in Q$ and

$$(\Phi^{-1}\Theta\Phi)_* v = \Phi_*^{-1}\Theta_*\Phi_*\Phi_*^{-1} w \in Q.$$

Hence because φ is a value, we have

$$\begin{aligned}\varphi'\Theta_* w &= \Phi_*\varphi\Phi_*^{-1}\Theta_*\Phi_*\Phi_*^{-1} w = \Phi_*\varphi(\Phi^{-1}\Theta\Phi)_* v \\ &= \Phi_*(\Phi^{-1}\Theta\Phi)_*\varphi v = \Phi_*\Phi_*^{-1}\Theta_*\Phi_*\varphi\Phi_*^{-1} w = \Theta_*\varphi' w,\end{aligned}$$

proving (6.5); the proof of (6.6) is trivial.

Let $v = f \circ \mu$. Then $v \in Q$ and so $\Phi_* v \in \Phi_* Q$. Now

$$(\Phi_* v)(S) = v(\Phi S) = f(\mu(\Phi S)) = f(\lambda(S)) = (f \circ \lambda)(S),$$

so $f \circ \lambda \in \Phi_* Q$. Since φ' is a value on $\Phi_* Q$, we may deduce from what we have already shown that $\varphi'(\Phi_* v) = \varphi'(f \circ \lambda) = \lambda$. Hence $\varphi v = \Phi_*^{-1} \varphi' \Phi_* v = \Phi_*^{-1} \lambda$; so

$$(\varphi v)(S) = (\Phi_*^{-1} \lambda)(S) = \lambda(\Phi^{-1} S) = \mu(\Phi \Phi^{-1} S) = \mu(S),$$

and the proof of our assertion is complete.

If one restricts oneself to subspaces of $bv'NA$, then Proposition 6.4 can also be generalized to the non-symmetric case.

7. The Value on pNA

In this section we shall prove Theorem B.

PROPOSITION 7.1. *Let* $\mu = (\mu_1, \ldots, \mu_n)$ *be a vector of measures in* NA, *and let* f *with* $f(0) = 0$ *be continuously differentiable on the range* R of μ. *Then* $f \circ \mu \in pNA$.

The proof of the proposition is accomplished in several stages. First, let f be a polynomial in n variables.

LEMMA 7.2. *Let* $\mu = (\mu_1, \ldots, \mu_n)$ *be a vector of measures in* NA, *and let* f *be a polynomial in* n *variables with* $f(0) = 0$. *Then* $f \circ \mu$ *is a linear combination of positive integer powers of* NA^+ *measures, and so in particular is in* pNA.

Proof. We first prove the formula

$$(7.3) \quad k! x_1 \cdots x_k = (x_1 + \cdots + x_k)^k - \sum_{1 \leq i \leq k} (x_1 + \cdots + x_k - x_i)^k + \sum_{1 \leq i < j \leq k} (x_1 + \cdots + x_k - (x_i + x_j))^k - + \cdots.$$

Indeed, the right side vanishes when $x_1 = 0$, and so it is divisible by x_1; similarly it is divisible by $x_2, \ldots, x_k$. Therefore, since it is of degree k, it must be a multiple of $x_1 \cdots x_k$. But the term $x_1 \cdots x_k$ only arises from the first term on the right side, and there its coefficient is $k!$, so (7.3) is proved.

From (7.3) it follows that every polynomial in n variables is a linear combination of powers of partial sums of the variables. Since every NA measure is the difference of two NA^+ measures, every polynomial in

NA measures is a polynomial in NA^+ measures. Hence every polynomial in NA measures is a linear combination of powers of NA^+ measures. This completes the proof of Lemma 7.2.

The proof of Proposition 7.1 proceeds by an argument involving approximation to a general f by polynomials. First, note that we may assume that R is of full dimension, i.e. that it has an interior in E^n. If not, let m be the true dimension of R; then because $(0, \ldots, 0) \in R$, we may find a linear transformation Ψ of E^n onto E^m which is $1-1$ on R; let $\Theta = \Psi|R$. Now define a vector measure ξ of dimension m by $\xi = \Theta_* \mu$, a function g on ΘR by $g(x) = f(\Theta^{-1}x)$, and a set function w by $w(S) = g(\xi(S))$; then $w = v$. Since the range of ξ is ΘR, which is of full dimension in E^m, the reduction is complete.

Now for continuous functions f on R, define

$$\|f\|_0 = \max_{x \in R} |f(x)|.$$

Let $C^1 = C^1(R)$ be the Banach space of continuously differentiable functions on R, with the norm

$$\|f\|_1 = \|f\|_0 + \sum_i \|f_i\|_0,$$

where $f_i = \partial f / \partial x_i$ in the interior of R, and is the appropriate continuous extension on the boundary.

LEMMA 7.4. *The polynomials are dense in* C^1.

Proof. This is proved in Courant and Hilbert (1953, p. 68) for the case in which R is an n-dimensional cube. The general case may be reduced to this one by imbedding R in a cube R' and extending f and its derivatives in an arbitrary way from R to R', so that the extended function is in $C^1(R')$. The existence of such an extension is well known (see, for example, Whitney, 1934).

To complete the proof of Proposition 7.1, let $\mu_i = \xi_i - \zeta_i$, where ξ_i and ζ_i are non-negative measures. Let g be any member of C^1, and let g' be the vector of its partial derivatives. Let $\emptyset = S_0 \subset S_1 \subset \cdots$ be a chain such that all the $\mu(S_i)$, except possibly the first and the last, are in the interior of R. Then by the mean value theorem,

$$\begin{aligned}\sum_j |g(\mu(S_{j+1})) - g(\mu(S_j))| &= \sum_j |\mu(S_{j+1}\backslash S_j)\cdot g'(\mu(S_j) + \theta_j\mu(S_{j+1}\backslash S_j))| \\ &\leqq \sum_j \sum_{i=1}^n |\mu_i(S_{j+1}\backslash S_j)|\ \|g\|_1 \\ &\leqq \|g\|_1 \sum_{i=1}^n [\xi_i(I) + \zeta_i(I)],\end{aligned}$$

where $0 \leqq \theta_j \leqq 1$. Hence if u is defined by $u(S) = g(\mu(S))$, then

$$\|u\| \leqq \|g\|_1 \sum_{i=1}^n [\xi_i(I) + \zeta_i(I)], \tag{7.5}$$

since the demand that $\mu(S_i)$ be in the interior of R cannot decrease the supremum of the expression defining the norm.

Now approximate to f in C^1 by a polynomial p, and let $w(S) = p(\mu(S))$. Then from (7.5) it follows that

$$\|w - v\| \leqq \|p - f\|_1 \sum_{i=1}^n [\xi_i(I) + \zeta_i(I)].$$

Since the right side can be made arbitrarily small, so can the left side, and the proof of Proposition 7.1 is complete.

PROPOSITION 7.6. *There is a value φ on pNA, with $\|\varphi\| = 1$. Furthermore, let v in pNA be such that there exist μ, f, and U as follows:*

(7.7) *μ is a vector of non-atomic measures with range R, f is a real-valued function defined on R and continuously differentiable there, U is a compact convex neighborhood in R of the diagonal* $[0, \mu(I)]$,

and

$$v(S) = f(\mu(S)) \text{ whenever } \mu(S) \in U.$$

Then

$$\varphi(v)(S) = \int_0^1 f_{\mu(S)}(t\mu(I))\, dt,$$

where $f_{\mu(S)}$ is the derivative of f in the direction $\mu(S)$.

Proof. Let v, μ, f, and U be as above. Define a signed measure ν by

$$\nu(S) = \int_0^1 f_{\mu(S)}(t\mu(I))\,dt.$$

Probably the easiest way to verify the countable additivity of ν is first to note that, when R has full dimension, countable additivity follows at once from the explicit formula in terms of partial derivatives (3.2), and then to reduce the general case to that of full dimension by arguments similar to those following (3.2); indeed, for g and ξ as defined in the proof of Proposition 7.1 it is easily verified that

$$\int_0^1 g_{\xi(S)}(t\xi(I))\,dt = \int_0^1 f_{\mu(S)}(t\mu(I))\,dt.$$

Let $I = S^+ \cup S^-$ be a Hahn decomposition (Halmos, 1950, p. 121) of I with respect to ν; that is, ν is non-negative on S^+ and its subsets, nonpositive on S^- and its subsets, and $S^+ \cap S^- = \emptyset$. Then

$$\|\nu\| = |\nu(S^+)| + |\nu(S^-)|.$$

Let m be an arbitrary positive integer. Because of Lyapunov's theorem, it is possible to partition S^+ into disjoint sets $S_1^+, \ldots, S_m^+$ such that $\mu(S_j^+) = \mu(S^+)/m$ for all j, and similarly to partition S^- into disjoint sets $S_1^-, \ldots, S_m^-$ such that $\mu(S_j^-) = \mu(S^-)/m$ for all j. Now define a chain (see Fig. 2)

$$S_0 \subset S_1 \subset \cdots \subset S_{2m}$$

by $S_0 = \emptyset$ and

$$S_{2j} = (S_1^+ \cup S_1^-) \cup \cdots \cup (S_j^+ \cup S_j^-),$$
$$S_{2j+1} = S_{2j} \cup S_{j+1}^+.$$

Let $y = \mu(S^+)$ and $b = \mu(I)$, so $\mu(S^-) = b - y$. Then

$$\mu(S_{2j}) = \frac{jb}{m},$$
$$\mu(S_{2j+1}) = \frac{jb + y}{m},$$

and hence we may choose m sufficiently large so that $\mu(S_l) \in U$ for all l. Hence

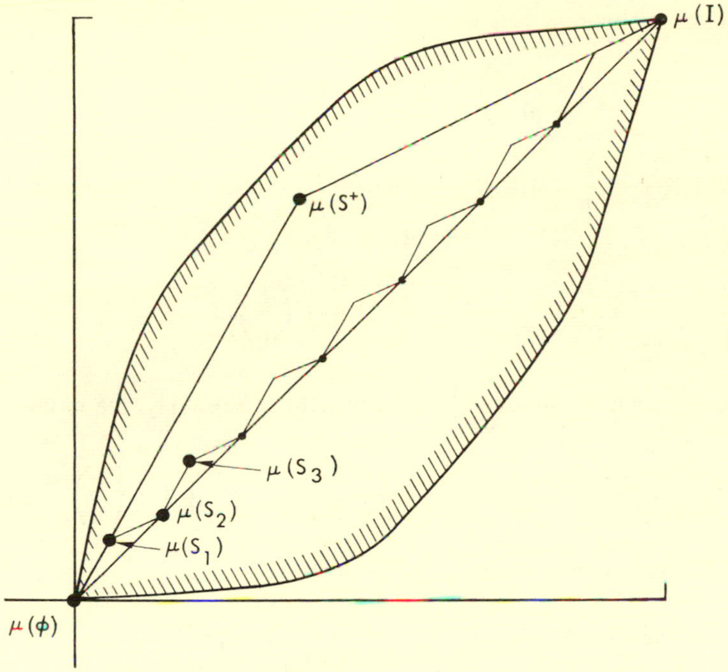

Figure 2. Construction of a chain near the diagonal

$$
\begin{aligned}
(7.8)\quad \|v\| &\geqq \sum_{l=1}^{2m} |f(\mu(S_l)) - f(\mu(S_{l-1}))| \\
&= \sum_{j=0}^{m-1} \left| f\left(\frac{jb+y}{m}\right) - f\left(\frac{jb}{m}\right) \right| \\
&\qquad + \sum_{j=0}^{m-1} \left| f\left(\frac{(j+1)b}{m}\right) - f\left(\frac{jb+y}{m}\right) \right| \\
&\geqq \left| \sum_{j=0}^{m-1} \left\{ f\left(\frac{jb+y}{m}\right) - f\left(\frac{jb}{m}\right) \right\} \right| \\
&\qquad + \left| \sum_{j=0}^{m-1} \left\{ f\left(\frac{(j+1)b}{m}\right) - f\left(\frac{jb+y}{m}\right) \right\} \right|.
\end{aligned}
$$

If we look at $f\left(\frac{jb}{m} + \theta y\right)$ as a function of θ, then an application of the mean value theorem yields

$$(7.9) \qquad f\left(\frac{jb+y}{m}\right) - f\left(\frac{jb}{m}\right) = \frac{1}{m} f_y\left(\frac{jb}{m} + \tau y\right)$$

where $0 \leqq \tau \leqq \frac{1}{m}$. Further, condition (7.7) implies that f_y is uniformly continuous in R; so the right side of (7.9) is

$$= \frac{1}{m} f_y\left(\frac{jb}{m}\right) + o\left(\frac{1}{m}\right)$$

as $m \to \infty$, where the $o\left(\frac{1}{m}\right)$ is uniform in j. Similarly, we have

$$f\left(\frac{(j+1)b}{m}\right) - f\left(\frac{jb+y}{m}\right) = \frac{1}{m} f_{b-y}\left(\frac{jb}{m}\right) + o\left(\frac{1}{m}\right).$$

Applying these remarks to (7.8), we get

$$\|v\| \geqq \left| \sum_{j=0}^{m-1} \frac{1}{m} f_y\left(\frac{jb}{m}\right) \right| + \left| \sum_{j=0}^{m-1} \frac{1}{m} f_{b-y}\left(\frac{jb}{m}\right) \right| + o(1).$$

Note that the sums are approximating sums to a Riemann integral; and since f_y is continuous near—and on—the diagonal, this integral exists. Hence by going to the limit we obtain

$$(7.10) \qquad \begin{aligned} \|v\| &\geqq \left| \int_0^1 f_y(tb)\,dt \right| + \left| \int_0^1 f_{b-y}(tb)\,dt \right| \\ &= |\nu(S^+)| + |\nu(S^-)| = \|\nu\|. \end{aligned}$$

Now define

$$\varphi v = \nu.$$

We shall prove that this is an admissible definition, i.e. that it does not depend on the choice of μ, f, and U in (7.7). Indeed, suppose ξ, g, and V satisfy (7.7), and that $v(S) = g(\xi(S))$ whenever $\xi(S) \in V$. Define a set function w by $w(S) = 0$ for all S, a vector measure ζ by $\zeta = (\mu, \xi)$, a function h by $h(x, y) = f(x) - g(y)$, and a compact convex neighborhood W of the diagonal $[0, \xi(I)]$ by $W = U \times V$. Then ζ, h, and W satisfy condition (7.7), and whenever $\zeta(S) \in W$, we have

$$h(\zeta(S)) = f(\mu(S)) - g(\xi(S)) = v(S) - v(S) = 0 = w(S).$$

Let

$$\eta(S) = \int_0^1 g_{\xi(S)}(t\xi(I))\,dt$$
$$\sigma(S) = \int_0^1 h_{\zeta(S)}(t\zeta(I))\,dt.$$

It is then easily verified that $\sigma = \nu - \eta$. Applying (7.10) to w, we deduce

$$\|\nu - \eta\| = \|\sigma\| \leqq \|w\| = \|0\| = 0.$$

So $\nu - \eta = 0$, i.e. $\nu = \eta$. This proves that φ is well-defined.

Let Q be the set of set functions v in pNA satisfying the second sentence of our proposition. Q contains all the linear combinations of powers of measures, so it is dense in pNA. We have already defined φ on Q; it is easily verified that Q is a linear subspace of pNA, and that φ is linear on Q. From (7.10) it follows that $\|\varphi v\| \leqq \|v\|$ for $v \in Q$, so φ is continuous on Q and $\|\varphi\| \leqq 1$; from $\varphi\lambda = \lambda$ we deduce $\|\varphi\| \geqq 1$, so $\|\varphi\| = 1$. From the completeness of FA, it then follows that φ can be uniquely extended to be a continuous linear operator from pNA to FA, and this extension will also have norm 1.

It remains to verify that φ is indeed a value, and for this we will use Proposition 6.4. Thus let $v = \mu^k$, where $\mu \in NA^1$. Then, using formula (3.2), we get

$$\begin{aligned}(\varphi v)(S) &= \mu(S) \int_0^1 \left(\frac{d}{dt}\,t^k\right) dt\\ &= \mu(S)[1^k - 0^k]\\ &= \mu(S),\end{aligned}$$

and the proof of Proposition 7.6 is complete.

Proposition 7.11. *The value on pNA is unique.*

Before proving this proposition, we require a number of preliminaries.

Let $[0, 1]^n$ denote the unit cube in dimension n. For any function f on $[0, 1]^n$, define a function $\hat{f}$ on $[0, 1]^n$ by

$$\hat{f}(x) = \sup \sum_{j=1}^{m} |f(x^j) - f(x^{j-1})|, \tag{7.12}$$

where the supremum is taken over all finite sequences

$$0 = x^0 \leqq x^1 \leqq \cdots \leqq x^m = x. \tag{7.13}$$

If $\hat{f}(x)$ is finite for all $x \in [0, 1]^n$—or equivalently, if $\hat{f}(e)$ is finite—then we shall say that f is of *bounded variation.*

LEMMA 7.14. *Let f be a continuously differentiable real valued function on $[0, 1]^n$. Then f is of bounded variation, and $\hat{f}(x)$ is continuous in x.*

Proof. From the fact that $f \in C^1([0, 1]^n)$, it follows that f satisfies a uniform Lipschitz condition and hence that it is of bounded variation.

Next, we prove that $\hat{f}$ is continuous. W.l.o.g., assume that in the sequence (7.13) over which the sup in (7.12) is taken, each x^j differs from x^{j-1} in exactly one coordinate (which may, of course, depend on j), and that $\|x^j - x^{j-1}\|$ is uniformly small—say $\|x^j - x^{j-1}\| < \gamma$, where γ will be determined. Such sequences will be called *admissible.*

We must show that if $\|y - x\|$ is small, so is $|\hat{f}(y) - \hat{f}(x)|$. W.l.o.g., let y differ from x in one coordinate only, say the first, and let $y_1 > x_1$. Assume that

$$y_1 - x_1 = \|y - x\| < \delta,$$

where δ will be determined later. For each admissible sequence $0 = y^0 \leqq y^1 \leqq \cdots \leqq y^m = y$, define an admissible sequence $0 = x^0 \leqq x^1 \leqq \cdots \leqq x^m = x$ by

$$x^j = (\min(x_1, y_1^j), y_2^j, \ldots, y_n^j);$$

then

$$\|y^j - x^j\| < \delta$$

for all j. Consider now a fixed j, and let $i = i(j)$ be such that y^j differs from y^{j-1}, if at all, only in the ith coordinate. Then also x^j differs from x^{j-1}, if at all, only in the ith coordinate; and we have

$$(7.15) \qquad \begin{cases} f(y^j) - f(y^{j-1}) = f_i(\bar{y}^j)(y_i^j - y_i^{j-1}) \\ f(x^j) - f(x^{j-1}) = f_i(\bar{x}^j)(x_i^j - x_i^{j-1}), \end{cases}$$

where $f_i = \partial f/\partial x_i$, $\bar{y}^j$ is on the segment joining y^{j-1} to y^j, and $\bar{x}^j$ is on the segment joining x^{j-1} to x^j. Hence

$$\|\bar{x}^j - \bar{y}^j\| \leqq \|\bar{x}^j - x^j\| + \|x^j - y^j\| + \|y^j - \bar{y}^j\| \leqq \delta + 2\gamma.$$

If $\epsilon > 0$ is given, then from the continuity of the f_i it follows that if δ and γ are sufficiently small, then

$$(7.16) \qquad |f_i(\bar{y}^j) - f_i(\bar{x}^j)| \leqq \frac{\epsilon}{2n}.$$

Now if $i > 1$, then $x_i^j = y_i^j$ and $x_i^{j-1} = y_i^{j-1}$, and so from (7.15) and (7.16) we obtain

$$|f(y^j) - f(y^{j-1})| - |f(x^j) - f(x^{j-1})| \leqq \frac{\epsilon}{2n}(y_i^j - y_i^{j-1}).$$

On the other hand, if $i = 1$, then if $y_1^j \leqq x_1$, we have

$$|f(y^j) - f(y^{j-1})| - |f(x^j) - f(x^{j-1})| = 0;$$

and if $y_1 \geqq y_1^j > x_1$, we have

$$\begin{aligned} |f(y^j) - f(y^{j-1})| - |f(x^j) - f(x^{j-1})| &\leqq |f(y^j) - f(y^{j-1})| \\ &\leqq M(y_1^j - y_1^{j-1}), \end{aligned}$$

where M is the maximum of $|f_1|$ over the whole square. Choosing δ and γ so that $M(\delta + \gamma) \leqq \epsilon/2$, and combining the last three displayed formulas, we obtain

$$\begin{aligned} &\sum_{j=1}^{m} (|f(y^j) - f(y^{j-1})| - |f(x^j) - f(x^{j-1})|) \\ &\qquad = \sum_{i(j)>1} + \sum_{i(j)=1, y_1{}^j \leqq x_1} + \sum_{i(j)=1, y_1 \geqq y_1{}^j > x_1} \\ &\qquad \leqq \frac{\epsilon}{2n} \sum_{j=1}^{m} (y_{i(j)}^j - y_{i(j)}^{j-1}) + 0 + M \sum_{y_1 \geqq y_1{}^j \geqq x_1} (y_1^j - y_1^{j-1}) \\ &\qquad \leqq \frac{\epsilon}{2} + M(y_1 - (x_1 - \gamma)) \leqq \frac{\epsilon}{2} + M(\delta + \gamma) \leqq \frac{\epsilon}{2} + \frac{\epsilon}{2} = \epsilon. \end{aligned}$$

Now the sequences $0 = y^0 \leqq y^1 \leqq \cdots \leqq y^m = y$ may be chosen so that $\Sigma |f(y^j) - f(y^{j-1})|$ is arbitrarily close to $\hat{f}(y)$. Hence it follows that if δ is sufficiently small, then $\hat{f}(x) \geqq \hat{f}(y) - \epsilon$. Since $\hat{f}(x) \leqq \hat{f}(y)$ follows from $x \leqq y$, the proof of Lemma 7.14 is complete.

We now return to the proof of Proposition 7.11. Assume that $(I, \mathcal{C}) = ([0, 1], \mathcal{B})$; by (2.1), this involves no loss of generality. For each $k > 0$

and each m with $1 \leqq m \leqq 2^k$, define a measure λ_m^k by

$$\lambda_m^k(S) = 2^k\lambda\left(S \cap \left[\frac{m-1}{2^k}, \frac{m}{2^k}\right]\right),$$

where λ is Lebesgue measure. Let λ^k be the vector measure, of dimension 2^k, defined by

$$\lambda^k = (\lambda_1^k, \ldots, \lambda_{2^k}^k);$$

the range of λ^k is the closed unit cube $R_k = [0, 1]^{2^k}$. Denote by A the set of all set functions of the form $f \circ \lambda^k$, where $k > 0$ and f is continuously differentiable on R_k.

LEMMA 7.17. *A is a subalgebra*[1] *of BV.*

Proof. Define a linear mapping Ψ_k: $E^{2^k} \to E^{2^{k-1}}$ by

$$\Psi_k(x_1, x_2, x_3, x_4, \ldots) = \left(\frac{x_1 + x_2}{2}, \frac{x_3 + x_4}{2}, \ldots\right).$$

Then Ψ_k takes R_k onto R_{k-1}, and $\Psi_k\lambda^k = \lambda^{k-1}$. It follows that when $k \geqq l$, there is a linear mapping Ψ_{kl}: $E^{2^k} \to E^{2^l}$ that takes R_k onto R_l, such that $\Psi_{kl}\lambda^k = \lambda^l$.

Suppose now that $v, w \in A$; let $v = f \circ \lambda^k$, $w = g \circ \lambda^l$, where $f \in C^1(R_k)$, $g \in C^1(R_l)$, and $k \geqq l$, say. Define a function h on R_k by $h = f + g \circ \Psi_{kl}$. Then $h \in C^1(R_k)$, and

$$v + w = f \circ \lambda^k + g \circ \Psi_{kl} \circ \lambda^k = h \circ \lambda^k \in A.$$

Similarly, it may be proved that $vw \in A$; for this it is necessary only to replace the addition sign in the definition of h by a multiplication sign. Therefore A is an algebra, and the proof of Lemma 7.17 is complete.

LEMMA 7.18. *A is internal.*

Proof. Represent $v \in A$ by $f \circ \lambda^k$, where $f \in C^1(R_k)$. Set $n = 2^k$, $R = R_k = [0, 1]^n$. Define f^+ on R by

$$f^+(x) = \sup \sum_{j=1}^{m} \max(0, f(x^j) - f(x^{j-1})),$$

[1] Linear subspace closed under multiplication.

taking the supremum over all finite sequences of the form (7.13). Define f^- similarly but with $\max(0, f(x^{j-1}) - f(x^j))$ for the summand. Then f^+ and f^- are non-decreasing, and we have

$$f^+ + f^- = \hat{f}$$
$$f^+ - f^- = f;$$

From Lemma 7.14 it then follows that f^+ and f^- are everywhere finite and are continuous. Moreover, we have

$$\|v\| = \hat{f}(e) = f^+(e) + f^-(e),$$

since the sequences (7.13) are exactly the sequences $\{\lambda^k(S_j)\}$ that arise from the chains $\emptyset = S_0 \subset \cdots \subset S_m = I$ that determine the variation norm. (Here the mutual singularity of the components of λ^k is crucial.) We are prevented from asserting similarly that $\|v\|_A = f^+(e) + f^-(e)$ because f^+ and f^- may fail to be differentiable.[2] Our object in the following will be to find suitably differentiable substitutes for f^+ and f^-; i.e. non-decreasing functions h, $k \in C^1(R)$ with $h - k = f$ and $h(0) = k(0) = 0$, and such that $h(e)$ and $k(e)$ are approximately equal to $f^+(e)$ and $f^-(e)$ respectively.[3]

Write f_i for $\partial f/\partial x_i$, and let $D = \max_i \max_x |f_i(x)|$. Fix $\epsilon > 0$, and let $\delta > 0$ be such that $\|x - y\| < \delta$ implies $\max_i |f_i(x) - f(y)| < \epsilon$, for all $x, y \in R$. We shall also require that $\delta < \epsilon/D$.

We now define a linear operator "#" on the continuous functions on R:

$$g^{\#}(x) = \int_{y \in R} g((1 - \delta)x + \delta y)\, dy,$$

or, equivalently,

$$g^{\#}(x) = \frac{1}{\delta^n} \int \cdots \int_{z_i = (1-\delta)x_i}^{(1-\delta)x_i + \delta} \cdots \int g(z)\, dz_1 \cdots dz_n.$$

(Note that the region over which this "moving average" is taken lies wholly within R.) From the second expression for $g^{\#}(x)$ it is apparent that $g^{\#} \in C^1(R)$, even if g is only continuous, and that if $g \in C^1(R)$

[2] A simple example for $n = 2$ is provided by $f(x) = x_1 + x_2 - 2x_1x_2$, in which case $f^+(x) = \max(x_1, x_2, f(x))$.

[3] Whether they can be made exactly equal, for all $f \in C^1([0, 1]^n)$, is an interesting open question. In the above example, we may take $h(x) = x_1 + x_2 - x_1x_2$ and $k(x) = x_1x_2$.

then $(g^{\#})_i = (1 - \delta)g_i^{\#}$. From the first expression it is apparent that if g is non-decreasing then so is $g^{\#}$, and moreover

$$g^{\#}(0) \geqq g(0) \text{ and } g^{\#}(e) \leqq g(e).$$

Finally, since we are averaging a continuous function, for every $x \in R$ there is a $y \in R$ such that $\|x - y\| < \delta$ and $g(y) = g^{\#}(x)$. Applying this last remark to the derivatives of our original function f we obtain

$$|f_i^{\#}(x) - f_i(x)| < \epsilon$$

for all x and i. Hence we have

$$\begin{aligned}\left|\frac{\partial(f^{\#} - f)(x)}{\partial x_i}\right| &\leqq |(f^{\#})_i(x) - f_i^{\#}(x)| + |f_i^{\#}(x) - f_i(x)| \\ &< \delta|f_i^{\#}(x)| + \epsilon \\ &\leqq \delta D + \epsilon \\ &\leqq 2\epsilon,\end{aligned}$$

which tells us that the function $f^{\#} - f + 2\epsilon u$ is non-decreasing, where u is defined by $u(x) \equiv \Sigma_1^n x_i$.

Now define

$$h = f^{+\#} - f^{+\#}(0) + 2\epsilon u.$$

Then $h \in C^1(R)$ and $h(0) = 0$; also h is non-decreasing, being the sum of non-decreasing functions. Next define

$$k = h - f.$$

Clearly, $k \in C^1(R)$ and $k(0) = 0$. Moreover, we can express k as a sum of non-decreasing functions:

$$k = (f^+ - f)^{\#} + (f^{\#} - f + 2\epsilon u) - f^{+\#}(0),$$

using the linearity of "#," so k too is non-decreasing. Thus, $h \circ \mu$ and $k \circ \mu$ are members of A^+, and so we have $\|v\|_A \leqq h(e) + k(e) = 2h(e) - f(e)$. Hence

$$\begin{aligned}\|v\| \leqq \|v\|_A &\leqq 2f^{+\#}(e) - 2f^{+\#}(0) + 4\epsilon u(e) - f(e) \\ &\leqq 2f^+(e) - f^+(0) + 4\epsilon n - f(e) \\ &= f^+(e) + f^-(e) + 4\epsilon n \\ &= \|v\| + 4\epsilon n.\end{aligned}$$

Since this holds for all $\epsilon > 0$, we have $\|v\| = \|v\|_A$, and the proof of Lemma 7.18 is complete.

PROPOSITION 7.19. *pNA is internal.*

Remark. A proposition related to and in a sense generalizing Proposition 7.19 will be proved in Appendix B.

Proof. Let L denote the space of all measures that are absolutely continuous w.r.t. Lebesgue measure λ. We claim that

$$L \subset \bar{A}. \tag{7.20}$$

Indeed, if $\mu \in L$, then by the Radon-Nikodym theorem we may find an integrable function f such that

$$\mu(S) = \int_S f(t)\, dt$$

for all $S \subset I$. Now f can be approximated by step functions in the L^1-norm, and hence by step functions in which each step is a dyadic interval; that is, f can be approximated in the L^1 norm by linear combinations of characteristic functions of dyadic intervals. In other words, we can find such a function g with

$$\int_I |f(t) - g(t)|\, dt < \epsilon.$$

Then if we define a measure ν by

$$\nu(S) = \int_S g(t)\, dt,$$

then $\nu \in A$ and $\|\mu - \nu\| < \epsilon$; thus μ can be approximated in variation by members of A, and this proves (7.20).

By Lemma 7.17, A is an algebra. Hence by Proposition 4.5 and the remark following it, $\bar{A}$ is also an algebra. From (7.20) it therefore follows that $\bar{A}$ contains all polynomials in measures in L, and hence also all limits of such polynomials.

Now let $v \in pNA$. Then v is the limit, in the variation norm, of a sequence $v_1, v_2, \ldots$, each of which is a polynomial in NA^1 measures. Since each v_i involves only finitely many such measures, there are only denumerably many such measures involved in the entire sequence; let them be $\mu_1, \mu_2, \ldots$. If we set

$$\mu = \sum_{i=1}^{\infty} \mu_i/2^i,$$

then $\mu_i \ll \mu$ for all i, and hence each v_i is a polynomial in measures that are $\ll\mu$. By Lemma 6.2, there is an automorphism Φ of $(I, \mathcal{C})$ such that $\Phi_*\mu = \lambda$. The $\Phi_* v_i$ are polynomials in the measures $\Phi_*\mu_i$, all of which are $\ll \Phi_*\mu = \lambda$. Hence $\Phi_* v_i \in \bar{A}$, and since $\|\Phi_* v_i - \Phi_* v\| = \|v_i - v\| \rightarrow 0$, it follows that

$$\Phi_* v \in \bar{A}. \tag{7.21}$$

From Proposition 4.12 and Lemma 7.18, it follows that $\bar{A}$ is internal. From this it follows easily that $\Phi_*^{-1}\bar{A}$ is internal as well, for each automorphism Φ. Hence by (7.21), every member of pNA is a member of some internal subspace of pNA. But from this it follows immediately that pNA is itself internal (if u and w in the definition of $\|v\|$ can be chosen to be in a subspace of pNA, they can a fortiori be chosen in pNA). This completes the proof of Proposition 7.19.

To prove Proposition 7.11, let φ be a value on pNA. Then by Propositions 4.7 and 7.19, φ must be continuous. On the other hand, by Proposition 6.1 we have $\varphi\mu^k = \mu$ for all μ in NA^1, and all positive integers k. This determines φ on a spanning subset of pNA and so, by linearity and continuity, on all of pNA. This completes the proof of Proposition 7.11.

Theorem B follows immediately from Propositions 7.6 and 7.11.

NOTES

1. Formula (7.3) is a folk lemma in algebra. It appears, for example, in Mitiagin, Rolewicz and Zelazko (1962) as Formula (8) on p. 293. There, however, the proof is more involved than the one in the text; the latter was communicated to us by S. Amitsur.

2. Let P be the space of all polynomials in NA measures, which by Lemma 7.2 is the same as the space of linear combinations of powers of NA^1 measures. Thus $pNA = \bar{P}$. As a "classroom note," we might mention that it is much easier to prove the existence of a unique value on P than on pNA.

The uniqueness, which constitutes the major difficulty in the proof of Theorem B, is of course in this case an immediate consequence of Proposition 6.1. But even the existence proof is considerably simpler for P than for

pNA. Indeed, let

$$v = \sum_{i=1}^{k} \alpha_i f_i \circ \mu_i, \tag{7.22}$$

where the α_i are real numbers, $f_i(x)$ is a positive integer power of x, and $\mu_i \in NA^1$. Let $\mu = (\mu_1, \ldots, \mu_k)$. Suppose $S \in \mathcal{C}$ and $0 < t < 1$; set $S' = I \backslash S$. Use Lyapunov's theorem to obtain subsets tS and tS' of S and S', respectively, such that $\mu(tS) = t\mu(S)$ and $\mu(tS') = t\mu(S')$. For $0 < \tau < 1 - t$, use Lyapunov's theorem again to obtain a subset τS of $S \backslash tS$ such that $\mu(\tau S) = \tau\mu(S)$; this is possible because $\mu(S\backslash tS) = (1 - t)\mu(S)$. Denote the set $tS \cup tS'$ by tI; then tI is disjoint from τS, and $\mu(tI) = t\mu(I) = (t, \ldots, t)$. Moreover, it is easily verified that

$$\lim_{\tau\to 0}(v(tI + \tau S) - v(tI))/\tau = \Sigma\alpha_i\mu_i(S)f_i'(t).$$

Integrating, and using

$$\int_0^1 f_i'(t)\, dt = 1,$$

we obtain

$$\int_0^1 (\lim_{\tau\to 0}(v(tI + \tau S) - v(tI))/\tau)\, dt = \Sigma\alpha_i\mu_i(S). \tag{7.23}$$

Now define φ on P by

$$\varphi v = \Sigma\alpha_i\mu_i. \tag{7.24}$$

To prove that this is an admissible definition, we must show that two different representations of the same v in the form (7.22) lead to the same φv. This is equivalent to showing that if the right side of (7.22) vanishes identically, then the right side of (7.24) vanishes identically; but that follows immediately from (7.23). Therefore φ is indeed well defined. Moreover it is easily verified that φ is linear and obeys the conditions of symmetry (2.2) and normalization (2.3). Thus it remains only to show the monotonicity, i.e. that φv is non-negative when v is monotonic; but this, too, follows immediately from (7.23).

3. It does not seem possible to weaken the hypothesis of Lemma 7.14 to any considerable extent and still get the continuity of $\hat{f}$. Already for $n = 1$, it is well known that there are functions (e.g. $x \sin(1/x)$) that are not of bounded variation, but that are continuous on the closed unit interval and continuously differentiable in its interior. There are even such functions (e.g. $x^2 \sin(1/x^2)$) that are differentiable (though not continuously) at the end points as well. When $n = 1$ the continuity of $\hat{f}$ *would* follow if we assume that f is continuous and of bounded variation; but already for $n = 2$ this is not enough. Indeed, we shall now describe a function f that satisfies a uniform Lipschitz condition on the closed unit square $[0, 1]^2$—and is a fortiori continuous and of bounded variation—and that is moreover differentiable on the closed square and continuously differentiable in its interior, but such that $\hat{f}$ is not continuous.[4]

[4] The example used at the beginning of Section 9 (as a counter-example to (9.1)) can be modified so as to satisfy all the above conditions except the Lipschitz condition.

Let g be a non-decreasing continuously differentiable function on [0, 1] such that $g(0) = 0$, $g(1) = 1$, and $g'(0) = g'(1) = 0$; for definiteness, take $g(t) = 3t^2 - 2t^3$. For each n, set

$$g_n(t) = g\left(\frac{t - 2^{-n}}{2^{-n+1} - 2^{-n}}\right) = g(2^n(t - 2^{-n})).$$

Denoting the points of $[0, 1]^2$ by (s, t), we note that $[0, 1]^2$ may be divided into vertical strips of the form $2^{-n} \leqq s \leqq 2^{-n+1}$. Define f on each such strip by

$$f(s,t) = s^2[(1 - g_n(s)) \sin(4^n t) + g_n(s) \sin(4^{n-1}t)].$$

The Lipschitz and differentiability conditions on f are easily verified. To see that f is not continuous, note that the variation of f over the vertical line $s = 2^{-n}$ tends to $2/\pi$ as $n \to \infty$; whereas on the line $s = 0$, f vanishes identically. Hence $\lim_{s\to 0} f(s,1) \geqq 2/\pi > 0 = f(0, 1)$.

8. The Value on *bv'NA*

In this section we will extend to all of *bv'NA* the value that we defined on *pNA* in the previous section, and prove that this extension is unique, thereby proving Theorem A. (The theorem is stated in Section 3.) For this purpose we must first recall some facts and definitions concerning functions of bounded variation.

It will be convenient to deal not only with functions defined on [0, 1] but with functions of bounded variation defined on any closed bounded interval. To avoid problems that would arise from adding functions having different domains, we will consider functions f of bounded variation on the entire real line, with the proviso that there exist real numbers c and d such that f is constant in $(-\infty, c]$ and in $[d, \infty)$; the interval $[c, d]$ is then called a *support* of f. The space of all such functions will be denoted bv^*, and the total variation of a bv^* function f will be denoted $\|f\|$. The real line $(-\infty, \infty)$ will be denoted E^1.

A bv^* function f is said to be a *left-continuous single-jump function* if there is a real number s such that

$$f(t) = \begin{cases} 1 & \text{for } t > s \\ 0 & \text{for } t \leqq s. \end{cases}$$

It is said to be a *right-continuous single-jump function* if there is a real number s such that

$$f(t) = \begin{cases} 1 & \text{for } t \geqq s \\ 0 & \text{for } t < s. \end{cases}$$

In either case it is said to have a *jump at s*. A bv^* function f is a *left-continuous* (*right-continuous*) *jump function* if it is of the form $\Sigma_i \alpha_i f_i$, where all the f_i are left-continuous (right-continuous) single-jump functions, and $\Sigma_i \alpha_i$ is either a finite sum or an absolutely convergent infinite sum of real numbers. If the jump of f_i is at s_i, then we may assume without loss of generality that the s_i are all different, and that none of the α_i vanish. Then the set $\{s_i\}$—the set of discontinuities of f—is called the *spectrum* of f and is denoted $\mathcal{S}(f)$; furthermore we have

$$\|f\| = \sum_i |\alpha_i|.$$

A function f in bv^* is said to be a *jump function* if

$$f = f^+ + f^-,$$

where f^+ is a right-continuous jump function and f^- is a left-continuous jump function; the decomposition is essentially[1] unique. In this case the spectrum $\mathcal{S}(f)$ is defined to be $\mathcal{S}(f^+) \cup \mathcal{S}(f^-)$, and we have

$$\|f\| = \|f^+\| + \|f^-\|.$$

Next, let f in bv^* be continuous. Then we may define a measure ν_f on E^1 by

$$\nu_f([s, t]) = f(t) - f(s),$$

whenever $s \leqq t$. If $\|\nu_f\|$ denotes the total variation of ν_f, then it may be verified that

$$\|f\| = \|\nu_f\|.$$

It is easily verified that every function f in bv^* may be essentially[2] uniquely written as

$$f = f^c + f^- + f^+$$

where f^c is continuous, f^- is a left-continuous jump-function, and f^+ is a right-continuous jump-function; furthermore, we have

$$\|f\| = \|f^c\| + \|f^-\| + \|f^+\|.$$

For the measure corresponding to the continuous component f^c of f, we will write ν_f rather than the more cumbersome ν_{f^c}.

[1] Up to an additive constant.

[2] Up to additive constants.

Two bv^* functions f and g are said to be *mutually singular* (written $f \perp g$) if ν_f and ν_g are mutually singular measures, $\mathcal{S}(f^+) \cap \mathcal{S}(g^+) = \emptyset$, and $\mathcal{S}(f^-) \cap \mathcal{S}(g^-) = \emptyset$. If $f \perp g$ then

$$\|f + g\| = \|f\| + \|g\|. \tag{8.1}$$

Clearly if $f \perp g$ then $f \perp \alpha g$ for all real α. Moreover

$$\text{if } f_1 \perp g \text{ and } f_2 \perp g, \text{ then } f_1 + f_2 \perp g; \tag{8.2}$$

this follows from the corresponding fact for mutually singular measures (cf. Halmos, 1950, p. 127, Exercise 10). Next, we have

Remark 8.3. If $g_0, \ldots, g_l$ are bv^* functions such that $g_i \perp g_j$ for all i and j, then

$$\|g_0 + \cdots + g_l\| = \|g_0\| + \cdots + \|g_l\|;$$

this remark follows at once from multiple applications of (8.1) and (8.2).

A bv^* function f is said to be *singular* if $\nu_f \perp \lambda$, where λ is Lebesgue measure; this is equivalent to saying that $f \perp g$ where g is the identity (i.e. $g(t) = t$) on a support of f.

A bv^* function f is said to be *absolutely continuous w.r.t.* a bv^* function g (written $f \ll g$) if $\mathcal{S}(f^+) \subset \mathcal{S}(g^+)$, $\mathcal{S}(f^-) \subset \mathcal{S}(g^-)$, and $\nu_f \ll \nu_g$ (in the sense of measures). A bv^* function f is said to be *absolutely continuous* if $\mathcal{S}(f^+) = \mathcal{S}(f^-) = \emptyset$ and $\nu_f \ll \lambda$; this is equivalent to saying that $f \ll g$, where g is the identity on a support of f.

If g is a fixed bv^* function, then every bv^* function f can be written uniquely in the form $f = f^{ac} + f^{\perp}$, where $f^{ac} \ll g$ and $f^{\perp} \perp g$; this follows easily from the corresponding fact for measures (Halmos, 1950, p. 134, Theorem C). In the particular case when g is the identity on a support of f, the component f^{ac} has a well-known explicit form (Saks, 1937, p. 119, Theorem 7.4). Indeed, the derivative f' of f exists a.e.[3] and is integrable w.r.t. Lebesgue measure, and for any real a we have

$$f^{ac}(t) = f^{ac}(a) + \int_a^t f'(s)\,ds. \tag{8.4}$$

We wish to obtain a similar explicit expression when g is not necessarily

[3] "Almost everywhere"—i.e. everywhere except possibly in a set of measure 0. If the measure in question is μ, we will write "a.e. w.r.t. μ"; the only exception to this rule is when μ is Lebesgue measure, as in this case, when we simply write "a.e."

the identity. For this purpose, define a function $f_{(g)} = df/dg$ on E^1 by

$$f_{(g)}(t) = \lim_{s \to t} \frac{f(s) - f(t)}{g(s) - g(t)}.$$

LEMMA 8.5. *Let f and g in bv^* be continuous and non-decreasing. Then $f_{(g)}$ exists a.e. w.r.t. ν_g and is integrable w.r.t. ν_g over E^1; and for any real a we have*

$$f^{ac}(t) = f^{ac}(a) + \int_a^t f_{(g)}(s)\, dg(s),$$

where $f = f^{ac} + f^{\perp}$ is the decomposition of f w.r.t. g.

The proof of this lemma proceeds by transforming the problem to one in which g is the identity on a support of f, applying (8.4), and then transforming back. It is not particularly difficult, but long and tedious. In order to avoid breaking the continuity of the presentation in this section, we therefore leave this proof to the reader.[4]

If $f \in bv^*$ and r is real, we may define a function $\Delta_r f \in bv^*$ by

$$(\Delta_r f)(t) = f(t + r).$$

Note that for all r,

$$f \text{ singular} \Rightarrow \Delta_r f \text{ singular};$$

$$f \text{ absolutely continuous} \Rightarrow \Delta_r f \text{ absolutely continuous};$$

$$f \perp g \Rightarrow \Delta_r f \perp \Delta_r g;$$

$$f \ll g \Rightarrow \Delta_r f \ll \Delta_r g.$$

LEMMA 8.6. *Let f, $g \in bv^*$, let g be singular, and let α and β be real numbers, $\alpha \neq \beta$. Then $\Delta_{r\alpha} f \perp \Delta_{r\beta} g$ for almost all r.*

Proof. Start out by assuming that $\alpha = 1$, $\beta = 0$, and f and g are continuous and non-decreasing; these assumptions will be removed later.

For each r, let

$$\Delta_r f = f_r^{ac} + f_r^{\perp} \tag{8.7}$$

be the decomposition of $\Delta_r f$ w.r.t. g. Our first claim is that for each t,

$$f_r^{ac}(t) \text{ is measurable in } r. \tag{8.8}$$

[4] The proof is carried out in detail in Appendix A of Aumann and Shapley (1968).

Indeed, let $[c, d]$ be a support of f; then $[c - r, d - r]$ is a support of $\Delta_r f$, and $f_r^{ac}(c - r) = (\Delta_r f)(c - r) = f(c)$. Hence from Lemma 8.5 it follows that

$$f_r^{ac}(t) = f(c) + \int_{c-r}^{t} \frac{d(\Delta_r f)}{dg}(s)\, dg(s).$$

Now it is easily verified that $(d(\Delta_r f)/dg)(s)$ is simultaneously measurable in r and s; and then it follows that its integral w.r.t. $g(s)$ is measurable in r. This proves (8.8). Incidentally, it is only for this purpose that Lemma 8.5 is needed.

Let $\Gamma = [\gamma, \delta]$ be a closed bounded interval, and write

$$\begin{aligned} F(t) &= \int_\Gamma (\Delta_r f)(t)\, dr \\ F^{ac}(t) &= \int_\Gamma f_r^{ac}(t)\, dr \qquad (8.9) \\ F^{\perp}(t) &= \int_\Gamma f_r^{\perp}(t)\, dr \end{aligned}$$

for all t. From (8.7) it follows that

$$F = F^{ac} + F^{\perp}.$$

Let

$$H(t) = \int_0^t f(t)\, dt.$$

Clearly H is absolutely continuous, and

$$F(t) = \int_\gamma^\delta f(t + r)\, dr = H(\delta + t) - H(\gamma + t).$$

Hence F, too, is absolutely continuous.

Let $\nu_r^{ac} = \nu_{f_r^{ac}}$, and $\nu^{ac} = \nu_{F^{ac}}$. If A is an interval of the form $[0, a]$, then from (8.9) it follows that

$$\nu^{ac}(A) = \int_\Gamma \nu_r^{ac}(A)\, dr.$$

Both sides of this equation are measures in A, so the equation must hold for all measurable A. From $f_r^{ac} \ll g$ it follows that $\nu_r^{ac} \ll \nu_g$. Hence if $U \subset E^1$ is such that $\nu_g(U) = 0$, then $\nu_r^{ac}(U) = 0$ for all r, and so

$$\nu^{ac}(U) = \int_\Gamma \nu_r^{ac}(U)\, dr = 0.$$

Since F is absolutely continuous and g singular, we have $\nu_F \perp \nu_g$, and so we may partition E^1 into disjoint sets U and V such that $\nu_g(U) = \nu_F(V) = 0$, so that in particular it follows that $\nu^{ac}(U) = 0$. Setting $\nu^{\perp} = \nu_{F^{\perp}}$, we deduce from $F = F^{ac} + F^{\perp}$ that

$$\nu_F = \nu^{ac} + \nu^{\perp}.$$

Now for any set $A \subset E^1$ we have

$$\begin{aligned}\nu^{ac}(A) + \nu^{\perp}(A) &= \nu_F(A) = \nu_F(A \cap U) + \nu_F(A \cap V)\\ &= \nu_F(A \cap U) = \nu^{ac}(A \cap U) + \nu^{\perp}(A \cap U) = \nu^{\perp}(A \cap U).\end{aligned}$$

On the other hand, since all measures involved are non-negative, we have $\nu^{\perp}(A) \geqq \nu^{\perp}(A \cap U)$; hence $\nu^{ac}(A) = 0$. Since A was chosen arbitrarily, it follows that ν^{ac} vanishes identically, and hence that F^{ac} vanishes identically. Hence f_r^{ac} vanishes identically for almost all r, as was to be proved. It remains only to remove the restrictions on α, β, f, and g.

First we get rid of the assumption that f and g are non-decreasing. Since f and g are in bv^* and are continuous, and g is singular, there are continuous non-decreasing f^1, f^2, g^1, g^2, in bv^* such that g^1 and g^2 are singular and $f = f^1 - f^2$, $g = g^1 - g^2$. Then from what we have already proved it follows that for almost all r we have $\Delta_r f^1 \perp g^1$, $\Delta_r f^1 \perp g^2$, $\Delta_r f^2 \perp g^1$, $\Delta_r f^2 \perp g^2$; and then it follows from (8.2) that

$$\Delta_r f = \Delta_r f^1 - \Delta_r f^2 \perp g^1 - g^2 = g.$$

Next, we allow f and g to have discontinuities. Then we may write $f = f^1 + f^2$, $g = g^1 + g^2$, where $f^1, f^2, g^1, g^2 \in bv^*$, f^1 and g^1 are continuous, g^1 is singular, and f^2 and g^2 are jump functions. Then for almost all r we have $\Delta_r f^1 \perp g^1$; and for all r we have $\Delta_r f^1 \perp g^2$ and $f^2 \perp g^1$. It remains only to prove that $\Delta_r f^2 \perp g^2$ for almost all r. Now $\mathcal{S}(f^2) - \mathcal{S}(g^2)$ (algebraic difference!) is denumerable, and so $r \notin \mathcal{S}(f^2) - \mathcal{S}(g^2)$ for almost all r. But for all such r we have $\mathcal{S}(\Delta_r f^2) \cap \mathcal{S}(g^2) = \emptyset$, and this in turn implies $\Delta_r f^2 \perp g^2$. So $\Delta_r f^2 \perp g^2$ a.e., and applying (8.2) we deduce $\Delta_r f \perp g$ a.e.

Finally, from $\alpha \neq \beta$ and what we have already proved, it follows that $\Delta_{(\alpha-\beta)r+\beta r} f \perp \Delta_{\beta r} g$ for almost all r, and the proof of Lemma 8.6 is complete.

COROLLARY 8.10. *Let f, $g \in bv^*$, g singular, α and β distinct real numbers. Then for almost all r, the functions $f((1 - r)t + r\alpha)$ and $g((1 - r)t + r\beta)$ of the argument t are mutually singular.*

Proof. For real s, define an operator Γ_s on bv^* by

$$(\Gamma_s h)(t) = h(st);$$

then $h_1 \perp h_2$ implies $\Gamma_s h_1 \perp \Gamma_s h_2$ for all s. From Lemma 8.6 we obtain $\Delta_{r\alpha} f \perp \Delta_{r\beta} g$ for almost all r; hence setting $s = 1 - r$ above, we obtain

$$\Gamma_{1-r}\Delta_{r\alpha} f \perp \Gamma_{1-r}\Delta_{r\beta} g$$

for almost all r. The latter two functions are precisely those appearing in the statement of the corollary, so the proof is complete.

If f is a function of bounded variation defined on a finite interval $[c, d]$, then f may be extended in a natural way to all of $(-\infty, \infty)$ so that the extended function f^* will be in bv^*. This is done by defining

$$f^*(s) = \begin{cases} f(c), & s \leqq c \\ f(s), & c \leqq s \leqq d \\ f(d), & d \leqq s. \end{cases}$$

The terminology introduced above for bv^* functions may then be naturally extended to functions of bounded variation on any interval; thus we shall say that f is absolutely continuous, singular, etc., if and only if f^* is absolutely continuous, singular, etc. The set of all singular functions in bv and in bv' will be denoted s and s', respectively.

PROPOSITION 8.11. *Let $g_1, \ldots, g_l \in s'$, let $\nu_1, \ldots, \nu_l$ be pairwise different measures in NA^1, and let $u \in AC$. Then*

$$\|u + g_1 \circ \nu_1 + \cdots + g_l \circ \nu_l\| \geqq |u(I)| + \|g_1\| + \cdots + \|g_l\|.$$

Proof. Let R denote the range of the vector measure $\nu = (\nu_1, \ldots, \nu_l)$. Then we claim that there is a point x^0 in R all of whose coordinates are different. Indeed, if not, then R is contained in the union of the $\binom{n}{2}$ hyperplanes H_{pq} determined by the conditions $x_p = x_q$. But by Lyapunov's theorem, R is convex; and any convex set contained in a finite union of hyperplanes must be contained in one of them. So

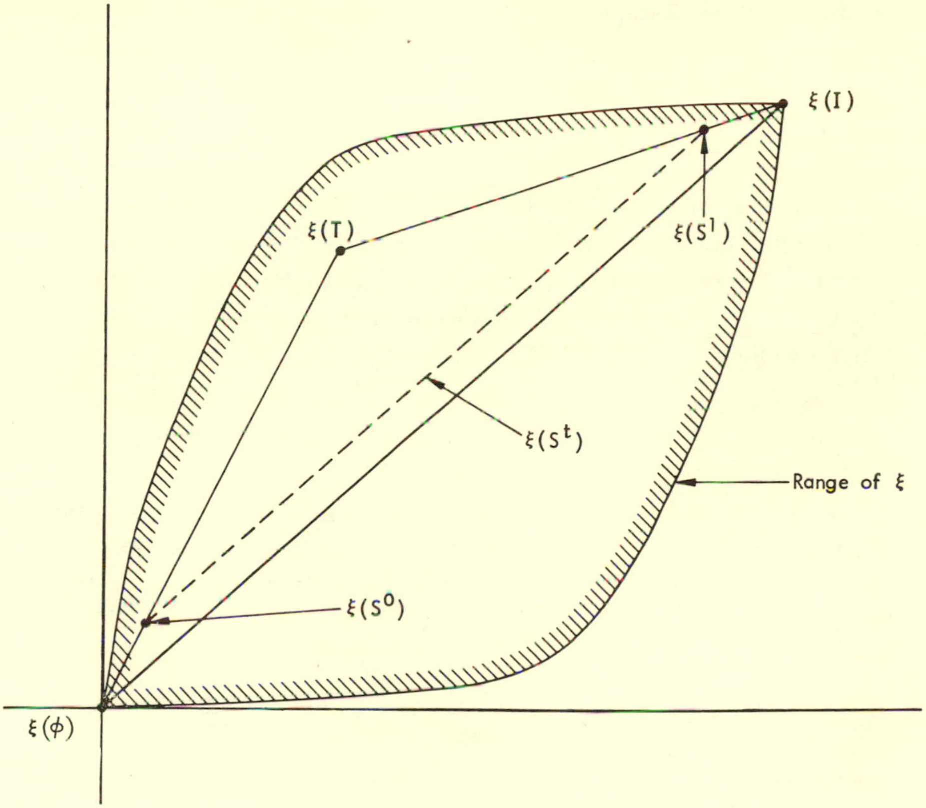

Figure 3. Construction of S^0 and S^1

$R \subset H_{pq}$ for some fixed p and q, i.e. $x_p = x_q$ for all $x \in R$; but this means that $\nu_p = \nu_q$, contrary to assumption.

Let T be such that $\nu(T) = x^0$, let $\nu_0 \in NA^+$ be such that $u \ll \nu_0$, and let ξ be the vector measure (ν_0, ν). By using Lyapunov's theorem, or more precisely Lemma 5.4, we can find, for each r in [0, 1], sets S^0 and S^1 such that $S^1 \supset T \supset S^0$ and

(8.12) $$\begin{cases} \xi(S^0) = r\xi(T) \\ \xi(S^1) = r\xi(T) + (1 - r)\xi(I) \end{cases}$$

(see Fig. 3). Apply Lemma 5.4 again, to assign to each t in (0, 1) a set S^t such that $t > s$ implies $S^t \supset S^s$, and

(8.13) $$\xi(S^t) = t\xi(S^1) + (1 - t)\xi(S^0).$$

Define functions $\tilde{u}$ and $\tilde{g}_1, \ldots, \tilde{g}_l$ in bv^* with support $[0, 1]$ as follows (for $t \in [0, 1]$):

$$(8.14) \qquad \begin{aligned} \tilde{u}(t) &= u(S^t) \\ \tilde{g}_p(t) &= g_p(\nu_p(S^t)) = g_p((1-r)t + r\nu_p(T)). \end{aligned}$$

From $u \ll \nu_0$ and $\nu_0(S^t) = (1-r)t\nu_0(I) + r\nu_0(T)$ it follows that $\tilde{u}$ is absolutely continuous, and hence $\tilde{u} \perp \tilde{g}_p$ for all p, since the $\tilde{g}_p$ are singular. Hence by Corollary 8.10, we may find arbitrarily small r such that $\tilde{g}_p \perp \tilde{g}_q$ for all p and q. Choosing such an r and applying Remark (8.3), we find

$$(8.15) \quad \left\| \tilde{u} + \sum_{p=1}^{l} \tilde{g}_p \right\| = \|\tilde{u}\| + \sum_{p=1}^{l} \|\tilde{g}_p\| \geqq |u(S^1) - u(S^0)| + \sum_{p=1}^{l} \|\tilde{g}_p\|.$$

Now since $g_p \in bv'$, it follows that $\|\tilde{g}_p\| \to \|g_p\|$ as $r \to 0$. Also, from $\nu_0(I \backslash S^1) = r\nu_0(I \backslash T)$, $\nu_0(S^0) = r\nu_0(T)$, and $u \ll \nu_0$, it follows that $u(S^1) \to u(I)$ and $u(S^0) \to u(\emptyset) = 0$ as $r \to 0$. Hence letting $r \to 0$ in (8.15), we obtain

$$\left\| u + \sum_{p=1}^{l} g_p \circ \nu_p \right\| \geqq \left\| \tilde{u} + \sum_{p=1}^{l} \tilde{g}_p \right\| \geqq |u(I)| + \sum_{p=1}^{l} \|g_p\|,$$

as was to be proved.

Proposition 8.16. *Let $v \in BV$, and let Ω be a chain $\emptyset = S^0 \subset \cdots \subset S^n = I$. For $i = 1, \ldots, n$, define*

$$I^i = S^i \backslash S^{i-1},$$

and let

$$\mathcal{C}^i = \{S \cap I^i : S \in \mathcal{C}\},$$

i.e. $\mathcal{C}^i$ is the set of measurable subsets of I^i. Let $v^i = v^i_\Omega$ be the set function defined on the underlying space $(I^i, \mathcal{C}^i)$ by

$$v^i(S) = v(S \cup S^{i-1}) - v(S^{i-1}).$$

Then v^i is of bounded variation for each i, and

$$\|v\| \geqq \sum_{i=1}^{n} \|v^i\|.$$

Proof. Let Ω_i be a chain

$$\emptyset = T_0^i \subset T_1^i \subset \cdots \subset I_i$$

at which $\|v^i\|$ is "almost" attained (i.e. such that $\|v^i\|_{\Omega_i} \geqq \|v^i\| - \epsilon$). From the Ω_i we construct the chain Ω, whose sets are of the form $S^{i-1} \cup T_j^i$; then

$$\|v\| \geqq \|v\|_\Omega \geqq \Sigma \|v^i\| - n\epsilon,$$

and letting $\epsilon \to 0$, we deduce our proposition.

PROPOSITION 8.17. *Let $g_1, \ldots, g_l \in s'$, let $\nu_1, \ldots, \nu_l$ be pairwise different measures in NA^1, and let $u \in AC$. Then*

$$\|u + g_1 \circ \nu_1 + \cdots + g_l \circ \nu_l\| = \|u\| + \|g_1\| + \cdots + \|g_l\|.$$

Proof. Let $v = u + \Sigma_{p=1}^l g_p \circ \nu_p$. For a given $\epsilon > 0$, let Ω be a chain $\emptyset = S^0 \subset S^1 \subset \cdots \subset S^n = I$ such that

$$\|u\|_\Omega > \|u\| - \epsilon. \tag{8.18}$$

Let I^i, $\mathcal{C}^i$, and v^i be as in Proposition 8.16. Define u^i on $(I^i, \mathcal{C}^i)$ by

$$u^i(S) = u(S \cup S^{i-1}) - u(S^{i-1}),$$

and define g_p^i on $[0, \nu_p(I^i)]$ by

$$g_p^i(t) = g_p(t + \nu_p(S^{i-1})) - g_p(\nu_p(S^{i-1}));$$

then

$$v^i = u^i + \Sigma_p g_p^i \circ \nu_p.$$

We wish to apply Proposition 8.11 to v^i. This is possible only if the g_p^i have jumps neither at 0 nor at $\nu_p(I^i)$; that is, we must have that for $p = 1, \ldots, l$ and $i = 0, \ldots, n$,

$$g_p \text{ does not have a jump at } \nu_p(S^i). \tag{8.19}$$

Now (8.19) certainly holds for $i = 0$ and $i = n$, since the g_p are in bv'. To make it hold for the other i may, however, necessitate some adjustment in the S^i, which we now proceed to make.

Let ν_0 be such that $u \ll \nu_0$, and let $\xi = (\nu_0, \nu_1, \ldots, \nu_l)$. By Lyapunov's theorem applied to I^i we may, for each t in [0, 1], find a set

S^{it} such that $S^{i-1} \subset S^{it} \subset S^i$ and

$$\xi(S^{it}) = t\xi(S^{i-1}) + (1 - t)\xi(S^i).$$

Then as $t \to 0$, we have, for each i, that $\nu_0(S^i \backslash S^{it}) \to 0$, and hence $|u(S^i) - u(S^{it})| \to 0$. So if we denote by Ω_t the chain obtained from Ω by replacing the S^i by S^{it}, then if t is chosen sufficiently small, we still have

$$\|u\|_{\Omega_t} > \|u\| - \epsilon. \tag{8.20}$$

On the other hand the g_p can have, between them, only denumerably many jumps; so by choosing t appropriately, we can see to it both that (8.20) holds and that the g_p have no jumps at the $\nu_p(S^{it})$. Thus we may as well assume that (8.19) holds for the chain Ω as originally defined.

We are now in a position to apply Proposition 8.11 to the v^i; this yields

$$\|v^i\| \geqq |u^i(I^i)| + \sum_{p=1}^{l} \|g_p^i\| = |u(S^i) - u(S^{i-1})| + \sum_{p=1}^{l} \|g_p^i\|.$$

Then by Proposition 8.16 and (8.18) we have

$$\begin{aligned} \|v\| \geqq \sum_{i=1}^{n} \|v^i\| &\geqq \sum_{i=1}^{n} |u(S^i) - u(S^{i-1})| + \sum_{p=1}^{l} \sum_{i=1}^{n} \|g_p^i\| \\ &= \|u\|_\Omega + \sum_{p=1}^{l} \|g_p\| > \|u\| - \epsilon + \sum_{p=1}^{l} \|g_p\|. \end{aligned}$$

Letting $\epsilon \to 0$, we deduce

$$\|v\| \geqq \|u\| + \sum_{p=1}^{l} \|g_p\|.$$

Since the opposite inequality is trivial, the proof of Proposition 8.17 is complete.

Let $s'NA$ be the subspace of BV spanned by all scalar measure functions $f \circ \mu$, where f in bv' is singular and $\mu \in NA^1$. From Proposition 8.17 we get

COROLLARY 8.21. *Let* $u \in AC$, $w \in s'NA$. *Then* $\|u + w\| = \|u\| + \|w\|$.

COROLLARY 8.22. $AC \cap s'NA = \{0\}$.

Proof. Let $v \in AC \cap s'NA$. Set $u = v$, $w = -v$, apply Corollary 8.21, and deduce

$$0 = \|v - v\| = \|v\| + \|-v\| = 2\|v\|.$$

Hence $v = 0$, as was to be proved.

Remark 8.23. From Theorem C and the fact that every bv' function can be decomposed into an absolutely continuous and a singular component in bv', it follows that

$$bv'NA = pNA + s'NA. \tag{8.24}$$

Corollary 8.22 shows that this sum is direct, i.e. that

$$pNA \cap s'NA = \{0\}. \tag{8.25}$$

Together with 8.25 and Corollary 5.3, Corollary 8.22 also shows that

$$AC \cap bv'NA = pNA. \tag{8.26}$$

We are now ready for the

Proof of Theorem A. First, we prove the existence of a value on $s'NA$. Let Q be the set of set functions v of the form

$$v = g_1 \circ \nu_1 + \cdots + g_l \circ \nu_l, \tag{8.27}$$

where the g_p are singular members of bv' and the ν_p are in NA^1. For each v in Q and each representation $\mathcal{R}$ of v in the form (8.27), we may define a measure $\theta = \theta_{v,\mathcal{R}}$ by

$$\theta = \sum_{p=1}^{l} g_p(1)\nu_p. \tag{8.28}$$

Then we claim that

$$\|\theta\| \leqq \|v\|. \tag{8.29}$$

To prove this, assume first that the ν_p are pairwise different. Then from Proposition 8.17 and the fact that the ν_p are in NA^1, it follows that

$$\|\theta\| \leqq \sum_p |g_p(1)|\,\|\nu_p\| = \sum_p |g_p(1)| \leqq \sum_p \|g_p\| = \|v\|,$$

establishing (8.29). If the ν_p are not necessarily pairwise different, we may group terms (e.g. if $\nu_1 = \nu_2$ we may write $(g_1 + g_2) \circ \nu_1$ instead of

$g_1 \circ \nu_1 + g_2 \circ \nu_2$). Although this leads to a different representation for v, it is easily seen that it does not change θ; hence since (8.29) holds for the "grouped" representation, it also holds for the original one.

Now let

$$v = \sum_{p=1}^{k} g'_p \circ \nu'_p \tag{8.30}$$

be a different representation of v in the form (8.27), which we denote $\mathcal{R}'$. Then $0 = v - v$ has a representation $\mathcal{P}$ given by

$$0 = \sum_{p=1}^{l} g_p \circ \nu_p - \sum_{p=1}^{k} g'_p \circ \nu'_p.$$

From (8.29) we obtain

$$\|\theta_{0,\mathcal{P}}\| \leqq \|0\| = 0,$$

so $\theta_{0,\mathcal{P}} = 0$. On the other hand it is easily verified that

$$\theta_{0,\mathcal{P}} = \theta_{v,\mathcal{R}} - \theta_{v,\mathcal{R}'};$$

hence

$$\theta_{v,\mathcal{R}} = \theta_{v,\mathcal{R}'}.$$

Thus $\theta_{v,\mathcal{R}}$ depends on the set function v only, and not on its representation $\mathcal{R}$; so we may define

$$\varphi' v = \theta_{v,\mathcal{R}},$$

using an arbitrary representation $\mathcal{R}$.

Clearly Q is dense in $s'NA$. We have already defined φ' on Q; it is easily verified that Q is a linear subspace of $s'NA$, and that φ' is linear on Q. From (8.29) it follows that $\|\varphi' v\| \leqq \|v\|$ for $v \in Q$, so φ' is continuous on Q and $\|\varphi'\| \leqq 1$. From the completeness of FA it then follows that φ' can be uniquely extended to be a continuous linear operator from $s'NA$ to FA, and this extension will also have norm $\leqq 1$. That the extension is indeed a value on $s'NA$ follows easily from (8.28) and Proposition 6.4.

Theorem B asserts the existence of a value φ_1 on pNA with $\|\varphi_1\| = 1$, and we have just proved that there is a value φ_2 on $s'NA$ with $\|\varphi_2\| \leqq 1$. Note that, by (8.24) and (8.25), each v in $bv'NA$ can be uniquely decomposed into a $u \in pNA$ and a $w \in s'NA$ such that $v = u + w$. We then define φ on $bv'NA$ by

$$\varphi v = \varphi_1 u + \varphi_2 w.$$

This φ is clearly linear; moreover from Corollary 8.21 it follows that

$$\|\varphi v\| = \|\varphi_1 u + \varphi_2 w\| \leqq \|\varphi_1 u\| + \|\varphi_2 w\| \\ \leqq \|u\| + \|w\| = \|u + w\| = \|v\|.$$

Hence $\|\varphi\| \leqq 1$, and so it follows from Proposition 6.4 (with F the union of s' and the positive integer powers) that φ is a value on $bv'NA$. That the range of φ is NA follows from Propositions 6.1 and 4.4.

To prove the uniqueness, let A be the set of all linear combinations of set functions $f \circ \mu$, where $f \in s'$ and $\mu \in NA^1$. Clearly A is reproducing. Furthermore, for every $g \in s'$ there are monotonic g^+ and g^- in s' such that $g = g^+ - g^-$ and

$$\|g\| = g^+(1) + g^-(1) = \|g^+\| + \|g^-\|.$$

From this and Proposition 8.17, applied with $u = 0$, we deduce that A is internal. Hence, by Proposition 4.12, $s'NA = \bar{A}$ is internal as well, and so, if φ is a value on $s'NA$ then, by Proposition 4.7, φ is continuous. On the other hand, Proposition 6.1 determines φ on a spanning subset of $s'NA$; so by the continuity, which we have just proved, φ is determined on all of $s'NA$.

Now if φ is a value on $bv'NA$, then $\varphi|pNA$ is a value on pNA and $\varphi|s'NA$ is a value on $s'NA$. The former is determined by Theorem B, and the latter by what we have just proved; this, with (8.24), completes the proof of Theorem A.

In Section 3 we asserted that, if in the definition of value we replace positivity by continuity, then Theorem A remains true as it stands. This assertion follows easily from the above line of proof.

From the internality of $s'NA$ (see above) and Proposition 7.19, we deduce (using (8.24) and Corollary 8.21)

PROPOSITION 8.31. *$bv'NA$ is internal.*

Finally, we wish to show that condition (3.3), which is weaker than the condition as stated in Theorem B, still implies $f \circ \mu \in pNA$ and formula (3.1). To this end, we first prove

PROPOSITION 8.32. *Let $v \in bv'NA$ be such that there is a positive integer n, an n-dimensional vector μ of non-atomic measures, and a compact convex neighborhood U in E^n of the diagonal $[0, \mu(I)]$ such that*

$$\mu(S) \in U \Rightarrow v(S) = 0.$$

Then $v \in pNA$.

Proof. Let $v = u + w$ be the decomposition, according to (8.24), (8.25), of v into set functions $u \in pNA$ and $w \in s'NA$. For each $\epsilon > 0$ we may find a set function

$$w^\epsilon = g_1 \circ \nu_1 + \cdots + g_l \circ \nu_l,$$

where the $g_p \in bv'$ are singular and the ν_p are in NA^1, such that

$$\|w - w^\epsilon\| < \epsilon.$$

Let $\nu = (\nu_1, \ldots, \nu_l)$, and let $\nu_0 \in NA$ be such that $u \ll \nu_0$. We now imitate the proof of Proposition 8.11, applying it to the $(n + l + 1)$-dimensional vector measure $\xi = (\mu, \nu_0, \nu)$ (rather than to $\xi = (\nu_0, \nu)$, as in Proposition 8.11). For each r in $[0, 1]$ and t in $[0, 1]$, we construct sets S^t so that (8.12) and (8.13) hold (see Fig. 3). For small r, then, we will have $\xi(S^0)$ close to 0 and $\xi(S^1)$ close to $\xi(I)$; hence $\mu(S^0)$ is close to 0, $\mu(S^1)$ is close to $\mu(I)$, and in particular we may choose r sufficiently small so that $\mu(S^0)$ and $\mu(S^1)$ are in U. From (8.13) it then follows that all the $\mu(S^t)$ are convex combinations of $\mu(S^0)$ and $\mu(S^1)$, and hence they are in U. Now define functions $\tilde{g}_1, \ldots, \tilde{g}_l, \tilde{u}, \tilde{w}$, and $\tilde{w}^\epsilon$ in bv^* with support $[0, 1]$ as in (8.14) and as follows (for $t \in [0, 1]$):

$$\tilde{w}(t) = w(S^t)$$

$$\tilde{w}^\epsilon(t) = w^\epsilon(S^t) = \sum_{p-1}^{l} \tilde{g}_p(t).$$

Then $\|\tilde{w} - \tilde{w}^\epsilon\| \leqq \|w - w^\epsilon\| < \epsilon$; and for $t \in [0, 1]$,

$$\tilde{u}(t) + \tilde{w}(t) = v(\mu(S^t)) = 0,$$

since $\mu(S^t) \in U$. But from (8.14) we get (8.15), and hence

$$0 = \|\tilde{u} + \tilde{w}\| = \|\tilde{u} + \tilde{w}^\epsilon + \tilde{w} - \tilde{w}^\epsilon\|$$
$$\geqq \|\tilde{u} + \sum_{p=1}^{l} \tilde{g}_p\| - \|\tilde{w} - \tilde{w}^\epsilon\| > \|\tilde{u}\| + \sum_{p=1}^{l} \|\tilde{g}_p\| - \epsilon.$$

In particular,

$$\sum_{p=1}^{l} \|\tilde{g}_p\| < \epsilon.$$

Now since $g_p \in bv'$, it follows that $\|\tilde{g}_p\| \to \|g_p\|$ as $r \to 0$. Hence, letting $r \to 0$, we obtain

$$\sum_{p=1}^{l} \|g_p\| < \epsilon.$$

Hence

$$\|w\| \leqq \sum_{p=1}^{l} \|g_p \circ \nu_p\| + \|w - w^\epsilon\| = \sum_{p=1}^{l} \|g_p\| + \|w - w^\epsilon\| < 2\epsilon.$$

Letting $\epsilon \to 0$, we obtain $w = 0$. Hence $v = u \in pNA$, and the proof of Proposition 8.32 is complete.

Now assume condition (3.3); that is, that $f \circ \mu \in bv'NA$ is such that μ is a vector of measures in NA with range R and f is continuously differentiable on $U \cap R$, where U is a convex neighborhood in E^n of the diagonal $[0, \mu(I)]$. Let f^* be a function that is continuously differentiable on all of R and coincides with f on $U \cap R$ (the existence of such a function is well known; see Whitney, 1934). Let $v = f \circ \mu - f^* \circ \mu$; then by Proposition 8.32, $v \in pNA$. But by Proposition 7.1, $f^* \circ \mu \in pNA$ as well; hence

$$f \circ \mu = v + f^* \circ \mu \in pNA.$$

Now apply Proposition 7.6, setting the v of that proposition equal to $f \circ \mu$ here, setting the f of that proposition equal to f^* here, and setting the U of that proposition equal to $U \cap R$ here (μ remains unchanged). Then since f and f^* are equal near the diagonal, we obtain (3.1). This completes the proof of the stronger form of Theorem B.

NOTES

1. It is tempting to try to simplify this section by defining the notion of "orthogonality" in BV by: $v \perp w$ if and only if $\|\alpha v + \beta w\| = |\alpha|\,\|v\| + |\beta|\,\|w\|$ for all real α and β (this of course is different from the ordinary notion of orthogonality in, say, a Hilbert space). If one could then prove that

(8.33) $$u \perp w \text{ and } v \perp w \text{ imply } u + v \perp w,$$

then the entire treatment would be greatly shortened and simplified. This notion of orthogonality can be defined in any Banach space, and one might have hoped that (8.33) is always true. Unfortunately, this is not the case. Consider, for example, the 3-dimensional space with norm $\| \quad \|'$ such that the unit ball is the convex hull of the 3 unit vectors e_1, e_2, e_3 and their negatives $-e_1$, $-e_2$, $-e_3$ and the 8 points $(\pm\frac{1}{2}, \pm\frac{1}{2}, \pm\frac{1}{2})$. Then $e_1 \perp e_2$ and $e_1 \perp e_3$ but $\|e_1 + e_2 + e_3\|' = 2 < \|e_1\|' + \|e_2 + e_3\|'$. As for BV, this does not enjoy property (8.33) either, but we will not quote the example here.

2. The space $bv'NA$ is not maximal among the subspaces of BV possessing a value and spanned by scalar measure games (compare Note 2 to Section 3). But one might say that it is "almost" maximal, in the sense that one can obtain a larger space with the requisite properties only by adding to the members of $bv'NA$ certain "degenerate" games whose value must vanish identically; and even this can be done in one and only one way. Let us clarify the situation.

Let

$$f_0(x) = \begin{cases} 0 & \text{if } x = 0 \\ 1 & \text{if } 0 < x < 1 \\ 0 & \text{if } x = 1. \end{cases}$$

Let R_0 be the space spanned by all the set functions $f_0 \circ \mu$, where $\mu \in NA^1$, and let

$$Q_0 = bv'NA + R_0.$$

Then we have

PROPOSITION 8.34.
(i) *Q_0 is symmetric and strictly includes $bv'NA$.*
(ii) *Q_0 is spanned by scalar measure games.*
(iii) *There is a value on Q_0.*
(iv) *Q_0 is the* only *subspace of BV satisfying (i), (ii), and (iii).*
(v) *The value on Q_0 is unique.*

Proof. Every v in Q_0 may be represented as

$$v = u + w, \tag{8.35}$$

where $u \in bv'NA$ and $w \in R_0$; then it may be seen that

$$\|v\| = \|u\| + \|w\|. \tag{8.36}$$

This yields

$$bv'NA \cap R_0 = \{0\} \tag{8.37}$$

(compare (8.25)). Since $R_0 \neq \{0\}$, it follows that Q_0 strictly includes $bv'NA$; since Q_0 is obviously symmetric, (i) is proved. Moreover, (8.36) yields that Q_0 is closed, and this proves (ii). To prove (iii), let the value vanish identically on R_0, and let it be as in Theorem A on $bv'NA$. This defines a unique operator on Q_0, since, by (8.37), the decomposition (8.35) of v is unique. Symmetry, $(\varphi v)(I) = v(I)$, and linearity are easily verified, and monotonicity follows from the fact that the only monotonic members of Q_0 are in $bv'NA$. This proves (iii).

Now any value φ on Q_0, when restricted to $bv'NA$, must be the unique value on $bv'NA$ provided by Theorem A (see Note 1 to Section 3). By Proposition 6.1, we have

$$\varphi(\mu + f_0 \circ \mu) = \mu = \varphi(\mu),$$

and hence $\varphi(f_0 \circ \mu) = 0$. Thus $\varphi | R_0 \equiv 0$; therefore φ is determined on R_0 and on $bv'NA$, and so on all of Q_0. This proves (v).

It remains to prove (iv). Suppose Q were another space with the stated properties. Then Q contains a scalar measure game $f \circ \mu \in Q \backslash bv'NA$, where $f \in bv$ and $\mu \in NA^1$. The function $f|(0, 1)$ has a unique extension f' to $[0, 1]$ that is continuous at 0 and at 1; set

$$f'' = f' - f'(0)$$

and

$$g = f - f''.$$

Since

$$f'' \circ \mu \in bv'NA \subset Q,$$

it follows that $g \circ \mu \in Q$. Now g is constant except for possible jumps at 0 and/or at 1. Unless these jumps are of equal magnitude but opposite sign—i.e. unless g is a constant multiple of f_0—we have $g(1) \neq 0$; and hence, reasoning as in the discussion of the μ-almost unanimity game in Section 3, we deduce that $g \circ \mu$ cannot have a value. Hence g is indeed a constant multiple of f_0. Therefore $f_0 \circ \mu \in Q$. Moreover since $f \circ \mu$ was an arbitrary scalar measure game in $Q \backslash bv'NA$, and since Q is spanned by scalar measure games, it follows that $Q \subset Q_0$. On the other hand since $f_0 \circ \mu \in Q$ for a particular μ, it follows from symmetry that $f_0 \circ \mu \in Q$ for all μ, and so, since Q is linear and closed (by (ii)), that $Q \supset Q_0$. Hence $Q = Q_0$ and the proof of Proposition 8.34 is complete.

9. Functions of a Signed Measure

Theorem C gives conditions for set functions of the form $f \circ \mu$ to be in pNA, where $\mu \in NA^+$. In this section we investigate set functions $f \circ \mu$ when μ is in NA but takes negative as well as positive values.

At the end of Section 5, we noted that in this situation the forward implication in Theorem C remains true, i.e. $f \circ \mu \in pNA$ implies the absolute continuity of f. However, the converse is no longer true. As we shall now see, absolutely continuous functions f exist for which $f \circ \mu \notin BV$, and so a fortiori $f \circ \mu \notin pNA$; they also exist for which $f \circ \mu$ *is* in BV, but is still not in pNA.

If R is any real interval containing the origin, we let $bv(R)$ denote the space of real-valued functions f of bounded variation on R that obey $f(0) = 0$; thus, we have $bv = bv([0, 1])$.

PROPOSITION 9.1. *Let* $v = f \circ \mu$, *where* $\mu \in NA$ *has range* $R = [-a,b]$, *with* $-a < 0 < b$. *Let*

$$g(x, y) = (x + a)(b - y)\frac{f(y) - f(x)}{y - x}.$$

Then $v \in BV$ *if and only if* $f \in bv(R)$ *and* $|g(x, y)|$ *is bounded in the domain* $-a < x < y < b$.

Remark. If f happens to be differentiable, then the condition on g says that the derivative of f must be bounded, except that near the boundary of R it is permitted to grow as fast as the reciprocal of the distance to the boundary. It is interesting that the conditions on f do not involve μ, except through its range R.

The underlying idea of the proof that follows is quite simple. Intuitively, we may think of the variation of f as being accumulated by a moving point, x, as it sweeps *once* across the domain R. For the variation of $f \circ \mu$, however, we must allow x to sweep *back and forth* within R, but with the proviso that the total distance traveled to the right (resp. left) must not exceed b (resp. a). Thus, if f becomes arbitrarily steep at some point in the interior of R (as in Example 9.3 below), then we can construct chains that oscillate in the neighborhood of that point and thereby accumulate an arbitrarily high variation for $f \circ \mu$. But if the steepness occurs only at the endpoints of R, then the moving point may not be able to remain in the neighborhood long enough to do any damage.

Proof of Proposition 9.1. First assume $f \circ \mu \in BV$. Given any increasing sequence $0 < x_1 < \cdots < x_p = b$, we can construct a chain (cf. Lemma 5.4) whose first p elements (after $\varnothing$) satisfy $\mu(S_i) = x_i$. This shows that $f \in bv([0, b])$. Similarly, $f \in bv([-a, 0])$; hence $f \in bv(R)$ as required.

To show that $g(x, y)$ is bounded, let k be a positive integer, and split $[x, y]$ into k equal intervals. At least one of them will have endpoints x', y' satisfying

$$|f(y') - f(x')| \geqq |f(y) - f(x)|/k,$$

as well as

$$x \leqq x' < y' \leqq y \text{ and } y' - x' = (y - x)/k.$$

If k is large enough, there will be a positive integer m such that

$$(y' - x')m \leqq \min(a, b, x' + a, b - x').$$

Using Lyapunov's theorem (drawing the increments $S_i \backslash S_{i-1}$ for $i > 1$ alternately from the positive and negative sides of a Hahn decomposition of I), we can construct a chain Ω: $\varnothing = S_0 \subset S_1 \subset \cdots \subset S_{2m+1} \subset I$ with the property

$$\mu(S_1) = \mu(S_3) = \cdots = \mu(S_{2m+1}) = x'$$
$$\mu(S_2) = \mu(S_4) = \cdots = \mu(S_{2m}) = y'.$$

(The conditions on m ensure that we have enough room for this much maneuvering.) Then

$$\begin{aligned} \|f \circ \mu\|_\Omega &= |f(x')| + 2m|f(y') - f(x')| + |f(x') - f(\mu(I))| \\ &\geqq \frac{2m}{k}|f(y) - f(x)| = \frac{2m(y - x)|g(x, y)|}{k(x + a)(b - y)}. \end{aligned}$$

Suppose now that m was chosen as large as possible. Then we have

$$\begin{aligned} m + 1 &> \frac{1}{y' - x'} \min(a, b, x' + a, b - x') \\ &\geqq \frac{k}{y - x} \min\left(a, b, \frac{(x' + a)(b - x')}{b + a}\right) \\ &= \frac{k(x' + a)(b - x')}{y - x} \min\left(\frac{\min(a, b)}{(x' + a)(b - x')}, \frac{1}{b + a}\right) \\ &\geqq \frac{k(x + a)(b - y)C}{y - x} \end{aligned}$$

where $C > 0$ is independent of x, y, and k. Hence

$$\|f \circ \mu\|_\Omega \geqq \frac{2mC}{m + 1}|g(x, y)| \geqq C|g(x, y)|.$$

Since $\|f \circ \mu\|_\Omega \leqq \|v\|$, this shows that g is bounded. This completes the proof in one direction.

For the other direction, let V denote the total variation of f on R; let G denote the supremum of $|g(x, y)|$ on $-a < x < y < b$; and let Ω be an arbitrary chain. We must show that $\|v\|_\Omega$ is bounded.

Let X denote the set of values assumed by $\mu(S)$ for $S \in \Omega$. First we wish to make precise the idea that the set X cannot get close to *both* endpoints of R. Let b' be the largest number in X and $-a'$ the smallest. Then we assert

$$\max(b - b', a - a') \geqq \min(a/2, b/2). \tag{9.2}$$

To prove this, let $S_i, S_j \in \Omega$ be such that $\mu(S_i) = b'$ and $\mu(S_j) = -a'$, and suppose $i < j$. Let Ω_0 be the chain $\{\emptyset, S_i, S_j, I\}$. Then, since $-a' \leqq \mu(I) = b - a$, we have

$$\|\mu\|_{\Omega_0} = b' + |-a' - b'| + |b - a + a'| = 2b' + 2a' - a + b.$$

But $\|\mu\|_{\Omega_0} \leqq \|\mu\| = a + b$; hence $a - a' \geqq b'$ and we have

$$\max(b - b', a - a') \geqq \max(b - b', b') \geqq b/2.$$

If $j < i$, a similar argument gives the estimate $a/2$ instead. This completes the proof of (9.2).

Next, we define an auxiliary chain as follows: For each link (S, T) of Ω for which the set[1] $Y = X \cap (\mu(S), \mu(T))$ is not empty, we introduce intermediate sets using Lyapunov's theorem, in such a way that each of the values in Y is assumed in its natural order (i.e. in increasing sequence if $\mu(S) < \mu(T)$, decreasing if $\mu(S) > \mu(T)$). The resulting, enlarged chain, denoted by Ω^*, has the property that the sequence of values $\mu(S_i^*)$ does not skip over any value that is assumed elsewhere in the sequence. By the triangle inequality, we have $\|f \circ \mu\|_{\Omega^*} \geqq \|f \circ \mu\|_\Omega$.

We now define two subchains of Ω^* (see Fig. 4). Let Λ_1 be the set of links $\{S_{i-1}^*, S_i^*\}$ such that $\mu(S_i^*) > 0$ and $\mu(S_j^*) \neq \mu(S_i^*)$ for all $j < i$. Let Λ_2 be the set of links $\{S_{i-1}^*, S_i^*\}$ such that $\mu(S_{i-1}^*) > b - a$ and $\mu(S_j^*) \neq \mu(S_{i-1}^*)$ for all $j > i - 1$. Then $\Lambda_1 \cap \Lambda_2 = \emptyset$, and we have

[1] We use $(\mu(S), \mu(T))$ here to denote the open interval whose end points are $\mu(S)$ and $\mu(T)$, even though it may happen that $\mu(S) > \mu(T)$.

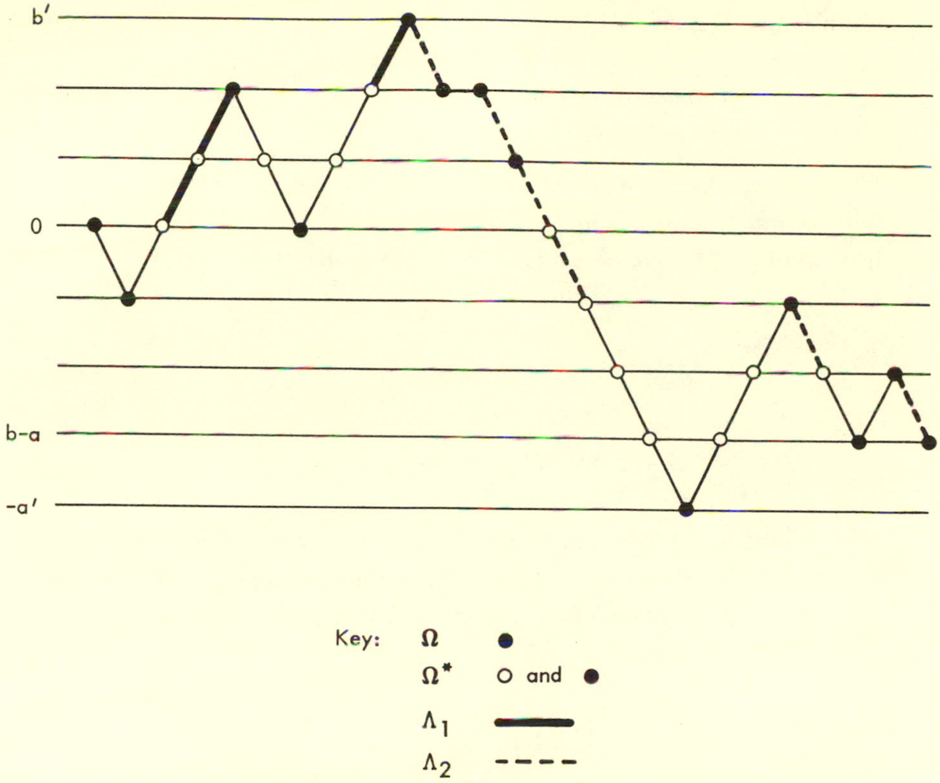

Figure 4. Relationship of Ω, Ω^*, Λ_1, Λ_2

$\|\mu\|_{\Lambda_1} = b'$ and $\|\mu\|_{\Lambda_2} = b' - (b - a)$. Since $\|\mu\|_{\Omega^*} \leqq \|\mu\| = a + b$, we have

$$\|\mu\|_{\Omega^* \backslash \Lambda_1 \backslash \Lambda_2} \leqq a + b - b' - (b' - (b - a)) = 2(b - b').$$

Hence

$$\|f \circ \mu\|_{\Omega^* \backslash \Lambda_1 \backslash \Lambda_2} \leqq \max_{\substack{i \\ \mu(S^*_i) \neq \mu(S^*_{i-1})}} \left| \frac{f(\mu(S_i^*)) - f(\mu(S_{i-1}^*))}{\mu(S_i^*) - \mu(S_{i-1}^*)} \right| \|\mu\|_{\Omega^* \backslash \Lambda_1 \backslash \Lambda_2}$$

$$\leqq 2(b - b') \max_{\substack{x,y \in X \\ x < y}} \frac{|f(y) - f(x)|}{y - x}.$$

On the other hand, $\|f \circ \mu\|_{\Lambda_1}$ and $\|f \circ \mu\|_{\Lambda_2}$ are each bounded by V, the variation of f. Hence

$$\|f \circ \mu\|_{\Omega^*} \leqq 2V + 2(b - b') \max_{\substack{x,y\in X \\ x<y}} \frac{|f(y) - f(x)|}{y - x}.$$

If $b = b'$ we therefore have $\|f \circ \mu\|_{\Omega^*} \leqq 2V$; the same holds if $a = a'$, by a symmetric argument that begins by redefining Λ_1 by $\mu(S_i^*) < 0$ instead of $\mu(S_i^*) > 0$. So we may continue under the assumption that both $b > b'$ and $a > a'$. Then

$$\begin{aligned}\|f \circ \mu\|_{\Omega^*} &\leqq 2V + 2(b - b') \max_{\substack{x,y\in X \\ x<y}} \frac{|g(x, y)|}{(x + a)(b - y)} \\ &\leqq 2V + 2(b - b') \frac{G}{(-a' + a)(b - b')} = 2V + \frac{2G}{a - a'}.\end{aligned}$$

This is still not the desired bound for $\|f \circ \mu\|_{\Omega^*}$, since a' depends on Ω. However the symmetric argument just mentioned can be carried out to give us $b - b'$ in place of $a - a'$ in the above, enabling us to apply (9.2). In fact, we have

$$\begin{aligned}\|f \circ \mu\|_{\Omega^*} &\leqq 2V + \min\left(\frac{2G}{a - a'}, \frac{2G}{b - b'}\right) \\ &= 2V + \frac{2G}{\max(a - a', b - b')} \leqq 2V + \frac{4G}{\min(a, b)}.\end{aligned}$$

Since $\|f \circ \mu\|_{\Omega} \leqq \|f \circ \mu\|_{\Omega^*}$, the proof of Proposition 9.1 is complete.

Example 9.3. Let the underlying space be the interval $[-1, 1]$, with its Borel subsets; let $\mu(S) = \int_S \operatorname{sgn} x \, dx$; and let $v(S) = \sqrt{|\mu(S)|}$. (See Fig. 5.) Then Proposition 9.1 tells us at once that $v \notin BV$, because of the behavior of f at 0.

To see this directly, merely define the chain $\Omega = \{\emptyset, S_1, \ldots, S_{2k}\}$ by $S_{2j-1} = [-(j-1)/k, j/k]$ and $S_{2j} = [-j/k, j/k]$. Then $\|v\|_{\Omega} = 2\sqrt{k}$, which is unbounded.

Example 9.4. Take μ as in Example 9.3, and let $v(S) = |\mu(S)|$. (See Fig. 5.) Clearly $v \ll \lambda$, where λ is Lebesgue measure; this shows that $v \in AC$, and hence that $v \in BV$. But we shall show that $v \notin pNA$ (and hence, by (8.26), that $v \notin bv'NA$).

The general idea of the proof that follows is to try to find a poly-

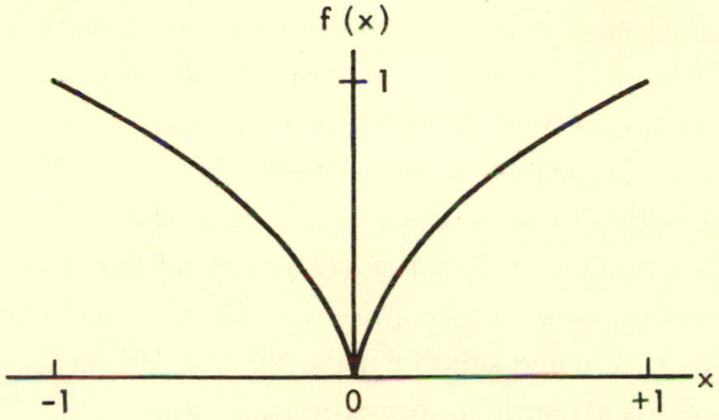

Example 9.3

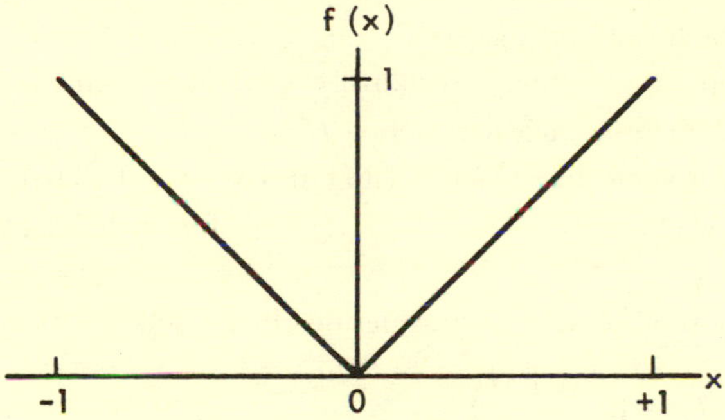

Example 9.4

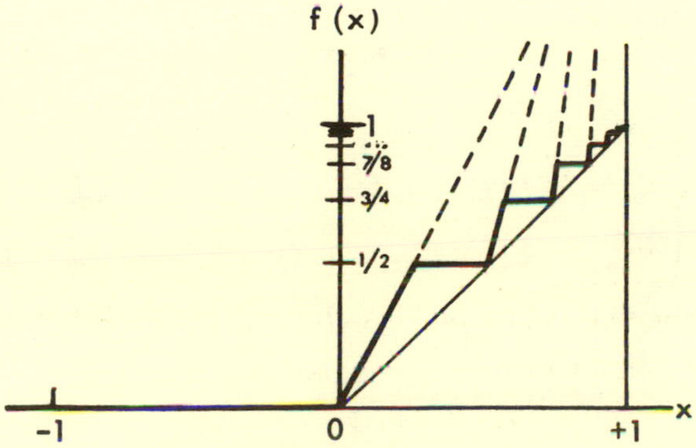

Example 9.5

Figure 5

nomial w in measures that approximates v in the variation norm. Such a w would behave "locally" like a measure plus a constant, just as a polynomial in several real variables behaves locally like a linear function in those variables (i.e. a homogeneous linear function plus a constant). On the other hand, in the neighborhood of sets $S_0 \subset I$ for which $\mu(S_0) = 0$, the function v does not behave at all like a measure plus a constant; this may be seen by considering, in such neighborhoods, both sets S with $\mu(S) > 0$ and sets S with $\mu(S) < 0$. Hence $\|v - w\|$ cannot be small, and the attempted approximation fails.

Proof[2] *for Example 9.4.* Let w be any polynomial in measures. Without loss of generality, $w(S) \equiv p(\nu(S))$ where p is a polynomial in n variables, for some n, and $\nu = (\nu_1, \ldots, \nu_n)$ is a vector of measures in NA^+. For each i, we may write $\nu_i = \xi_i + \zeta_i$, where ξ_i and ζ_i are both in NA^+ and are respectively absolutely continuous and singular, with respect to Lebesgue measure λ. Let $I^* = I \backslash (I_1 \cup \cdots \cup I_n)$, where each $I_i \subset I$ is such that $\zeta_i(I_i) = \zeta_i(I)$ and $\lambda(I_i) = 0$. Then for all Borel sets $S \subset I$ we have $\lambda(S \cap I^*) = \lambda(S)$ and, for each i, $\zeta_i(S \cap I^*) = 0$ and $\nu_i(S \cap I^*) = \xi_i(S)$.

Now choose a number k, and define the $2k$ sets

$$\begin{cases} T_j = \left(\dfrac{j-1}{k}, \dfrac{j}{k}\right) \cap I^*, & j = 1, \ldots, k, \\ U_j = \left(-\dfrac{j}{k}, -\dfrac{j-1}{k}\right) \cap I^*, & j = 1, \ldots, k. \end{cases}$$

These are pairwise disjoint, and partition I^* (and I) except for a set of measure zero. Next define a chain $\Omega = \{S_0, \ldots, S_{2k}, I\}$ by

$$\begin{cases} S_0 = \varnothing \\ S_{2j} = \bigcup_{i=1}^{j} (T_i \cup U_i), & j = 1, \ldots, k, \\ S_{2j+1} = S_{2j} \cup T_{j+1}, & j = 0, \ldots, k-1. \end{cases}$$

Note that the values of v on this chain alternate between 0 and $1/k$. Accordingly, we have

$$\begin{aligned} \|w - v\| &\geqq \|w - v\|_\Omega \\ &\geqq \sum_{j=1}^{k} \left| w(S_{2j-1}) - w(S_{2j-2}) - \frac{1}{k} \right| + \sum_{j=1}^{k} \left| w(S_{2j}) - w(S_{2j-1}) + \frac{1}{k} \right|. \end{aligned}$$

[2] Shorter (but deeper) proofs will be given in Sections 16, 23, and 27.

Setting $\xi = (\xi_1, \ldots, \xi_n)$, we have by the mean value theorem that

$$w(S_{2j-1}) - w(S_{2j-2}) = \xi(T_j) \cdot q(x_j),$$

where q denotes the gradient of p and x_j is some point on the line between $\xi(S_{2j-1})$ and $\xi(S_{2j-2})$. Similarly, we have

$$w(S_{2j}) - w(S_{2j-1}) = \xi(U_j) \cdot q(y_j),$$

where y_j is between $\xi(S_{2j})$ and $\xi(S_{2j-1})$. Hence

$$\begin{aligned}\|w - v\| &\geqq \sum \left| \xi(T_j) \cdot q(x_j) - \frac{1}{k} \right| + \sum \left| \xi(U_j) \cdot q(y_j) + \frac{1}{k} \right| \\ &\geqq \sum \left(-\xi(T_j) \cdot q(x_j) + \frac{1}{k} \right) + \sum \left(\xi(U_j) \cdot q(y_j) + \frac{1}{k} \right) \\ &= 2 - \sum \xi(T_j) \cdot q(x_j) + \sum \xi(U_j) \cdot q(y_j).\end{aligned}$$

If we begin instead with the chain $\Omega' = \{S'_0, \ldots, S'_{2k}, I\}$, defined by

$$\begin{cases} S'_{2j} = S_{2j}, & j = 0, \ldots, k, \\ S'_{2j+1} = S_{2j} \cup U_{j+1}, & j = 0, \ldots, k-1, \end{cases}$$

the exactly analogous argument yields

$$\|w - v\| \geqq 2 - \sum \xi(U_j) \cdot q(y'_j) + \sum \xi(T_j) \cdot q(x'_j).$$

Here x'_j is between $\xi(S'_{2j})$ and $\xi(S'_{2j-1})$ and y'_j is between $\xi(S'_{2j-1})$ and $\xi(S'_{2j-2})$. Combining the two estimates, we obtain

$$\begin{aligned}2\|w - v\| &\geqq 4 - \sum \xi(T_j) \cdot [q(x_j) - q(x'_j)] - \sum \xi(U_j) \cdot [q(y'_j) - q(y_j)] \\ &\geqq 4 - n\|\xi(I^*)\|Q,\end{aligned}$$

where Q is the maximum of the norms of all the expressions in square brackets.

To complete the argument, suppose k is large. Then by the absolute continuity of the ξ_i, $\|\xi(T_j)\|$ and $\|\xi(U_j)\|$ will be small (uniformly in j). The distances $\|x_j - x'_j\|$ and $\|y'_j - y_j\|$ are also small, as each is bounded by $\|\xi(T_j)\| + \|\xi(U_j)\|$ (this may be seen by applying the triangle inequality to the parallelogram with vertices $\xi(S_{2j})$, $\xi(S_{2j-1})$, $\xi(S_{2j-2})$, and $\xi(S'_{2j-1})$). Thus, finally, applying the continuity of q (i.e. the continuous differentiability of p), we see that Q itself can be made arbitrarily small. Hence $\|w - v\| \geqq 2$, and v is not in pNA.

Example 9.5. Take μ as in Example 9.3, and take f to be a monotonic, piecewise linear function as illustrated in Fig. 5. If the rising segments are given slopes 2, 8, 24, . . . , $k2^k$, . . . , then $|g|$ is unbounded. (Graphically, this means that the extensions of these segments (dashed lines) would cross the vertical $x = 1$ at arbitrarily high points.) Thus $f \circ \mu$ is *not* in BV, by Proposition 9.1. Note that f could be made continuously differentiable in $[-1, 1)$, by "rounding off the corners" of the graph, without affecting this conclusion.

On the other hand, we could give the rising segments slopes of 2, 4, 8, . . . , 2^k, . . . and $|g|$ would be bounded (the dashed lines would all pass below the point (1, 3)), despite the unbounded derivative of f. In this case, therefore, $f \circ \mu$ is in BV. It is not in pNA, however, because of the points at which f is not differentiable (compare Example 9.4). It appears reasonable to think that if the corners were rounded, as above, then the result would be in pNA; but Kohlberg (1973) has shown that this is not the case.

More generally, it is shown by Kohlberg that a necessary and sufficient condition that $f \circ \mu$ be in pNA (when μ is allowed to take both positive and negative values) is that f be continuously differentiable at each point in the interior of the range of μ, and that $g(x, y) \to 0$ as $x \to -a$ or $y \to b$ (where g is as in Proposition 9.1).[3]

10. A Variation on a Theme of Theorem B

In this section we prove a theorem that relaxes somewhat the differentiability conditions on f under which it may be concluded that $f \circ \mu \in pNA$. In particular, consider the case in which μ is a vector of measures in NA^+, f is defined on the non-negative orthant of E^n, and for each i, the partial derivative $f_i(x) = \partial f/\partial x_i$ exists and is continuous whenever it is defined as a two-sided derivative, i.e. whenever $x_i > 0$. Thus f is continuously differentiable in the interior of the orthant; and on each of the faces (of various dimensions) of the orthant, some of the

[3] The multidimensional case is discussed at the end of Section 10.

partial derivatives, but not all, must exist and be continuous. At the origin none of the partial derivatives need exist.

Under these conditions it can still be proved that $f \circ \mu$ is in pNA, if it is also assumed that f is increasing in each of its variables. The precise statement is Proposition 10.17 below.

Lemma 7.4 is useless in this context. Instead of working with the C^1 norm $\| \quad \|_1$, we define and work with a variation norm for functions of several variables. Because the range of μ is necessarily compact, it is natural to work with the cube rather than with the entire orthant.

At the end of the section we will briefly discuss a possible generalization from the orthant to arbitrary convex closed sets in E^n containing the origin.

For x and y in E^n we will write $x \geqq y$ if $x_i \geqq y_i$ for all i. A real function f of n real variables is said to be *non-decreasing* if $x \geqq y$ implies $f(x) \geqq f(y)$ for all x and y for which f is defined.

Suppose f is a non-decreasing continuous real function on the unit square $[0, 1]^2$. Let m be even, and consider a "staircase" sequence of points $x^0 \leqq \cdots \leqq x^m$ in the square, i.e. a sequence in which x^{2j+1} differs from x^{2j}, if at all, only in the first coordinate; and x^{2j} differs from x^{2j-1}, if at all, only in the second coordinate. Then the total increment $\Delta = f(x^m) - f(x^0)$ of the function f over the sequence can be split into two parts: the increment

$$\Delta_1 = \sum_j (f(x^{2j+1}) - f(x^{2j}))$$

over the horizontal segments, and the increment

$$\Delta_2 = \sum_j (f(x^{2j}) - f(x^{2j-1}))$$

over the vertical segments.

For $\delta \geqq 0$, let $\Delta_1(\delta)$ denote the supremum of Δ_1 over all staircase sequences involving only points x for which $x_1 \leqq \delta$ (see Fig. 6). Clearly if $\delta = 0$ the horizontal segments all disappear, and we necessarily have $\Delta_1(0) = 0$. Because of the continuity of f, it is reasonable to conjecture that

$$\Delta_1(\delta) \to 0 \text{ as } \delta \to 0. \tag{10.1}$$

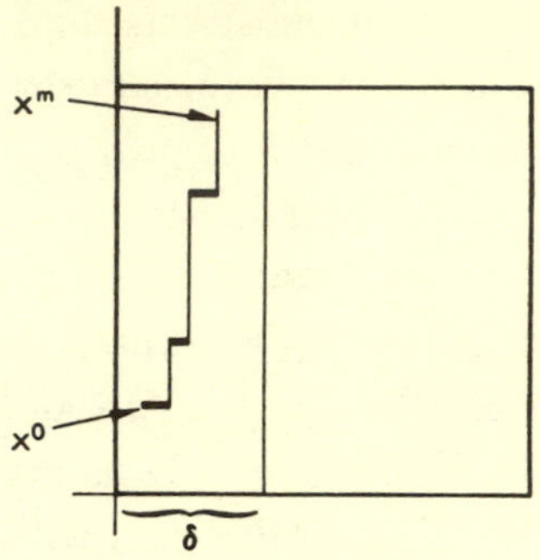

Figure 6. Staircase sequence with $x_1 \leqq \delta$

But in fact, under the conditions we have stated, (10.1) is false! Indeed, for each $k \geqq 2$ let $A_k \subset [0, 1]^2$ be the parallelogram whose vertices are $(2^{-k}, 0)$, $(2^{-k} + 4^{-k}, 0)$, $(2^{-k+1} + 4^{-k}, 1)$, and $(2^{-k+1}, 1)$ (see Fig. 7). Then we may find a non-decreasing continuous function f on the square such that for $x \in A_k$,

$$f(x) = f(x_1, x_2) = 2^k x_1 + 2^{-k+1} - 1. \tag{10.2}$$

For example, for x between A_k and A_{k-1} we may define

$$f(x) = 2^{-k+1} + x_2 + (x_1 - 2 \cdot 4^{-k})/(1 - 2^{-k} + x_2);$$

for x to the right of A_2 we may define f by the same formula that defines f on A_2, i.e. $f(x) = 4x_1 - \frac{1}{2}$; and for $x_1 = 0$ we may define

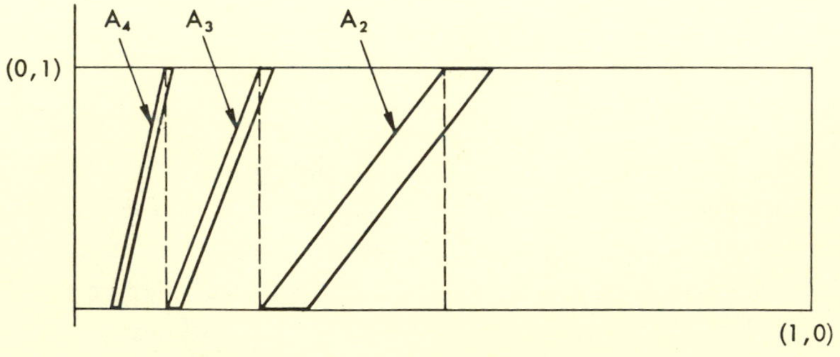

Figure 7. Counter-example to (10.1)

$f(x) = x_2$. The monotonicity and continuity of f are easily verified. If

$$(2^{-k}, 0) = x^0 \leqq \cdots \leqq x^m = (2^{-k+1}, 1)$$

is a staircase sequence all of whose points are in A_k, then from (10.2) it follows at once that the vertical increment Δ_2 vanishes, so the total increment Δ equals the horizontal increment Δ_1. But

$$\Delta = f(2^{-k+1}, 1) - f(2^{-k}, 0) = 1.$$

As $k \to \infty$, we have $x_1^m = 2^{-k+1} \to 0$, but Δ_1 tends to (in fact equals) 1, not 0.

Note that the example can easily be modified so that f is continuously differentiable whenever $x_1 > 0$ and $x_2 > 0$.

In order to prove (10.1), it is of course sufficient to assume that f satisfies a uniform Lipschitz condition in x_1, or in particular that the horizontal derivative $\partial f/\partial x_1$ exists and is continuous on the entire square. But this is not necessary; rather surprisingly, it is sufficient to assume the existence and continuity of the *vertical* derivative $\partial f/\partial x_2$ for $x_2 > 0$. We first sketch the proof intuitively. Because of the continuity of the vertical derivative, the vertical increment of a sequence $\{x^j\}$ with $x_1^m \leqq \delta$, δ small, can be approximated by the vertical increment of a nearby sequence $\{y^j\}$ with all $y_1^j = 0$. But for the latter, the vertical increment equals the total increment, and the total increment is clearly continuous (as a function of the endpoints of the sequence). So for small δ, the vertical increment Δ_2 must be close to the total increment $\Delta = \Delta_1 + \Delta_2$; hence Δ_1 is necessarily small. We now give the complete proof, for arbitrary n (rather than $n = 2$).

PROPOSITION 10.3. *Let f be a continuous non-decreasing real function on the unit cube $[0, 1]^n$ such that for each i, the partial derivative $f_i = \partial f/\partial x_i$ exists*[1] *and is continuous whenever $x_i > 0$. Then for every $\epsilon > 0$ there is a $\delta > 0$ such that for all i in $\{1, \ldots, n\}$, the following implication holds:*

IF $x^0 \leqq \cdots \leqq x^m$ is a sequence of points in $[0, 1]^n$ such that for each j, x^j differs from x^{j-1} in at most one coordinate, and such that $x_i^m < \delta$,

THEN

$$\sum_{j \in M(i)} (f(x^j) - f(x^{j-1})) < \epsilon,$$

[1] When $x_i = 1$ the derivative is one-sided.

where $M(i)$ is the set of j in $\{1, \ldots, m\}$ for which $x_i^j > x_i^{j-1}$.

Proof. The proof is by induction on the dimension n. Let $\mathcal{P}_n$ denote the proposition as stated. $\mathcal{P}_1$ follows at once from the continuity of f. Assume $\mathcal{P}_{n-1}$; we wish to prove $\mathcal{P}_n$. Let $N = \{1, \ldots, n\}$ and $M = \{1, \ldots, m\}$.

In proving $\mathcal{P}_n$, we may restrict attention to a single fixed i; for if we have found $\delta(i)$ for each i, we may define $\delta = \min_i \delta(i)$. Without loss of generality, we may let this fixed i be n. We will make use of the first $n - 1$ derivatives only. Define a function f^* on $[0, 1]^{n-1}$ by

$$f^*(x_1, \ldots, x_{n-1}) = f(x_1, \ldots, x_{n-1}, 0).$$

f^* satisfies the hypotheses of $\mathcal{P}_{n-1}$; so for a given $\epsilon > 0$, we may choose a $\delta_1 > 0$ that corresponds to f^* and to $\epsilon/3n$ in accordance with the conclusions of $\mathcal{P}_{n-1}$. Next, choose $\delta_2 > 0$ so that for all $r \in \{1, \ldots, n - 1\}$ and all x and y in $[0, 1]^n$, if $\|x - y\| \leqq \delta_2$ and $x_r \geqq \delta_1/2$, $y_r \geqq \delta_1/2$, then

$$|f_r(x) - f_r(y)| < \epsilon/3n.$$

This is possible because of the uniform continuity of f_r in $\{x \in [0, 1]^n: x_r \geqq \delta_1/2\}$.

Let δ_3 be such that $\|x - y\| \leqq \delta_3$ implies $|f(x) - f(y)| < \epsilon/3$; this is possible because of the uniform continuity of f in $[0, 1]^n$. Define

$$\delta = \delta(n) = \min(\delta_3, \delta_2).$$

Let $x^0, \ldots, x^m$ be given as in the statement of the proposition. Without loss of generality, assume that

$$\|x^j - x^{j-1}\| < \min(\delta_1/2, \delta_2) \tag{10.4}$$

for all j; if this is not already the case, we may insert additional points x^j so that it becomes true, without effecting any change in the statements appearing in the proposition.

For each $r \in N$, let $M(r) = \{j \in M: x_r^j > x_r^{j-1}\}$ and let $M_0 = \{j \in M: x^j = x^{j-1}\}$. The $M(r)$ are disjoint, because x^j can differ from x^{j-1} in at most one coordinate. For $r \in \{1, \ldots, n - 1\}$, let

$$K(r) = \{j \in M(r): x_r^j < \delta_1\}$$
$$L(r) = \{j \in M(r): x_r^j \geqq \delta_1\} = M(r)\backslash K(r).$$

Define a sequence $y^0, \ldots, y^m$ by

$$y^j = (x_1^j, \ldots, x_{n-1}^j, 0).$$

Now

$$\begin{aligned} f(y^m) - f(y^0) &= \sum_{j=1}^{m} (f(y^j) - f(y^{j-1})) \\ &= \sum_{r=1}^{n} \sum_{j \in M(r)} (f(y^j) - f(y^{j-1})) \\ &= \sum_{r=1}^{n-1} \Big[\sum_{j \in K(r)} + \sum_{j \in L(r)} \Big] (f(y^j) - f(y^{j-1})). \end{aligned}$$

The terms with $j \in M_0$ contribute nothing because then $x^j = x^{j-1}$, hence $y^j = y^{j-1}$; and when $j \in M(n)$ then $y^j = y^{j-1}$, so the terms with $r = n$ contribute nothing either.

For $r < n$, it follows from the definitions of $K(r)$ and δ_1 that

$$\sum_{j \in K(r)} (f(y^j) - f(y^{j-1})) < \epsilon/3n.$$

Hence

$$\sum_{r=1}^{n-1} \sum_{j \in K(r)} (f(y^j) - f(y^{j-1})) < \epsilon/3.$$

It follows that

$$\sum_{r=1}^{n-1} \sum_{j \in L(r)} (f(y^j) - f(y^{j-1})) > f(y^m) - f(y^0) - \epsilon/3. \tag{10.5}$$

Let e^r be the rth unit vector $(0, \ldots, 0, 1, 0, \ldots, 0)$. Then for $r < n$ and $j \in L(r)$ the mean value theorem yields

$$f(y^j) - f(y^{j-1}) = f_r(y^j - \theta_r^j \Delta_r^j e^r) \Delta_r^j$$

where $0 \leqq \theta_r^j \leqq 1$ and $\Delta_r^j = y_r^j - y_r^{j-1}$. Similarly

$$f(x^j) - f(x^{j-1}) = f_r(x^j - \psi_r^j \Delta_r^j e^r) \Delta_r^j,$$

where $0 \leqq \psi_r^j \leqq 1$. Now since $j \in L(r)$, we have $x_r^j \geqq \delta_1$; from (10.4) it then follows that $x_r^j - \Delta_r^j > \delta_1/2$. Hence

$$\begin{cases} (x^j - \psi_r^j \Delta_r^j e^r)_r \geqq x_r^j - \Delta_r^j > \delta_1/2, \text{ and} \\ (y^j - \theta_r^j \Delta_r^j e^r)_r \geqq y_r^j - \Delta_r^j = x_r^j - \Delta_r^j > \delta_1/2. \end{cases} \tag{10.6}$$

Furthermore, we have

$$\|(x^j - \psi_r^j \Delta_r^j e^r) - (y^j - \theta_r^j \Delta_r^j e^r)\| = \|x_n^j e_n + (\theta_r^j - \psi_r^j) \Delta_r^j e^r\|$$
$$\leqq \max(x_n^j, \Delta_r^j) \leqq \delta_2;$$

indeed $\Delta_r^j \leqq \delta_2$ because of (10.4), and $x_n^j \leqq x_n^m < \delta \leqq \delta_2$. Combining this with (10.6) and the definition of δ_2, we deduce

$$|[f(y^j) - f(y^{j-1})] - [f(x^j) - f(x^{j-1})]|$$
$$= |f_r(y^j - \theta_r^j \Delta_r^j e^r) - f_r(x^j - \psi_r^j \Delta_r^j e^r)| \Delta_r^j < \frac{\epsilon}{3n} \Delta_r^j.$$

From this it follows that

$$\begin{aligned} f(x^m) - f(x^0) &= \sum_{j=1}^{m} (f(x^j) - f(x^{j-1})) = \sum_{r=1}^{n} \sum_{j \in M(r)} (f(x^j) - f(x^{j-1})) \\ &\geqq \sum_{r=1}^{n-1} \sum_{j \in L(r)} (f(x^j) - f(x^{j-1})) + \sum_{j \in M(n)} (f(x^j) - f(x^{j-1})) \\ &\geqq \sum_{r=1}^{n-1} \sum_{j \in L(r)} (f(y^j) - f(y^{j-1})) \\ &\qquad - \frac{\epsilon}{3n} \sum_{r=1}^{n-1} \sum_{j \in L(r)} \Delta_r^j + \sum_{j \in M(n)} (f(x^j) - f(x^{j-1})). \end{aligned}$$

Now $\Sigma_{j \in L(r)} \Delta_r^j \leqq \Sigma_{j \in M(r)} \Delta_r^j = x_r^m - x_r^0 \leqq 1$; applying this and (9.5) to the above string of inequalities, we deduce

$$f(x^m) - f(x^0) > f(y^m) - f(y^0) - \frac{\epsilon}{3} - \frac{\epsilon}{3} + \sum_{j \in M(n)} (f(x^j) - f(x^{j-1})).$$

Since $x^0 \geqq y^0$, it follows that

$$\sum_{j \in M(n)} (f(x^j) - f(x^{j-1})) < f(x^m) - f(y^m) - (f(x^0) - f(y^0)) + 2\epsilon/3$$
$$< f(x^m) - f(y^m) + 2\epsilon/3.$$

Now $\|x^m - y^m\| = x_n^m < \delta$; since $\delta \leqq \delta_3$, it follows that $f(x^m) - f(y^m) < \epsilon/3$, and $\mathcal{P}_n$ follows. This completes the proof of Proposition 10.3.

If f is a real function defined on the unit cube $[0, 1]^n$, with $f(0) = 0$, then the *variation* $\|f\|$ of f is defined to be $\hat{f}(e)$ (see (7.12)). The space of functions of bounded variation on $[0, 1]^n$ is a linear space, and the variation is a norm for this space. Denote the space by bv^n.

If f is a real function defined on $[0, 1]^n$ and $\delta > 0$, define $f^\delta(x)$ for $x \in [0, 1]^n$ by

$$f^\delta(x) = f\left(\frac{x + \delta e}{1 + 2\delta}\right) - f\left(\frac{\delta e}{1 + 2\delta}\right),$$

where $e = (1, \ldots, 1)$.

PROPOSITION 10.7. *Let f be a continuous non-decreasing real function on the unit cube $[0, 1]^n$ such that for each i, the partial derivative $f_i = \partial f/\partial x_i$ exists*[2] *and is continuous whenever $x_i > 0$. Then for each $\delta > 0$, f^δ and f are in bv^n, and*

$$\|f^\delta - f\| \to 0 \text{ as } \delta \to 0.$$

Proof. From the fact that f is non-decreasing, it follows that f^δ also is, and hence at once that both are in bv^n.

For a given ϵ, let δ_1 correspond to $\epsilon/3n$ in accordance with Proposition 10.3; we may assume that $\delta_1 \leqq 1$. For each i, the derivative f_i is uniformly continuous in $\{x \in [0, 1]^n: x_i \geqq \delta_1/2\}$; therefore we may choose $\delta_2 > 0$ so that for all i and all x and y in the cube with $\|x - y\| < \delta_2$ and $x_i \geqq \delta_1/2$, $y_i \geqq \delta_1/2$, we have

$$|f_i(y) - f_i(x)| < \epsilon/6n. \tag{10.8}$$

Let β be a bound for $f_i(x)$ for all i and all x in the cube with $x_i \geqq \delta_1/2$.

Now choose $\delta > 0$ to satisfy the following conditions:

$$\delta \leqq \epsilon/6\beta n \tag{10.9}$$

$$\delta \leqq \delta_2 \tag{10.10}$$

$$\delta \leqq \delta_1/4. \tag{10.11}$$

Let $g = f^\delta - f$. For an arbitrary i and x in the cube with $x_i \geqq \delta_1/2$, and $\gamma \geqq 0$ such that $x + \gamma e^i \in [0, 1]^n$, we have from the mean value theorem that

$$|g(x + \gamma e^i) - g(x)| = |g_i(x + \theta\gamma e^i)|\gamma,$$

where $0 \leqq \theta \leqq 1$. Setting $y = x + \theta\gamma e^i$, and $z = (y + \delta e)/(1 + 2\delta)$,

[2] When $x_i = 1$, the derivative is one-sided.

we obtain $y_i \geqq x_i \geqq \delta_1/2$. Further, if $y_i \leqq \frac{1}{2}$, then $\delta \geqq 2\delta y_i$, and hence

$$z_i = (y_i + \delta)/(1 + 2\delta) \geqq y_i \geqq \delta_1/2; \tag{10.12}$$

and if $y_i \geqq \frac{1}{2}$, then

$$z_i \geqq (\tfrac{1}{2} + \delta)/(1 + 2\delta) = \tfrac{1}{2} \geqq \delta_1/2,$$

because $\delta_1 \leqq 1$; so that (10.12) is in any case established. From (10.10) it follows that

$$\|z - y\| = \|e - 2y\| \frac{\delta}{1 + 2\delta} \leqq \frac{\delta}{1 + 2\delta} \leqq \delta \leqq \delta_2.$$

From these observations and (10.8) and(10.9) it follows that

$$\begin{aligned} |g(x + \gamma e^i) - g(x)| &= |g_i(y)|\gamma = \left| \frac{1}{1 + 2\delta} f_i\left(\frac{y + \delta e}{1 + 2\delta}\right) - f_i(y) \right| \gamma \\ &= \left| (f_i(z) - f_i(y)) - \frac{2\delta}{1 + 2\delta} f_i(z) \right| \gamma \\ &\leqq (|f_i(z) - f_i(y)| + 2\delta\beta)\gamma \leqq \left(\frac{\epsilon}{6n} + \frac{\epsilon}{6n}\right)\gamma \\ &= \frac{\epsilon}{3n}\gamma. \end{aligned} \tag{10.13}$$

Suppose now that

$$0 = x^0 \leqq \cdots \leqq x^k = e; \tag{10.14}$$

we wish to prove that

$$\sum_{j=1}^{k} |g(x^j) - g(x^{j-1})| < \epsilon. \tag{10.15}$$

Without loss of generality we may assume that

$$\|x^j - x^{j-1}\| < \delta_1/4$$

for all j; for, if this is not already the case, it can be made so by inserting additional points into the sequence (10.14), without decreasing the left side of (10.15); so that, if (10.15) holds for the new sequence, it certainly holds for the original sequence.

For each i in $\{0, 1, \ldots, n\}$ and each j in $\{1, \ldots, k\}$, let

$$x^{ji} = (x_1^j, \ldots, x_i^j, x_{i+1}^{j-1}, \ldots, x_n^{j-1}).$$

Note that $x^{j0} = x^{j-1,n} = x^{j-1}$; and that x^{ji} and $x^{j,i-1}$ differ in the ith coordinate only. Then

$$g(x^j) - g(x^{j-1}) = g(x^{jn}) - g(x^{j0}) = \sum_{i=1}^{n} (g(x^{ji}) - g(x^{j-1,i})),$$

and hence

$$\text{(10.16)} \qquad \sum_{j=1}^{k} |g(x^j) - g(x^{j-1})| \leqq \sum_{i=1}^{n} \sum_{j=1}^{k} |g(x^{ji}) - g(x^{j-1,i})|.$$

Consider now an arbitrary but fixed i. We have

$$0 = x_i^0 \leqq x_i^1 \leqq \cdots \leqq x_i^k = 1.$$

Let $m = m(i)$ be such that

$$\tfrac{1}{2}\delta_1 \leqq x_i^m \leqq \tfrac{3}{4}\delta_1 \leqq \delta_1;$$

the existence of such an m follows from $\|x^j - x^{j-1}\| < \delta_1/4$. Since x^{ji} and $x^{j,i-1}$ differ in the ith coordinate only, and $x_i^{mi} = x_i^m \leqq \delta_1$, it follows from the definition of δ_1 (i.e. from Proposition 10.3) that

$$\sum_{j=1}^{m} |f(x^{ji}) - f(x^{j,i-1})| < \epsilon/3n.$$

Furthermore, from $x_i^m \leqq \frac{3}{4}\delta_1$ and (10.11) it follows that

$$(x^{mi} + \delta e)_i/(1 + 2\delta) \leqq \delta_1,$$

so again using the definition of δ_1, we deduce

$$\sum_{j=1}^{m} |f^\delta(x^{ji}) - f^\delta(x^{j,i-1})| < \epsilon/3n.$$

Hence

$$\sum_{j=1}^{m} |g(x^{ji}) - g(x^{j,i-1})| < 2\epsilon/3n.$$

Since $x_i^{m+1,i-1} = x_i^m \geqq \delta_1/2$, we have from (10.13) that

$$\sum_{j=m+1}^{k} |g(x^{ji}) - g(x^{j,i-1})| \leqq \frac{\epsilon}{3n} \sum_{j=m+1}^{k} (x_i^j - x_i^{j-1}) \leqq \frac{\epsilon}{3n}.$$

Summing up, we obtain

$$\sum_{i=1}^{n} \sum_{j=1}^{k} |g(x^{ji}) - g(x^{j-1,i})| \leqq \sum_{i=1}^{n} \epsilon/n \leqq \epsilon,$$

and, by applying (10.16), we complete the proof of Proposition 10.7.

PROPOSITION 10.17. *Let f be a continuous non-decreasing real function on the non-negative orthant of E^n, such that for each i, the partial derivative $\partial f/\partial x_i$ exists and is continuous whenever $x_i > 0$. Let μ be an n-dimensional vector of measures in NA^+. Then $f \circ \mu \in pNA$.*

Proof. Since the range R of μ is compact by Lyapunov's theorem (Proposition 1.3), we may w.l.o.g. assume that $R \subset [0, 1]^n$ (otherwise replace $f(x)$ by $f(x/k)$ and μ by $k\mu$). For any function $g \in bv^n$, we then have $g \circ \mu \in BV$, and $\|g \circ \mu\| \leqq \|g\|$. Since $f^\delta \circ \mu - f \circ \mu = (f^\delta - f) \circ \mu$, it follows that $\|f^\delta \circ \mu - f \circ \mu\| \to 0$ as $\delta \to 0$. But $f^\delta \circ \mu \in pNA$ because of Theorem B; hence also $f \circ \mu \in pNA$, because pNA is closed. This completes the proof.

It is not difficult to extend the proof of Proposition 10.7 so that it holds when the condition $x_i > 0$ is replaced by $0 < x_i < 1$. In other words, we may prove the proposition for continuous non-decreasing functions defined on the unit cube whose partial derivatives exist whenever they are defined as two-sided derivatives. Proposition 10.17 therefore also holds for such functions, provided of course that the range of μ is in the unit cube. An appropriate analogue of Proposition 10.3 also holds.

More generally, let X be a closed convex subset of E^n containing the origin, and let f be a real continuous non-decreasing function defined on X such that f is continuously differentiable on Rel Int X (the relative interior of X). If x is in the relative boundary of X, and z is X-admissible, then f is said to be *continuously differentiable at x in the direction z* if it is continuously differentiable on $\{x\} \cup$ Rel Int X in the direction z. Next, z is called *tangent* to X at x if $x + z$ is in the intersection of all the hyperplanes through x that support X. It seems reasonable to conjecture the following generalization of Proposition 10.17:

Assume that f is continuously differentiable on Rel Int X, and that for each x in the relative boundary of X and for each X-admissible z tangent to X at x, f is continuously differentiable at x in the direction z. Let μ be a vector of non-negative measures whose range is included in X. Then $f \circ \mu \in pNA$.

Chapter II. The Random Order Approach

11. Introduction

Shapley's original paper (1953a) contained two equivalent approaches to the valuation of games with finitely many players: one based on axioms, and one based on determining the expected marginal payoff to a player when the players are brought into coalition in a random order (see Appendix A). Chapter I was devoted to a generalization of the first of these approaches to games with a continuum of players. The present chapter will be devoted to investigating analogues of the second approach.

It turns out that a direct generalization of the random order approach is impossible; this will be shown in Sections 12 and 13. A way to circumvent this impossibility is through the use of what are called mixing transformations; these enable us, in a sense, to "shuffle" the players and achieve, by a limiting process, much the same effect as a random order. In Section 14 we define the concept of "mixing value" and state the basic theorem (Theorem E), which relates it to the value defined in Chapter I; the theorem is proved in Section 15. Section 16 contains an application. Further properties of the mixing value will be developed throughout the book, in particular the "diagonal property" in Chapter VII.

Another way to circumvent the above impossibility is to return to the finite model, where random orders are of course always possible, and to define the value for a game with a continuum of players as the limit of the values of an approximating sequence of finite games. This "asymptotic value," due to Kannai (1966), will be discussed in Chapter III. It proves to have much in common with the mixing value, because of the dominant role that random order methods play in investigations of the values of finite games.

An important difference between the axiomatic approach and the mixing and asymptotic approaches is that the latter two work with individual games, whereas the former requires a linear space of games.

Another difference is that the mixing and asymptotic values will be expressed directly in terms of the set function v, while the axiomatic approach gave us expressions for the value only in terms of some representation of v—for example, as a sum of scalar measure functions or as a C^1 function of a vector measure.

12. The Random Order Impossibility Principle

The generalization to games with a continuum of players of the random order approach to value would, if it were possible, proceed as follows:[1] first one defines (and possibly restricts) in an appropriate manner the notion of an order $\mathcal{R}$ on the player space. Then with each characteristic function v and each $\mathcal{R}$, one associates a measure $\varphi^{\mathcal{R}}v$ on the player space; intuitively $(\varphi^{\mathcal{R}}v)(S)$ is the marginal contribution of the coalition S to the payoff if the players "enter the scene" according to the order $\mathcal{R}$. Next, one defines the notion of measurability, and a probability measure ω, on the space Ω of orders; and finally, one defines the value φv to be the expectation of $\varphi^{\mathcal{R}}v$, i.e.

$$(12.1)\qquad (\varphi v)(S) = \int_\Omega (\varphi^{\mathcal{R}}v)(S)\, d\omega(\mathcal{R}).$$

Let us see how far we can get in carrying out this program.

An *order* on the underlying space I is a relation $\mathcal{R}$ on I that is transitive, irreflexive, and complete.[2] For each order $\mathcal{R}$ and each $s \in I$, define the *initial segment* $I(s; \mathcal{R})$ by

$$I(s; \mathcal{R}) = \{t \in I : s \,\mathcal{R}\, t\}.$$

The entire space I and the empty set $\emptyset$ will also be considered initial segments, and as such will be denoted $I(\infty; \mathcal{R})$ and $I(-\infty; \mathcal{R})$, respectively; it will be understood that $\infty \,\mathcal{R}\, s \,\mathcal{R}\, (-\infty)$ for all $s \in I$, and we will denote[3] $\{-\infty\} \cup I \cup \{\infty\}$ by $\bar{I}$. The intuitive requirement on the measure $\varphi^{\mathcal{R}}v$ that was given in the previous paragraph is then mathe-

[1] Compare the discussion in Appendix A.

[2] I.e. for all $s, t \in I$, one and only one of the three statements $s \,\mathcal{R}\, t$, $t \,\mathcal{R}\, s$, $s = t$ holds. We shall interpret "$s \,\mathcal{R}\, t$" as "s comes after t" (or "s is greater than t").

[3] Formally, we extend $\mathcal{R}$ to $\bar{I}$ by the condition $\infty \,\mathcal{R}\, s \,\mathcal{R}\, (-\infty)$ for all $s \in I$. This however is a notational device only; we are not adding anything to the underlying space, and all set functions, measures, etc., continue to be defined on subsets of I only. Note that if I has an "$\mathcal{R}$-first" element t, then $I(t; \mathcal{R}) = \emptyset = I(-\infty; \mathcal{R})$.

matically expressed as follows:

$$(\varphi^{\mathcal{R}}v)(I(s;\mathcal{R})) = v(I(s;\mathcal{R})) \tag{12.2}$$

for all $s \in \bar{I}$. For this equation to be meaningful, $I(s;\mathcal{R})$ must be measurable for all s; thus if $F(\mathcal{R})$ is the σ-field generated by all the initial segments $I(s;\mathcal{R})$, then we must have $\mathcal{C} \supset F(\mathcal{R})$. If moreover

$$\mathcal{C} = F(\mathcal{R}), \tag{12.3}$$

i.e. if the $I(s;\mathcal{R})$ generate $\mathcal{C}$, then (12.2) "determines" $\varphi^{\mathcal{R}}v$, in the sense that there is then at most one measure $\varphi^{\mathcal{R}}v$ on $\mathcal{C}$ satisfying (12.2). Orders $\mathcal{R}$ obeying (12.3) will be called *measurable*, and we will henceforth restrict the discussion to measurable orders (see Note 1).

Although the measurability condition (12.3) guarantees that there is at most one $\varphi^{\mathcal{R}}v$ satisfying (12.2), it does not guarantee that there is at least one. For a simple example using the underlying space $([0, 1], \mathcal{B})$, let the order $\mathcal{R}$ be the usual order $>$ (which is clearly measurable); in this case $I(s;\mathcal{R}) = [0, s)$. Let $v = f \circ \lambda$, where

$$f(x) = \begin{cases} 0 & \text{for } x < \frac{1}{2} \\ 1 & \text{for } x \geqq \frac{1}{2}. \end{cases}$$

Then there is no measure $\varphi^{\mathcal{R}}v$ satisfying (12.2).

It might be thought that what makes this example possible is the discontinuity in the function f; but this is not the case. Indeed, suppose again that the underlying space is $([0, 1], \mathcal{B})$, and let the order be defined as follows: for two points s and t that are either both in the Cantor set, or both outside of the Cantor set, $s \mathcal{R} t$ if and only if $s > t$; and if t is in the Cantor set whereas s is not, then $s \mathcal{R} t$ (i.e. the entire Cantor set comes "before" its complement). Let $v = f \circ \lambda$, where f is the Cantor function. Then there is no measure $\varphi^{\mathcal{R}}v$ satisfying (12.2).

If, however, f is *absolutely* continuous, then a counterexample of this kind is impossible. Indeed, we shall show below (Proposition 12.8) that, if $v \in AC$, then, for each measurable order $\mathcal{R}$, there is a unique measure $\varphi^{\mathcal{R}}v$ satisfying (12.2). Thus, though we must drastically limit the domain of set functions in which we work, at this point there might still be hope that the program outlined at the beginning of this section would work out for $v \in AC$, and so for $v \in pNA$ in particular (cf. Corollary 5.3).

Unfortunately, the program does *not* work out, even if we restrict consideration to pNA. Let Ω denote the space of all measurable orders on $(I, \mathcal{C})$. Then we have

THEOREM D. *It is impossible to define measurability and a probability measure* ω *on* Ω, *so that* (12.1) *defines a value* φ *on* pNA.

A comment on this theorem is in order. For (12.1) to define a value on pNA, its right side must be meaningful; in particular, we must have that for all v in pNA,

(12.4) $(\varphi^{\mathcal{R}}v)(S)$ is measurable as a function of $\mathcal{R}$, for each fixed measurable $S \subset I$.

Now if we define measurability on Ω so that (12.4) holds, and if ω is any probability measure on the resulting measurable space, then it is easily verified that (12.1) defines an operator φ that satisfies all the conditions in the definition of value[4] except possibly symmetry (2.2). Thus Theorem D may be interpreted as saying that if there are sufficiently many measurable sets in Ω to make the integral in (12.1) meaningful, then it is impossible to define a probability measure on Ω that is sufficiently sensitive to the symmetries of the situation to yield (2.2).

We now proceed to prove Theorem D.

Given an order $\mathcal{R}$ on I, we will let "$s \overset{\mathcal{R}}{=} t$" denote "$s \mathcal{R} t$ or $s = t$." Then a subset R of I will be called $\mathcal{R}$-*dense* if for all s and t in I such that $s \mathcal{R} t$, there is a member r of R such that $s \overset{\mathcal{R}}{=} r \overset{\mathcal{R}}{=} t$.

LEMMA 12.5. *If* $\mathcal{R}$ *is a measurable order on* I, *then* I *has a denumerable* $\mathcal{R}$-*dense subset.*

Proof. For each family $\mathcal{S}$ of subsets of I, let $F(\mathcal{S})$ denote the σ-field generated by $\mathcal{S}$, i.e. the smallest σ-field that includes $\mathcal{S}$. From (12.3) we know that every measurable subset S of I is in $F(\mathcal{R})$, which by definition is the σ-field generated by all the initial segments. Our first claim is that to generate a given measurable S, only a denumerable number of initial segments are needed; more precisely, for each $S \in \mathcal{C}$ there are initial segments $I_1, I_2, \ldots$ such that $S \in F(\{I_1, I_2, \ldots\})$. To show this, note first that it is trivially true when S is itself an initial

[4] The monotonicity follows from $\|\varphi^{\mathcal{R}}v\| \leqq \|v\|$ (see Proposition 12.8) and Proposition 4.6.

segment. Next, if it is true for a given S then it is also true for its complement, and if it is true for a denumerable sequence of S_i then it is also true for their union. Hence it is true for every S in the smallest σ-field containing all the initial segments, and as we noted above, this is precisely $\mathcal{C}$. Since the player space is isomorphic to $([0, 1], \mathcal{B})$, it follows that $\mathcal{C}$ is countably generated; that is, there is a countable family $S^1, S^2, \ldots$ of measurable sets such that $\mathcal{C} = F(\{S^1, S^2, \ldots\})$. For each S^i there are initial segments $I_1^i, I_2^i, \ldots$ such that $S^i \in F(\{I_1^i, I_2^i, \ldots\})$. Then if we renumber the I_j^i so that they form a sequence $I_1, I_2, \ldots$, we obtain

$$\mathcal{C} = F(\{I_1, I_2, \ldots\}). \tag{12.6}$$

Define $r_1, r_2, \ldots$ by $I_i = I(r_i; \mathcal{R})$ for all i.

Suppose $s \mathcal{R} t$; we claim that there is an r_i such that $s \overset{\mathcal{R}}{=} r_i \mathcal{R} t$. Indeed, if not, then each of the initial segments I_i either contains both s and t or neither one. Then from (12.6) it follows that every measurable S either contains both s and t or neither one; this contradicts the fact that $(I, \mathcal{C})$ is isomorphic to $([0, 1], \mathcal{B})$. Thus $\{r_1, r_2, \ldots\}$ is a denumerable $\mathcal{R}$-dense subset of I, and the proof of the lemma is complete.

The following corollary is of some interest, though it will not be used in the sequel.

COROLLARY 12.7. *If $\mathcal{R}$ is a measurable order on I, then $\{(s, t): s \mathcal{R} t\}$ is a measurable subset of $I \times I$.*

Proof. Sets containing a single point are measurable; hence each set of the form $\{s: s \mathcal{R} t\}$ is also measurable, since

$$\{s: s \mathcal{R} t\} = I \backslash (I(t; \mathcal{R}) \cup \{t\}).$$

Now if $\{r_1, r_2, \ldots\}$ is an $\mathcal{R}$-dense subset, then

$$\begin{aligned}\{(s, t): s \mathcal{R} t\} = &\bigcup_{i=1}^{\infty} \{s: s \mathcal{R} r_i\} \times \{t: r_i \mathcal{R} t\} \\ &\cup \bigcup_{i=1}^{\infty} \{r_i\} \times \{t: r_i \mathcal{R} t\} \\ &\cup \bigcup_{i=1}^{\infty} \{s: s \mathcal{R} r_i\} \times \{r_i\}.\end{aligned}$$

This completes the proof.

The converse of Corollary 12.7 is false. Take the unit square with its Borel subsets as the underlying space. If $\mathcal{R}$ is the lexicographic order, then $\mathcal{R}$ is not measurable, but $\{(s, t): s \,\mathcal{R}\, t\}$ is a measurable subset of $I \times I$ (see Note 1).

We saw above that although there is always at most one measure $\varphi^{\mathcal{R}}v$ satisfying (12.2), there may be none. Let us define v to be *orderable* if $\varphi^{\mathcal{R}}v$ does exist for all measurable $\mathcal{R}$, i.e. if for each measurable $\mathcal{R}$, there is a measure $\varphi^{\mathcal{R}}v$ on $\mathcal{C}$ satisfying (12.2) for all $s \in I$. Denote by ORD the linear space of all orderable v in BV.

PROPOSITION 12.8. *If $v \in ORD$, then for all measurable $\mathcal{R}$,*

$$\|\varphi^{\mathcal{R}}v\| \leqq \|v\|.$$

Furthermore, $AC \subset ORD$. Finally, if v is in AC and $\mu \in NA^1$ is such that $v \ll \mu$, then $\varphi^{\mathcal{R}}v \ll \mu$, uniformly[5] *for all measurable $\mathcal{R}$; in particular, $\varphi^{\mathcal{R}}v$ is non-atomic.*

Remark. $\varphi^{\mathcal{R}}v \ll \mu$ uniformly for all $\mathcal{R}$ means that the δ appearing in the definition of absolute continuity (see Sec. 5) depends on ϵ only and not on $\mathcal{R}$. If $|\varphi^{\mathcal{R}}v|$ is the "total variation" of $\varphi^{\mathcal{R}}v$ (Halmos, 1950, p. 122), this may be restated: For each $\epsilon > 0$ there is a $\delta > 0$—depending on ϵ and v only, and not on $\mathcal{R}$—such that $\mu(S) \leqq \delta$ implies $|\varphi^{\mathcal{R}}v|(S) \leqq \epsilon$. Of course, this condition of uniform absolute continuity implies the ordinary absolute continuity, i.e. it implies that $\varphi^{\mathcal{R}}v \ll \mu$ in the ordinary sense, for each $\mathcal{R}$ separately.

Proof. Let $H(\mathcal{R})$ be the field (not σ-field) generated by the initial segments $I(s; \mathcal{R})$; if we define a *half-open $\mathcal{R}$-interval* (or simply an *$\mathcal{R}$-interval* for short) to be the set difference between two initial segments, then the members of $H(\mathcal{R})$ are precisely the finite unions of disjoint $\mathcal{R}$-intervals. Whether or not $v \in ORD$, there is always a unique finitely additive measure ν on $H(\mathcal{R})$ such that

$$\nu(I(s; \mathcal{R})) = v(I(s; \mathcal{R})) \tag{12.9}$$

for all s.

Let $U \in H(\mathcal{R})$. Then we may write

$$U = \bigcup_{i=1}^{k} [t_i, s_i)_{\mathcal{R}}, \tag{12.10}$$

[5] See the remark below.

where the right side contains a self-explanatory notation for $\mathcal{R}$-intervals, and

(12.11) $$s_k \mathcal{R} t_k \mathcal{R} s_{k-1} \mathcal{R} \cdots \mathcal{R} s_1 \mathcal{R} t_1.$$

Let

$$S_i = I(s_i; \mathcal{R}),$$
$$T_i = I(t_i; \mathcal{R}),$$

let Ω be the chain

$$\varnothing \subset T_1 \subset S_1 \subset \cdots \subset T_k \subset S_k \subset I,$$

and let Λ be the subchain whose links are the $\{T_i, S_i\}$; we will say that Ω and Λ are *associated* with U. Then

(12.12) $$\|v\|_\Lambda = \sum_{i=1}^{k} |v(S_i) - v(T_i)| = \sum_{i=1}^{k} |\nu(S_i \backslash T_i)| \geqq \left| \sum_{i=1}^{k} \nu(S_i \backslash T_i) \right| = |\nu(U)|.$$

If $v \in ORD$, then ν may be uniquely extended from $H(\mathcal{R})$ to the measure $\varphi^{\mathcal{R}}v$ on $\mathcal{C}$; and in that case we may rewrite (12.12) as

$$\|v\|_\Lambda \geqq |(\varphi^{\mathcal{R}}v)(U)|.$$

To prove $\|\varphi^{\mathcal{R}}v\| \leqq \|v\|$, let $I = S^+ \cup S^-$ be a Hahn decomposition of I w.r.t. $\varphi^{\mathcal{R}}v$ (Halmos, 1950, p. 121). By a standard approximation theorem,[6] every measurable S can be approximated by members of $H(\mathcal{R})$ w.r.t. $\varphi^{\mathcal{R}}v$; that is, there is a $U^+ \in H(\mathcal{R})$ such that

$$|(\varphi^{\mathcal{R}}v)(S^+) - (\varphi^{\mathcal{R}}v)(U^+)| < \epsilon;$$

setting $U^- = I \backslash U^+$, we deduce easily that

$$|(\varphi^{\mathcal{R}}v)(S^-) - (\varphi^{\mathcal{R}}v)(U^-)| < \epsilon.$$

Now the chain Ω associated with U^+ is also associated with U^-; denote by Λ_+ and Λ_- the subchains associated with U^+ and U^-, respectively. Then

$$\|v\| \geqq \|v\|_\Omega = \|v\|_{\Lambda_+} + \|v\|_{\Lambda_-} \geqq |(\varphi^{\mathcal{R}}v)(U^+)| + |(\varphi^{\mathcal{R}}v)(U^-)| \geqq |(\varphi^{\mathcal{R}}v)(S^+)| - \epsilon + |(\varphi^{\mathcal{R}}v)(S^-)| - \epsilon = \|\varphi^{\mathcal{R}}v\| - 2\epsilon.$$

Letting $\epsilon \to 0$, we obtain $\|v\| \geqq \|\varphi^{\mathcal{R}}v\|$.

[6] One uses Theorem D of Halmos (1950, p. 56) on the "total variation" $|\varphi^{\mathcal{R}}v|$.

Suppose now that $v \in AC$; we wish to prove that $v \in ORD$, i.e. that ν can be extended to a countably additive measure $\varphi^{\mathcal{R}}v$ on $F(\mathcal{R}) = \mathcal{C}$. Now, by a well-known theorem,[7] in order to do this it is sufficient to prove that ν is bounded and that it is countably additive on $H(\mathcal{R})$; that is, if $U_1, U_2, \ldots$ is an infinite sequence of pairwise disjoint sets in $H(\mathcal{R})$ whose union U is also in $H(\mathcal{R})$, we must show that

$$\nu(U) = \sum_{i=1}^{\infty} \nu(U_i). \tag{12.13}$$

To prove that ν is bounded, let $U \in H(\mathcal{R})$, and let Λ be the subchain associated with U; then

$$|\nu(U)| \leqq \|v\|_\Lambda \leqq \|v\|.$$

To prove (12.13) let us for the moment fix k, and let $W_k = U \backslash \bigcup_{i=1}^k U_i$. Then $W_k \in H(\mathcal{R})$, so there is a subchain Λ associated with W_k; for this Λ we have

$$\|v\|_\Lambda \geqq |\nu(W_k)|.$$

Now let us choose a measure μ in NA^1 with $v \ll \mu$. Let $\epsilon > 0$ be given, and choose $\delta > 0$ to correspond to ϵ in accordance with the definition of $v \ll \mu$ (see Section 5). If k is sufficiently large, then $\|\mu\|_\Lambda = \mu(W_k) = \mu(U \backslash \bigcup_{i=1}^k U_i) < \delta$; this follows from the countable additivity of μ and $U = \bigcup_{i=1}^\infty U_i$. Then by (12.12),

$$|\nu(U) - \sum_{i=1}^{k} \nu(U_i)| = |\nu(W_k)| \leqq \|v\|_\Lambda < \epsilon;$$

it follows that $\nu(U) = \Sigma_{i=1}^\infty \nu(U_i)$. This completes the proof that $AC \subset ORD$.

Finally, to prove the uniform absolute continuity, let $\epsilon > 0$ be given, and let $\delta > 0$ correspond to $\epsilon/2$ in accordance with the definition of $v \ll \mu$. Let $\mathcal{R}$ be a measurable order, let $U \in H(\mathcal{R})$, and let the subchain Λ be associated with U; then

$$\mu(U) = \|\mu\|_\Lambda \leqq \delta \Rightarrow |(\varphi^{\mathcal{R}}v)(U)| = \|v\|_\Lambda \leqq \epsilon/2.$$

Now by a standard approximation theorem,[8] every measurable S can

[7] See, for example, Dunford and Schwartz (1958), III.5.9, p. 136.
[8] One uses Theorem D of Halmos (1950, p. 56) on the measure $\mu + |\varphi^{\mathcal{R}}v|$.

be approximated by members of $H(\mathcal{R})$, simultaneously w.r.t. μ and w.r.t. $\varphi^{\mathcal{R}}v$; hence

$$\mu(S) < \delta \Rightarrow |(\varphi^{\mathcal{R}}v)(S)| \leqq \epsilon/2,$$

for every measurable S. If $I = S^+ \cup S^-$ is a Hahn decomposition, then it follows that

$$\mu(S) < \delta \Rightarrow \mu(S \cap S^+) < \delta \Rightarrow |(\varphi^{\mathcal{R}}v)(S \cap S^+)| \leqq \frac{\epsilon}{2},$$

and similarly

$$\mu(S) < \delta \Rightarrow |(\varphi^{\mathcal{R}}v)(S \cap S^-)| \leqq \frac{\epsilon}{2}.$$

Hence

$$\mu(S) \leqq \frac{\delta}{2} < \delta \Rightarrow |\varphi^{\mathcal{R}}v|(S) = |(\varphi^{\mathcal{R}}v)(S \cap S^+)| + |(\varphi^{\mathcal{R}}v)(S \cap S^-)| \leqq \epsilon,$$

and Proposition 12.8 is proved.

A subset J of I will be called an *initial set* (to be distinguished from an initial segment!) if

$$s \in J,\ s\,\mathcal{R}\,s' \Rightarrow s' \in J.$$

Of course, any initial segment is an initial set, but not conversely.

LEMMA 12.14. *Let $\mathcal{R}$ be a measurable order, J an initial set; then J is measurable. Furthermore, if $v \in AC^+$, then setting $\bar{J} = J \cup \{-\infty\}$, we have*

$$(\varphi^{\mathcal{R}}v)(J) = v(J) = \sup_{s \in \bar{J}} v(I(s; \mathcal{R})) = \inf_{t \in \bar{I} \setminus \bar{J}} v(I(t; \mathcal{R})).$$

Proof. Let R be a denumerable $\mathcal{R}$-dense subset of I (Lemma 12.5), let $\bar{R} = \{-\infty\} \cup R \cup \{\infty\}$, and let

$$J_1 = \bigcap_{q \in \bar{R} \setminus \bar{J}} I(q; \mathcal{R}),$$

$$J_0 = \bigcup_{q \in \bar{R} \cap \bar{J}} I(q; \mathcal{R}).$$

Then $J_1 \supset J \supset J_0$, and because R is $\mathcal{R}$-dense and each $q \in \bar{R}$ must be either in $\bar{R} \cap \bar{J}$ or in $\bar{R} \setminus \bar{J}$, it follows that $J_1 \setminus J_0$ can contain at most two points. Since the intersection and union defining J_1 and J_0 respectively are denumerable, it follows that J_1 and J_0, and therefore also J, are measurable.

Now let $v \in AC^+$. Since the $I(q; \mathcal{R})$ are linearly ordered under inclusion, each finite intersection of the $I(q; \mathcal{R})$ is equal to one of them; hence

$$J_1 = \bigcap_i I(q_1^i; \mathcal{R}),$$

where $\{q_1^i\}$ is a sequence of points in $\bar{R}\backslash\bar{J}$ such that

$$q_1^1 \overset{\mathcal{R}}{\geqq} q_1^2 \overset{\mathcal{R}}{\geqq} \cdots,$$

i.e.,

$$I(q_1^1; \mathcal{R}) \supset I(q_1^2; \mathcal{R}) \supset \cdots.$$

Similarly

$$J_0 = \bigcup_i I(q_0^i; \mathcal{R}),$$

where $\{q_0^i\}$ is a sequence of points in $\bar{R} \cap \bar{J}$ such that

$$\cdots \overset{\mathcal{R}}{\geqq} q_0^2 \overset{\mathcal{R}}{\geqq} q_0^1,$$

i.e.,

$$I(q_0^1; \mathcal{R}) \subset I(q_0^2; \mathcal{R}) \subset \cdots.$$

Now since $\varphi^{\mathcal{R}}v$ is totally finite, we have

$$(\varphi^{\mathcal{R}}v)(J_1) = \lim_{i\to\infty} (\varphi^{\mathcal{R}}v)(I(q_1^i; \mathcal{R}))$$
$$(\varphi^{\mathcal{R}}v)(J_0) = \lim_{i\to\infty} (\varphi^{\mathcal{R}}v)(I(q_0^i; \mathcal{R})).$$

But since $J_1\backslash J_0$ consists of at most two points, and $\varphi^{\mathcal{R}}v$ is non-atomic (Proposition 12.8), it follows that $(\varphi^{\mathcal{R}}v)(J_1) = (\varphi^{\mathcal{R}}v)(J_0)$. Hence

$$v(I(q_1^i; \mathcal{R})) - v(I(q_0^i; \mathcal{R})) = (\varphi^{\mathcal{R}}v)(I(q_1^i; \mathcal{R})) - (\varphi^{\mathcal{R}}v)(I(q_0^i; \mathcal{R})) \to 0$$

as $i \to \infty$. But clearly, for each i we have

$$v(I(q_1^i; \mathcal{R})) \geqq \inf_{t\in\bar{I}\backslash\bar{J}} v(I(t; \mathcal{R})) \geqq \sup_{s\in\bar{J}} v(I(s; \mathcal{R})) \geqq v(I(q_0^i; \mathcal{R}));$$

hence if we let $i \to \infty$ we deduce that

$$\inf_{t\in\bar{I}\backslash\bar{J}} v(I(t; \mathcal{R})) = \sup_{s\in\bar{J}} v(I(s; \mathcal{R})).$$

Now note that, because v is monotonic, $(\varphi^{\mathcal{R}}v)(S)$ is non-negative whenever S is a half-open $\mathcal{R}$-interval, and so by (12.2) for all S. Since for $t \in \bar{I} \backslash \bar{J}$ and $s \in \bar{J}$ we have

$$I(t; \mathcal{R}) \supset J \supset I(s; \mathcal{R}),$$

it follows that

$$v(I(t; \mathcal{R})) = (\varphi^{\mathcal{R}}v)(I(t; \mathcal{R})) \geqq (\varphi^{\mathcal{R}}v)(J) \geqq (\varphi^{\mathcal{R}}v)(I(s; \mathcal{R})) = v(I(s; \mathcal{R}));$$

since v is monotonic, it also follows that

$$v(I(t; \mathcal{R})) \geqq v(J) \geqq v(I(s; \mathcal{R})).$$

Taking the inf w.r.t. t and the sup w.r.t. s and using the fact that they are equal, we deduce the conclusion of the lemma.

Now if $\mathcal{R}$ is a measurable order and μ an NA^1 measure on the underlying space, and if $\alpha \in [0, 1]$, let $J(\alpha; \mu, \mathcal{R})$ denote the intersection of all initial segments of μ-measure $> \alpha$.

Lemma 12.15. *$J(\alpha; \mu, \mathcal{R})$ is measurable, and $\mu(J(\alpha; \mu, \mathcal{R})) = \alpha$.*

Proof. The case $\alpha = 1$ is trivial, so w.l.o.g. we may exclude it. Denote $J = J(\alpha; \mu, \mathcal{R})$. Clearly J is an initial set; therefore by the previous lemma it is measurable. Set $v = \mu$. If $t \notin \bar{J}$, then there is an initial segment $I(t'; \mathcal{R})$ of μ-measure $> \alpha$ not containing t; hence $I(t; \mathcal{R}) \supset I(t'; \mathcal{R})$, so $\mu(I(t; \mathcal{R})) \geqq \mu(I(t'; \mathcal{R})) > \alpha$. Hence the inf appearing in the previous lemma is $\geqq \alpha$. If $s \in \bar{J}$, then we must have $\mu(I(s; \mathcal{R})) \leqq \alpha$; for if $\mu(I(s; \mathcal{R})) > \alpha$, then by definition $J \subset I(s; \mathcal{R})$, so $s \in I(s; \mathcal{R})$, a contradiction. Hence the sup appearing in the previous lemma is $\leqq \alpha$. Since the inf and the sup are equal, they both $= \alpha$, and the result follows from the previous lemma.

Note that this proof demonstrates

Corollary 12.16.

$$J(\alpha; \mu, \mathcal{R}) = \{s: \mu(I(s; \mathcal{R})) \leqq \alpha\}. \tag{12.17}$$

Lemma 12.18. *Let $f \in bv$ be non-decreasing and absolutely continuous, and let $\mu \in NA^1$. Let $\mathcal{R}$ be a measurable order and let $S \in \mathcal{C}$. Then*

$$\varphi^{\mathcal{R}}(f \circ \mu)(S) = \int_S f'(\mu(I(s; \mathcal{R}))) \, d\mu.$$

Proof. Define a transformation (not necessarily 1–1 or onto)

$$\Phi: (I, \mathcal{C}) \to ([0, 1], \mathcal{B})$$

by

$$\Phi(s) = \mu(I(s; \mathcal{R})).$$

Then for each α, Corollary 12.16 yields

$$\Phi^{-1}[0, \alpha] = \{s: \mu(I(s; \mathcal{R})) \leqq \alpha\} = J(\alpha; \mu, \mathcal{R}).$$

By Lemma 12.15, the right side of this is measurable; hence Φ is a measurable transformation. Again applying Lemma 12.15 we find that $\mu\Phi^{-1}[0, \alpha] = \alpha$; hence the measure $\mu\Phi^{-1}$ coincides with Lebesgue measure λ. Hence, by the change-of-variables formula (Halmos, 1950, p. 163, Theorem C), we have

$$\begin{aligned}(12.19)\quad \int_{J(\alpha;\mu,\mathcal{R})} f'(\mu(I(s; \mathcal{R}))) \, d\mu &= \int_{\Phi^{-1}[0,\alpha]} f'(\Phi(s)) \, d\mu \\ &= \int_0^\alpha f'(t) \, d\mu\Phi^{-1} = \int_0^\alpha f'(t) \, d\lambda = f(\alpha) \\ &= f(\mu(J(\alpha; \mu, \mathcal{R}))) = (f \circ \mu)(J(\alpha; \mu, \mathcal{R})) \\ &= \varphi^{\mathcal{R}}(f \circ \mu)(J(\alpha; \mu, \mathcal{R})),\end{aligned}$$

where the fourth equality sign in this sequence follows from the absolute continuity of f (see (8.4)) and the last one from Lemma 12.14. This proves the assertion of the lemma when S is of the form $J(\alpha; \mu, \mathcal{R})$. When S is an initial segment, let $\alpha = \mu(S)$. Then $S \subset J(\alpha; \mu, \mathcal{R})$ (by Corollary 12.16), and so $\mu(J(\alpha; \mu, \mathcal{R})\backslash S) = \alpha - \alpha = 0$ (by Lemma 12.15). Hence

$$\int_S f'(\mu(I(s; \mathcal{R}))) \, d\mu = \int_{J(\alpha;\mu,\mathcal{R})} f'(\mu(I(s; \mathcal{R}))) \, d\mu.$$

Moreover from Lemma 12.14 (and the fact that an initial segment is an initial set) it follows that

$$\varphi^{\mathcal{R}}(f \circ \mu)(S) = f(\mu(S)) = f(\alpha).$$

Hence it follows from (12.19) that

$$\int_S f'(\mu(I(s; \mathcal{R}))) \, d\mu = \varphi^{\mathcal{R}}(f \circ \mu)(S).$$

But both sides of this are measures in S, and since the initial segments generate $\mathcal{C}$ (by (12.3)), the proof of Lemma 12.18 is complete.

Proof of Theorem D. Suppose it were possible to define measurability and the measure ω on Ω; define φ by (12.1). Let $\mu \in NA^1$ and let $v = \mu^3$. Since by assumption φ is a value, Proposition 6.1 yields $\varphi v = \mu$. Hence from (12.1) and Lemma 12.18 we get

$$\mu(S) = \int_\Omega \varphi^{\mathcal{R}}(\mu^3)(S)\, d\omega = \int_\Omega d\omega \int_S 3\mu^2(I(s;\mathcal{R}))\, d\mu \tag{12.20}$$

for all $S \in \mathcal{C}$.

Now let V and T be disjoint subsets of I, both of positive μ-measure. Define measures μ_T, ν, and ξ by

$$\begin{aligned} \mu_T(S) &= \mu(T \cap S)/\mu(T), \\ \nu &= (\mu + \mu_T)/2, \\ \xi &= (\mu + \nu)/2 = \tfrac{3}{4}\mu + \tfrac{1}{4}\mu_T; \end{aligned}$$

note that for $S \subset V$,

$$\mu(S) = 2\nu(S) = \tfrac{4}{3}\xi(S).$$

Now set $S = V$ and apply (12.20) three times—as it stands, and with ν and then ξ substituted for μ. This yields

$$\int_\Omega d\omega \int_V 3\mu^2(I(s;\mathcal{R}))\, d\mu = \mu(V),$$

$$\int_\Omega d\omega \int_V 3\nu^2(I(s;\mathcal{R}))\, d\mu = \int_\Omega d\omega \int_V 3\nu^2(I(s;\mathcal{R}))\, 2d\nu = 2\nu(V) = \mu(V),$$

and

$$\int_\Omega d\omega \int_V 3\xi^2(I(s;\mathcal{R}))\, d\mu = \int_\Omega d\omega \int_V 3\xi^2(I(s;\mathcal{R}))\, \tfrac{4}{3}d\xi = \tfrac{4}{3}\xi(V) = \mu(V).$$

Using $(x-y)^2 = 2x^2 + 2y^2 - 4((x+y)/2)^2$, we deduce that

$$\begin{aligned} \int_\Omega d\omega \int_V 3(\mu(I(s;\mathcal{R})) - \nu(I(s;\mathcal{R})))^2\, d\mu &= 2\int_\Omega d\omega \int_V 3\mu^2(I(s;\mathcal{R}))\, d\mu \\ &\quad + 2\int_\Omega d\omega \int_V 3\nu^2(I(s;\mathcal{R}))\, d\mu - 4\int_\Omega d\omega \int_V 3\xi^2(I(s;\mathcal{R}))\, d\mu \\ &= 2\mu(V) + 2\mu(V) - 4\mu(V) = 0. \end{aligned}$$

Therefore ω-a.e. in Ω, it is true that μ-a.e. in V, $\mu(I(s;\mathcal{R})) = \nu(I(s;\mathcal{R}))$, which yields $\mu(I(s;\mathcal{R})) = \mu_T(I(s;\mathcal{R}))$, and hence

$$\mu(T)\mu(I(s;\mathcal{R})) - \mu(I(s;\mathcal{R}) \cap T) = 0. \tag{12.21}$$

Now if we fix V with $0 < \mu(V) < 1$, this will be true for all subsets T of $I \backslash V$; in particular, if $\mathfrak{F}$ is a denumerable family of subsets of $I \backslash V$ that generates[9] all the measurable subsets of $I \backslash V$, then for each T in $\mathfrak{F}$, the previous sentence is true. But then it follows that ω-a.e. in Ω, μ-a.e. in V, for all T in $\mathfrak{F}$, (12.21) holds. Now for fixed $\mathcal{R}$ and s, the left side of (12.21) is a measure in T; since it vanishes for all T in the generating family $\mathfrak{F}$, it must vanish for all subsets T of $I \backslash V$. Thus we have shown that ω-a.e. in Ω, μ-a.e. in V, for all $T \subset I \backslash V$, (12.21) holds.

Now pick an $\mathcal{R}$ in Ω for which the previous sentence is true, i.e. which does not belong to the exceptional set of ω-measure 0. For this $\mathcal{R}$, Lemma 12.15 yields

$$\mu(J(\tfrac{1}{2}; \mu, \mathcal{R}) \backslash J(0; \mu, \mathcal{R})) = \tfrac{1}{2} - 0 > 0.$$

Hence there is an s in $J(\frac{1}{2}; \mu, \mathcal{R}) \backslash J(0; \mu, \mathcal{R})$ that is not exceptional, i.e., such that for all $T \subset I \backslash V$, (12.21) holds. If we choose such an s and let

$$J = I(s; \mathcal{R}),$$

then from Corollary 12.16 it follows that

$$0 < \mu(J) \leqq \tfrac{1}{2}.$$

Setting $T = I \backslash V$ in (12.21), we deduce

$$\mu(J \backslash V) = \mu(J \cap (I \backslash V)) = \mu(J)\mu(I \backslash V) > 0. \tag{12.22}$$

Then setting $T = J \backslash V$ in (12.21), we get $\mu(J \backslash V) = \mu(T \cap J) = \mu(T)\mu(J) \leqq \frac{1}{2}\mu(T) = \frac{1}{2}\mu(J \backslash V)$; hence $\mu(J \backslash V) = 0$, in contradiction to (12.22). This completes the proof of Theorem D.

Remark. Note that we have proved a stronger form of Theorem D. It is even impossible to define measurability and a probability measure ω on Ω, so that (12.1) defines a value φ on the space[10] P of polynomials in NA measures.[11] In the proof, we did not even use all such polynomials; we used cubes, i.e. third-degree polynomials only.

[9] I.e., the smallest σ-field containing $\mathfrak{F}$ consists of all measurable subsets of $I \backslash V$.

[10] See Note 2 to Section 7.

[11] Moreover, since we used only measures absolutely continuous w.r.t. μ, it follows that ω cannot be defined even if one restricts oneself to measures absolutely continuous w.r.t a given measure.

NOTES

1. The measurability condition $\mathcal{C} = F(\mathcal{R})$ for orders says that any coalition can be constructed in denumerably many steps from initial segments $I(s; \mathcal{R})$ by using the operations of complementation and denumerable union. In finite games, every individual is the difference of initial segments, and so every coalition can be constructed from the initial segments. It is this property that we wish to capture in the definition of measurability.

An example of a non-measurable order is the lexicographic order $\mathcal{R}$ on the unit square $[0, 1]^2$, defined by

$$x \,\mathcal{R}\, y \Leftrightarrow x_1 > y_1 \quad \text{or} \quad (x_1 = y_1 \text{ and } x_2 > y_2).$$

In this case a set is in $F(\mathcal{R})$ if and only if it is of the form

$$(S_0 \times [0, 1]) \cup \bigcup_{i=1}^{\infty} (\{s_i\} \times S_i),$$

where $S_0, S_1, S_2, \ldots$ are Borel subsets of $[0, 1]$ and $s_1, s_2, \ldots$ are members of $[0, 1]$; that is, the union of a "cylinder set" whose base is a Borel set on the horizontal axis, and denumerably many Borel sets sitting on vertical lines. The bottom half $\{x: x_2 \leqq \frac{1}{2}\}$ of I, though it is a Borel set, is not in $F(\mathcal{R})$; thus the inclusion $\mathcal{C} \supset F(\mathcal{R})$ is strict, so that $\mathcal{R}$ is not measurable.

Condition (12.2) defines $\varphi^{\mathcal{R}}v$ only on the initial segments. If $\mathcal{R}$ is measurable, then $\varphi^{\mathcal{R}}v$ is determined on $\mathcal{C}$, in the sense that there is at most one measure on $\mathcal{C}$ satisfying (12.2). If, on the other hand, the inclusion $\mathcal{C} \supset F(\mathcal{R})$ is strict, then $\varphi^{\mathcal{R}}v$ is in general not determined. Thus let $\mathcal{R}$ be the lexicographic order discussed above and let v be (2-dimensional) Lebesgue measure; then there are many measures $\varphi^{\mathcal{R}}v$ on $\mathcal{C}$ obeying (12.2), for example Lebesgue measure or the measure μ defined by

$$\mu(S) = \iint_S 2x_2 \, dx_1 \, dx_2.$$

It is perhaps worthwhile to mention that an order $\mathcal{R}$ is measurable if and only if it has a denumerable $\mathcal{R}$-dense subset and all the initial segments are measurable. The necessity is Lemma 12.15. For the sufficiency, note that from the existence of a denumerable $\mathcal{R}$-dense subset it follows that $F(\mathcal{R})$ is countably generated. Moreover, since $\mathcal{R}$ is an order, $F(\mathcal{R})$ is separating; i.e. for each s and t there is a member of $F(\mathcal{R})$ containing s but not t. Since $F(\mathcal{R}) \subset \mathcal{C}$ and a countably generated separating σ-field of Borel sets in the unit interval must be the whole Borel σ-field [Mackey, 1957, Theorem 3.3, p. 139], our assertion follows.

13. The Impossibility Principle: An Alternative Formulation

As stated, Theorem D shows that the random order approach cannot yield a value in the axiomatic sense of Section 2. However, we might take a slightly different view of the random order approach.

Let's start out by recalling that in n-player games (n finite) there is a "natural" probability measure on the set of all orderings of the players, namely the uniform measure that assigns probability $1/n!$ to each of the $n!$ possible orderings. Thus one might look for a probability measure ω on Ω that is "natural" or "uniform" and then use (12.1) to define a value. One would expect some connection between such an approach and the notion of symmetry as expressed in (2.2); in particular, one would be surprised if a "natural" ω yields a φ that does not obey (2.2). Nevertheless, conceptually this view is slightly different from that expressed in Theorem D, and it deserves at least to be explored.

Obviously, there are many ways to impose a probability measure on Ω; for a trivial example, one can concentrate all the probability on one order $\mathcal{R}$. This is unsatisfactory because of its arbitrariness; why one order rather than another order? The situation as given consists of the underlying space $(I, \mathcal{C})$. Singling out a specific order $\mathcal{R}$ on which to concentrate all of the probability adds a new, non-intrinsic element to the situation, which cannot be derived from what is originally given.

Let $I = (I, \mathcal{C})$ and $I' = (I', \mathcal{C}')$ be two underlying spaces, Ω and Ω' the spaces of all measurable orders on I and I' respectively, Θ an isomorphism from I onto I'. Then to each order $\mathcal{R}$ on I, there corresponds an order $\Theta\mathcal{R}$ on I', defined by

$$\Theta s \; \Theta\mathcal{R} \; \Theta t \Leftrightarrow s \; \mathcal{R} \; t.$$

Note that $\Theta\mathcal{R}$ is measurable if and only if $\mathcal{R}$ is; thus Θ may be considered a one-one mapping from Ω onto Ω', in addition to being an isomorphism from I onto I'. Furthermore, if on the space Ω there is defined a σ-field of measurable subsets and a probability measure ω, then we may define a corresponding σ-field and a corresponding probability measure ω' on Ω', as follows: the measurable subsets of Ω' are precisely those of the form $\Theta\Gamma$, where Γ is measurable in Ω; and their probability $\omega'(\Theta\Gamma)$ is given by

$$\omega'(\Theta\Gamma) = \omega(\Gamma).$$

Now set $I' = I$, whence $\Omega' = \Omega$. Then Θ transforms the given situation back into itself; it does not change anything—it is a symmetry of the situation. Therefore it should not change ω either, i.e. we should have $\omega' = \omega$. This means that for each automorphism Θ of the underlying space and each $\Gamma \subset \Omega$, we should have

(13.1) $\Theta\Gamma$ *is measurable in* Ω *if and only if* Γ *is measurable,*

and

(13.2) $$\omega\Theta = \omega.$$

Together with (12.4), (13.1) and (13.2) spell out what one must expect from a "natural" probability measure[1] on Ω. As we might by now expect, it turns out that there is no such measure; we have

PROPOSITION 13.3. *It is impossible to define measurability and a probability measure* ω *on* Ω, *so that* (12.4), (13.1), *and* (13.2) *hold for each automorphism* Θ *of the underlying space and each measurable* $\Gamma \subset \Omega$.

Proposition 13.3 is an immediate consequence of Theorem D; the proof of this assertion points up the close connection between the notion of a "natural" measure on Ω and the symmetry axiom (2.2). Indeed, if φ is defined by (12.1), then (13.1) and (13.2) imply (2.2); this can be seen by means of a straightforward computation (using a change of variables[2] in integration). However, the reverse implication (i.e., that Proposition 13.3 implies Theorem D) is by no means clear; it appears that the "naturalness" conditions ((13.1) and (13.2)) on ω are stronger than (2.2), and so in a certain sense more difficult to satisfy. This feeling gains support from the fact that one can prove a stronger form of Proposition 13.3—namely Proposition 13.4 below—whereas it is not clear whether a corresponding strengthening of Theorem D is true; moreover, the proof of this stronger result is independent of Theorem D and is comparatively simple.

Before stating Proposition 13.4, let us discuss its motivation. Both in Theorem D and in Proposition 13.3 it was tacitly assumed that the probability distribution ω on Ω should depend on I and $\mathcal{C}$ only, and not on any specific set function v. It would, indeed, be curious, and contrary to the spirit of the discussion, if the definition of "ran-

[1] In case of a finite underlying space, those conditions uniquely determine the uniform probability measure discussed above.

[2] See Theorem C of Halmos (1950, p. 163).

dom order" changed from game to game. For the purposes of Proposition 13.3 this was, however, not a crucial point; our conclusions remain unchanged if we fix v and even Θ.

A *symmetry* of a set function v is an automorphism Θ such that $\Theta_* v = v$.

PROPOSITION 13.4. *There is a v in pNA and a symmetry of v such that there are no σ-field of measurable sets and probability measure ω on Ω satisfying* (12.4), (13.1), *and* (13.2). *Moreover, v can be chosen to be the square of a single NA^1 measure.*

Proposition 13.4 shows that the impossibility asserted in Proposition 13.3 is not of a "global" nature, i.e. it is not caused by some kind of incompatibility between various set functions in *pNA* or even between various automorphisms of the underlying space. Rather, it holds even if one restricts attention entirely to a single v, and what is perhaps even more remarkable, to a single automorphism[3] Θ. Thus the measurability properties on Ω asserted by (12.4) are asserted only for one specific v; and (13.1) and (13.2) now assert invariance only under one specific Θ, which is moreover a symmetry of v.

Before going on to the proof of Proposition 13.4, we remark that the impossibility theorems of this and the previous section depend basically on symmetry considerations, i.e. on the demand that the situation remain essentially invariant under any automorphism of the underlying space. Such a demand is unavoidable in the context of this book. In a different context, however—if more "structure" were imposed on the game—it might be avoided; or if not avoided, at least rendered harmless. We could, for example, view the underlying space as having a topological as well as a measurable structure. That, of course, would change our whole outlook. We would restrict Θ to be a homeomorphism of the topological space as well as an automorphism of the measurable space; this restriction would be applied to the invariance conditions

[3] Intuitively, fixing Θ does not appear to add much strength to the result; (13.1) and (13.2) are no less convincing for all Θ simultaneously than for any specific Θ. Mathematically, though, it is interesting that ω cannot be defined even if we allow it to depend on Θ as well as on v. Incidentally, if we allowed ω to depend on v but not on Θ, we could deduce the result fairly easily from the above proof of Theorem D. For the slightly stronger result stated here, we will, however, require a separate proof.

(13.1) and (13.2) and to the symmetry condition (2.2). The outcome would be a value that is not the analog of the ordinary value for finite games as first treated by Shapley (1953a), but rather of variants in which the symmetry axiom is relaxed in one way or another. We have not followed this direction any further (cf. Shapley, 1953b, and Owen, 1968, 1971, 1972).

The following lemma will be needed in the sequel:

LEMMA 13.5. *If* Θ *is any automorphism of the underlying space, then*

$$\Theta J(\alpha; \mu, \mathcal{R}) = J(\alpha; \Theta_*^{-1}\mu, \Theta\mathcal{R}).$$

Proof: This is a straightforward consequence of $\Theta I(s; \mathcal{R}) = I(\Theta s; \Theta\mathcal{R})$; the details are omitted.

We now return to the

Proof of Proposition 13.4. W.l.o.g. let $(I, \mathcal{C}) = ([0, 1], \mathcal{B})$. Define $v = f \circ \lambda$, where

$$f(x) = \begin{cases} x, & \text{when } x \leqq \frac{1}{2}, \\ \frac{1}{2}, & \text{when } x \geqq \frac{1}{2}. \end{cases}$$

For each measurable order $\mathcal{R}$, set $J = J(\mathcal{R}) = J(\frac{1}{2}; \lambda, \mathcal{R})$. Then for each $S \subset I$,

$$\varphi^{\mathcal{R}}(v)(S) = \lambda(J \cap S); \tag{13.6}$$

this follows from Lemma 12.18 and Corollary 12.16.

We now apply a trick due to Furstenberg (cf. Aumann, 1967b). Let Θ be a Lebesgue-measure-preserving automorphism of the underlying space, which is moreover (strongly) mixing.[4] For each fixed measurable $S \subset I$, define a sequence of random variables $\mathbf{x}_1, \mathbf{x}_2, \ldots$ on the probability space Ω by

$$\mathbf{x}_n = \mathbf{x}_n^{\mathcal{R}} = \lambda(J(\Theta^n\mathcal{R}) \cap S);$$

from (12.4), (13.1), and (13.6) it follows[5] that the $\mathbf{x}_n$ are indeed random variables, i.e., measurable in $\mathcal{R}$. From (13.2) it follows that all the $\mathbf{x}_n$ have the same distribution. On the other hand, since Θ^n is measure-

[4] Readers not familiar with this term may refer to the definition, which is given in the next section.

[5] Note that Θ and hence all the Θ^n are not only automorphisms of the under-lying space but also symmetries of v.

preserving, Lemma 13.5 yields $J(\Theta^n \mathcal{R}) = \Theta^n J(\mathcal{R})$. Therefore, since Θ is mixing, it follows that

$$\mathbf{x}_n = \lambda(\Theta^n J(\mathcal{R}) \cap S) \to \lambda(J(\mathcal{R}))\lambda(S) = \tfrac{1}{2}\lambda(S).$$

So $\mathbf{x}_n$ tends to a constant (independent of $\mathcal{R}$); but since all the $\mathbf{x}_n$ have the same distribution, it follows that for each n, $\mathbf{x}_n = \frac{1}{2}\lambda(S)$ with probability 1. In particular, setting $n = 0$, we deduce

$$\lambda(J(\mathcal{R}) \cap S) = \tfrac{1}{2}\lambda(S) \tag{13.7}$$

with probability 1. In particular, this is true whenever S is a finite union of intervals with rational endpoints; therefore, since there are only denumerably many S of this kind, (13.7) holds simultaneously for *all* such S with probability 1. But from this, a simple approximation argument leads to the conclusion that, with probability 1, (13.7) holds simultaneously for *all* measurable S. If we then take an $\mathcal{R}$ for which (13.7) does in fact hold for all S, and set $S = J(\mathcal{R})$, we obtain

$$\tfrac{1}{2} = \lambda(J(\mathcal{R})) = \lambda(J(\mathcal{R}) \cap S) = \tfrac{1}{2}\lambda(S) = \tfrac{1}{2}\lambda(J(\mathcal{R})) = \tfrac{1}{4},$$

an absurdity. This shows that v and Θ indeed have the property asserted in the proposition.

We could, alternatively, have chosen $v = \lambda^2$. As above, let Θ be a mixing transformation. For fixed S, define random variables $\mathbf{x}_n$ on Ω by

$$\mathbf{x}_n = (\varphi^{\Theta^n \mathcal{R}} v)(S) = \int_S 2\lambda(I(s; \Theta^n \mathcal{R}))\, d\lambda$$

(see Lemma 12.18); from (13.2) it then follows that all the $\mathbf{x}_n$ have the same distribution. On the other hand, it can be shown by using Lemma 15.12 below that $\mathbf{x}_n \to \lambda(S)$ for all $\mathcal{R}$. Hence, as above, $x_0 = \lambda(S)$ with probability 1, and, again as above, we may deduce that with probability 1, we have for all S that

$$\int_S 2\lambda(I(s; \mathcal{R}))\, d\lambda = \lambda(S).$$

Hence with probability 1, λ-a.e. in s,

$$\lambda(I(s; \mathcal{R})) = \tfrac{1}{2},$$

which easily leads to a contradiction (cf. Lemma 12.15 and Corollary 12.16). This completes the proof of Proposition 13.4.

Rather than paralleling the proof in Aumann (1967b), as we have done here, it is also possible to use the theorem of that paper directly in order to prove Proposition 13.4; but little if any space would be saved in the process.

14. The Mixing Value: Statement of Results

The impossibility theorems in the previous sections may be intuitively understood as follows: Suppose that there is a large finite number n of players, rather than a continuum. Let S be a coalition of these players which is "not too small"; to fix ideas, let $|S| = [n/4]$ (the greatest integer in $n/4$). For each ordering $\mathcal{R}$ of the players let $J(\mathcal{R})$ be the coalition consisting of the first $[n/2]$ players. If we assign probability $1/n!$ to each of the $n!$ possible orderings $\mathcal{R}$ of the player set, then it may be seen that $|J(\mathcal{R}) \cap S|/n$ will be close to $\frac{1}{2} \cdot \frac{1}{4}$ with high probability; more precisely, for each $\epsilon > 0$,

$$\text{Prob}\left(\left|\frac{|J(\mathcal{R}) \cap S|}{n} - \tfrac{1}{4} \cdot \tfrac{1}{2}\right| > \epsilon\right) \to 0$$

as $n \to \infty$. Thus if we shuffle a deck of cards and then cut it in half, the proportion of hearts in the top half will with high probability be close to $\frac{1}{4}$.

More generally, if we replace $\frac{1}{4}$ by any other fixed number θ such that $0 < \theta \leqq 1$, i.e., if we take $|S| = [\theta n]$, then $|J(\mathcal{R}) \cap S|/n$ will be close to $\theta/2$ with high probability.

If we now "pass to the limit," i.e. use a continuum of players rather than a finite number, then we might expect to replace the phrase "close to," by "equal to," and "high probability" by "probability 1." An appropriate analogue for the relative number of players in a coalition T (i.e., the quantity $|T|/n$) might be the measure $\mu(T)$. Of course it is not quite clear which measure μ is the correct one to use, and presumably this depends on just how one "passes to the limit"; but this is a technical point that need not concern us for the present, since the whole discussion is meant only to be suggestive. The upshot is that we would expect, for every S, that

$$\mu(J(\mathcal{R}) \cap S) = \tfrac{1}{2}\mu(S)$$

with probability 1; and this leads to a contradiction as at (13.7).

We saw in the previous section that matters are not really as simple as all this. Nevertheless, it will be useful to keep in mind the principle behind the impossibility theorem: If it were possible to define the notion of random order, then the first "half" of the players would with probability 1 be evenly mixed among all the players, i.e. the first half would intersect each coalition S in the same proportion that it intersects the whole player space; and such evenly mixed sets do not exist.[1]

This idea of "even mixing" reminds us of the notion of mixing transformation from ergodic theory. Let μ be a probability measure on the underlying space $(I, \mathcal{C})$. Recall that a *mixing transformation* of the measure space $(I, \mathcal{C}, \mu)$ is a μ-measure preserving automorphism Θ of $(I, \mathcal{C})$ such that for all $S, T \in \mathcal{C}$, we have

$$\mu(S \cap \Theta^n T) \to \mu(S)\mu(T) \tag{14.1}$$

as $n \to \infty$. Roughly speaking, the sequence $(\Theta, \Theta^2, \ldots)$ accomplishes in the long run the mixing job that we would have liked to (but cannot) accomplish "in one fell swoop" with a single automorphism. We will now see how such a sequence can be used as an approach to the value concept.

First let us generalize slightly the notion of mixing transformation. All that really interests us is the property (14.1) of the sequence $\{\Theta, \Theta^2, \ldots\}$; the fact that it is a sequence of powers is irrelevant. W.r.t. a fixed non-atomic probability measure μ on $(I, \mathcal{C})$, we therefore define a *μ-mixing sequence* to be a sequence $\{\Theta_1, \Theta_2, \ldots\}$ of μ-measure-preserving automorphisms of $(I, \mathcal{C})$, such that for all $S, T \in \mathcal{C}$, we have

$$\mu(S \cap \Theta_n T) \to \mu(S)\mu(T).$$

Now let $\mathcal{R}$ be a fixed measurable order (cf. Sec. 12) on $(I, \mathcal{C})$. Then for each automorphism Θ of $(I,\mathcal{C})$, $\Theta\mathcal{R}$ is also a measurable order. We shall be interested in measures of the form $\varphi^{\Theta\mathcal{R}}v$ (cf. (12.2)); since this

[1] Cf. (12.22) ff. or (13.7) ff.

notation will occasionally be cumbersome, we will sometimes write $\varphi(v; \Theta\mathcal{R})$ rather than $\varphi^{\Theta\mathcal{R}}v$.

DEFINITION. Let $v \in ORD$. A set function[2] φv is said to be the *mixing value of* v if there is a measure μ_v in NA^1 such that for all μ in NA^1 with $\mu_v \ll \mu$,

(14.2) for all μ-mixing sequences $\{\Theta_1, \Theta_2, \ldots\}$, for all measurable orders $\mathcal{R}$, and for all coalitions S, we have

$$\varphi(v; \Theta_n\mathcal{R})(S) \to (\varphi v)(S) \text{ as } n \to \infty.$$

The mixing value, if it exists, is clearly[3] unique. The set of all members of ORD that have mixing values will be denoted MIX.

Let us review the idea behind this definition: We saw before that a random order (if there were such a thing) would be "thoroughly mixing" with probability 1. So rather than averaging the $\varphi^{\mathcal{R}}v$ over all $\mathcal{R}$ as we tried to do in the previous sections, we could let $\mathcal{R}$ be a single "thoroughly mixing" order, and define the value to be $\varphi^{\mathcal{R}}v$ for this $\mathcal{R}$. This would be especially convincing if $\varphi^{\mathcal{R}}v$ were the same for all "thoroughly mixing" $\mathcal{R}$. Unfortunately, as we have seen, there are no "thoroughly mixing" orders. But there *are* orders that are "approximately thoroughly mixing," namely the orders $\Theta_n\mathcal{R}$, where $\{\Theta_1, \Theta_2, \ldots\}$ is a μ-mixing sequence and $\mathcal{R}$ is arbitrary. So the measures $\varphi(v; \Theta_n\mathcal{R})$ may be considered "approximate values"; and if their limit exists and is independent of the various choices that must be made, we may feel justified in calling it a value.

Let us now devote a few words to the meaning of the measure μ_v. Let us think of a measure μ as being "more sensitive" than a measure ν whenever $\nu \ll \mu$ but not $\mu \ll \nu$. Then the above definition says that φv is the mixing value of v if and only if (14.2) holds for all "sufficiently sensitive" μ. The need for such a condition may be illustrated as follows: If $(I, \mathcal{C}) = ([0, 1], \mathcal{B})$, $v = \lambda^2$, and μ is defined by $\mu(S) = \lambda([0, \frac{1}{2}] \cap S)$, then a sequence whose behavior on $[\frac{1}{2}, 1]$ is perfectly arbitrary may be μ-mixing, and clearly this cannot always produce convergence in (14.2). In Appendix C we shall show that when $v \in AC$,

[2] We will see below that the mixing value, if it exists, must necessarily be in FA.

[3] On condition that there are μ-mixing sequences. See Proposition 14.3.

replacing the condition $\mu_v \ll \mu$ by the somewhat similar condition $v \ll \mu$ leads to an equivalent definition of mixing value.

Finally, we remark that the concept of μ-mixing sequence on which the definition is based is not a vacuous one. Indeed, we have

PROPOSITION 14.3. *For every μ in NA^1 there is a μ-mixing sequence.*

Proof. W.l.o.g. let $(I, \mathcal{C}) = ([0, 1], \mathcal{B})$. Let Θ be a Lebesgue-measure-preserving mixing transformation, e.g., the bilateral 1-shift.[4] If $\mu = \lambda$ then $\{\Theta, \Theta^2, \ldots\}$ is a μ-mixing sequence. For general μ, let Φ be an automorphism such that $\Phi_* \mu = \lambda$ (Lemma 6.2); then $\{\Phi\Theta\Phi^{-1}, \Phi\Theta^2\Phi^{-1}, \ldots\}$ is a μ-mixing sequence. This completes the proof of Proposition 14.3.

We are now ready for the statement of the main result.

THEOREM E. *MIX is a closed symmetric linear subspace of BV which contains pNA, and the operator φ that associates to each v its mixing value φv is a value on MIX.*

It follows from this that the mixing value exists for all set functions in pNA and coincides with the unique value on pNA as determined in Chapter I (Propositions 7.6 and 7.11). Our proof (in Sec. 15) will be independent of the proof of Theorems A and B given in Chapter I, and indeed it will enable us to give an alternative proof of Proposition 7.6 (the existence of a value on pNA). Also, we shall show (in Section 16) that the set function of Example 9.4 is not in MIX. Combined with the fact that $pNA \subset MIX$, this provides an independent proof that this set function is not in pNA; it also shows that MIX does not comprise all of ORD, nor even all of AC.

15. The Mixing Value: Proof of Theorem E

LEMMA 15.1. *Let $\mathcal{R}$ be a measurable order. Then $\varphi^{\mathcal{R}}$ is a linear operator on ORD.*

Proof. Follows easily from (12.2).

LEMMA 15.2. *Let $v \in MIX$, with mixing value φv. Then $\varphi v \in FA$, and*

$$\|\varphi v\| \leqq \|v\|.$$

[4] Halmos (1956), pp. 9 and 37, or see Section 16 below.

Proof. The finite additivity of φv follows from the definition in a straightforward manner. Next, let S and T be disjoint sets whose union is I. Then by Proposition 12.8,

$$|\varphi(v; \Theta_n\mathcal{R})(S)| + |\varphi(v; \Theta_n\mathcal{R})(T)| \leqq \|\varphi(v; \Theta_n\mathcal{R})\| \leqq \|v\|. \tag{15.3}$$

Hence

$$|(\varphi v)(S)| + |(\varphi v)(T)| = \lim_{n\to\infty} (|\varphi(v; \Theta_n\mathcal{R})(S)| + |\varphi(v; \Theta_n\mathcal{R})(T)|) \leqq \|v\|.$$

Since $\|\varphi v\| = \sup(|(\varphi v)(S)| + |(\varphi v)(T)|)$, where the sup is taken over all choices of S and T, the lemma is proved.

PROPOSITION 15.4. *MIX is a closed linear subspace of BV.*

Proof. That $v \in MIX \Rightarrow \alpha v \in MIX$ for all scalar α is immediate. Suppose that $v, w \in MIX$ and that μ_v and μ_w correspond to v and w respectively in accordance with the definition of mixing value. Then $(\mu_v + \mu_w)/2$ corresponds to $v + w$. This proves that MIX is a subspace.

To prove that MIX is closed, let $v_i \to v$, where $v_i \in MIX$. Set

$$\mu_v = \sum_{i=1}^{\infty} \mu_{v_i}/2^i.$$

Let $\mu \gg \mu_v$, let $\{\Theta_1, \Theta_2, \ldots\}$ be a μ-mixing sequence, and let $\mathcal{R}$ be a measurable order. Let φv_i be the mixing value of v_i. From $v_i \to v$ and Lemma 15.2 it follows that φv_i is a Cauchy sequence (in the variation norm); denote its limit by φv. Since $\mu_{v_i} \ll \mu_v \ll \mu$, it follows that for all $S \subset I$,

$$\varphi(v_i; \Theta_k\mathcal{R})(S) \to (\varphi v_i)(S)$$

as $k \to \infty$. We wish to show that

$$\varphi(v; \Theta_k\mathcal{R})(S) \to (\varphi v)(S) \tag{15.5}$$

for all $S \subset I$. To this end, let $S \subset I$ and $\epsilon > 0$ be given. Then there is a number $j = j(\epsilon)$ such that

$$\|v_j - v\| \leqq \epsilon/3 \text{ and } \|\varphi v_j - \varphi v\| \leqq \epsilon/3.$$

For this j we may find a number $k_0 = k_0(\epsilon)$ such that for $k \geqq k_0$,

$$|\varphi(v_j; \Theta_k\mathcal{R})(S) - (\varphi v_j)(S)| \leqq \epsilon/3.$$

For $k \geqq k_0$ it then follows from Lemma 15.1 and Proposition 12.8 that

$$\begin{aligned}|\varphi(v; \Theta_k\mathcal{R})(S) - (\varphi v)(S)| &\leqq |\varphi(v; \Theta_k\mathcal{R})(S) - \varphi(v_j; \Theta_k\mathcal{R})(S)| \\ &\quad + |\varphi(v_j; \Theta_k\mathcal{R})(S) - (\varphi v_j)(S)| + |(\varphi v_j)(S) - (\varphi v)(S)| \\ &\leqq |\varphi(v - v_j; \Theta_k\mathcal{R})(S)| + \epsilon/3 + |(\varphi v_j - \varphi v)(S)| \\ &\leqq \|\varphi(v - v_j; \Theta_k\mathcal{R})\| + \epsilon/3 + \|\varphi v_j - \varphi v\| \\ &\leqq \|v - v_j\| + 2\epsilon/3 \leqq 3\epsilon/3 = \epsilon.\end{aligned}$$

This proves (15.5), and it follows that $v \in MIX$. The proof of Proposition 15.4 is now complete.

PROPOSITION 15.6. *MIX is symmetric, and the operator φ that associates with each v in MIX its mixing value φv is a value on MIX (in the sense of Section 2).*

Proof. The linearity of φ follows easily from (14.2) and the linearity of $\varphi^{\mathcal{R}}$ (Lemma 15.1). The normalization condition (2.3) follows at once from (14.2) and (12.2). The positivity also follows from (14.2) and (12.2); alternatively, it follows from Lemma 15.2 and Proposition 4.6.

The proof that MIX is symmetric and that φ satisfies the symmetry condition (2.2) is a straightforward computation, which the reader can supply for himself. This completes the proof of Proposition 15.6.

The proof of Theorem E will be complete if we show that $pNA \subset MIX$. Since MIX is a closed subspace, it would be sufficient to show that $\nu^k \in MIX$, for each $\nu \in NA^1$ and each positive integer k. As it turns out, it is just as easy to prove that $f \circ \nu \in MIX$, when f is non-decreasing and continuously differentiable on the range $[0, 1]$ of ν; and, notationally, it makes the proof somewhat more transparent.

From now until the end of this section, f will denote a non-decreasing, continuously differentiable function on $[0, 1]$ with $f(0) = 0$ and $f(1) = 1$, ν and μ will denote measures in NA^1, and $\mathcal{R}$ will denote a measurable order; f and ν will be fixed throughout the discussion but μ and $\mathcal{R}$ may vary as indicated in each context. The set function $f \circ \nu$ will be denoted v; note that[1] $v \in ORD$. $J(\alpha; \mu, \mathcal{R})$ will be as in Sec. 12, i.e., the intersection of all initial segments of μ-measure $> \alpha$. The difference $J(\beta; \mu, \mathcal{R}) \backslash J(\alpha; \mu, \mathcal{R})$ will be denoted $(\alpha, \beta]^{\mu}_{\mathcal{R}}$, or, when no confusion

[1] Follows from Theorem C, Corollary 5.3, and Proposition 12.8.

can result, simply $(\alpha, \beta]^{\mu}$. Parentheses will not be repeated around $(\alpha, \beta]^{\mu}$; thus $\nu(\alpha, \beta]^{\mu}$ means $\nu((\alpha, \beta]^{\mu})$, and so on. The derivative of f will be denoted f'. Since f' is continuous on a closed interval, it is uniformly continuous there; its modulus of uniform continuity will be denoted by $\eta(\epsilon)$.

LEMMA 15.7. *Let μ and $\mathfrak{R}$ be arbitrary, and let $G(\mu, \mathfrak{R})$ denote the σ-field generated by all the sets $J(\alpha; \mu, \mathfrak{R})$ for $\alpha \in [0, 1]$. Then for each measurable T there is a set $T' \in G(\mu, \mathfrak{R})$ such that*[2] $\mu(T' + T) = 0$.

Proof. First let T be an initial segment. Set $\alpha = \mu(T)$, and let $J = J(\alpha; \mu, \mathfrak{R})$. Then $J \supset T$ and by Lemma 12.15, $\mu(J) = \alpha = \mu(T)$; since $J \in G(\mu, \mathfrak{R})$, the lemma is proved in this case. Now if F is the family of all sets T satisfying the conclusion of the lemma, then F is a σ-field; since it contains all the $I(s, \mathfrak{R})$, it follows from (12.3) that it contains $\mathcal{C}$, and the lemma is proved.

LEMMA 15.8. *Let μ be such that $\nu \ll \mu$. Then for all $\epsilon > 0$, all $\mathfrak{R}$, and all α, β such that $0 \leqq \alpha < \beta \leqq 1$ and $\nu(\alpha, \beta]^{\mu}_{\mathfrak{R}} < \eta(\epsilon)$, we have*

$$|(\varphi^{\mathfrak{R}}v)(\alpha, \beta]^{\mu}_{\mathfrak{R}} - \nu(\alpha, \beta]^{\mu}_{\mathfrak{R}} f'(\nu(J(\alpha; \mu, \mathfrak{R})))| \leqq \epsilon\nu(\alpha, \beta]^{\mu}_{\mathfrak{R}}.$$

Proof. Set $x = \nu(J(\alpha; \mu, \mathfrak{R}))$, $x + \Delta x = \nu(J(\beta; \mu, \mathfrak{R}))$, and hence $\Delta x = \nu(\alpha, \beta]^{\mu} \geqq 0$. Then by Lemma 12.14 and the mean value theorem,

$$\begin{aligned}(\varphi^{\mathfrak{R}}v)(\alpha, \beta]^{\mu} &= (\varphi^{\mathfrak{R}}v)(J(\beta; \mu, \mathfrak{R})) - (\varphi^{\mathfrak{R}}v)(J(\alpha; \mu, \mathfrak{R}))\\ &= v(J(\beta; \mu, \mathfrak{R})) - v(J(\alpha; \mu, \mathfrak{R})) = f(x + \Delta x) - f(x)\\ &= \Delta x f'(x + \theta\, \Delta x) = \Delta x f'(x) + \Delta x(f'(x + \theta\, \Delta x) - f'(x))\\ &= \nu(\alpha, \beta]^{\mu} f'(\nu(J(\alpha; \mu, \mathfrak{R})))\\ &\qquad + \nu(\alpha, \beta]^{\mu}(f'(x + \theta\nu(\alpha, \beta]^{\mu}) - f'(x)),\end{aligned}$$

where $0 \leqq \theta \leqq 1$. The lemma now follows easily.

LEMMA 15.9. *Let μ be such that $\nu \ll \mu$. Then for each $\epsilon > 0$, there is a $\delta > 0$ such that for all $\mathfrak{R}$, all α, β with $0 \leqq \alpha < \beta \leqq 1$ and $\beta - \alpha < \delta$, and all $T \subset (\alpha, \beta]^{\mu}_{\mathfrak{R}}$, we have*

$$|(\varphi^{\mathfrak{R}}v)(T) - \nu(T)f'(\nu(J(\alpha; \mu, \mathfrak{R})))| \leqq \epsilon\nu(\alpha, \beta]^{\mu}_{\mathfrak{R}}. \tag{15.10}$$

[2] Here $+$ denotes the symmetric difference, i.e. $T' + T = (T' \backslash T) \cup (T \backslash T')$.

Proof. We shall prove a somewhat stronger statement, namely the one obtained by replacing the right side of (15.10) by $\epsilon\nu(T)$. Choose $\delta < 1$ such that $\mu(S) < \delta$ implies $\nu(S) < \eta(\epsilon/2)$; this is possible because $\nu \ll \mu$. Assume $0 < \beta - \alpha < \delta$.

First, let T be of the form $(\gamma, \zeta]^\mu$. Since $T \subset (\alpha, \beta]^\mu$, it follows that $\alpha \leqq \gamma \leqq \zeta \leqq \beta$; hence

$$0 \leqq \mu(\gamma, \zeta]^\mu = \zeta - \gamma \leqq \beta - \alpha < \delta,$$

and hence $\nu(\gamma, \zeta]^\mu < \eta(\epsilon/2)$, and similarly $\nu(\alpha, \gamma]^\mu < \eta(\epsilon/2)$. Therefore by Lemma 15.8,

$$|(\varphi^{\mathcal{R}}v)(T) - \nu(T)f'(\nu(J(\gamma; \mu, \mathcal{R})))| \leqq \frac{\epsilon}{2}\nu(T),$$

and by the definition of $\eta(\epsilon/2)$,

$$|f'(\nu(J(\gamma; \mu, \mathcal{R}))) - f'(\nu(J(\alpha; \mu, \mathcal{R})))| \leqq \frac{\epsilon}{2}.$$

Combining these, we obtain

$$(15.11) \qquad |(\varphi^{\mathcal{R}}v)(T) - \nu(T)f'(\nu(J(\alpha; \mu, \mathcal{R})))| \leqq \epsilon\nu(T).$$

Now think of α and β as fixed but of T as varying over $\mathcal{C}$; denote by $\xi(T)$ what is inside the absolute value symbol in (15.11). From $v = f \circ \nu$ and the conditions on f it follows that $v \ll \nu$; hence also $\varphi^{\mathcal{R}}v \ll \nu$ (Proposition 12.8), and so $\xi \ll \nu$. What (15.11) says is that when T is an $\mathcal{R}$-interval included in $(\alpha, \beta]^\mu$, then $\xi(T) < \epsilon\nu(T)$ and $-\xi(T) < \epsilon\nu(T)$. But then the same inequalities follow for arbitrary finite unions of $\mathcal{R}$-intervals included in $(\alpha, \beta]^\mu$, and then, by a standard approximation theorem,[3] for all T in $G(\mu, \mathcal{R})$ (with $T \subset (\alpha, \beta]^\mu$, of course). Now if $T \subset (\alpha, \beta]^\mu$ is an arbitrary measurable set, we may find T' in $G(\mu, \mathcal{R})$ with $\mu(T' + T) = 0$, and we may suppose w.l.o.g. that $T' \subset (\alpha, \beta]^\mu$ (otherwise intersect it with $(\alpha, \beta]^\mu$). From $\xi \ll \nu \ll \mu$ it then follows that $\xi(T' + T) = \nu(T' + T) = 0$; hence $\xi(T') = \xi(T)$ and $\nu(T') = \nu(T)$, and so (15.11) is proved for all measurable sets included in $(\alpha, \beta]^\mu$. This proves the lemma.

[3] Theorem D of Halmos (1950), p. 56.

LEMMA 15.12. *For given μ, let ξ be a nonnegative measure with $\xi \ll \mu$, and let $\{\Theta_n\}$ be a μ-mixing sequence. Then*

$$\xi(\Theta_n T) \rightarrow \mu(T)\xi(I)$$

as $n \rightarrow \infty$, for all measurable T.

Proof. The Radon-Nikodym theorem gives $\xi(T) = \int_T g(t)\, d\mu(t)$ for some $g \geqq 0$. Suppose first that g is a characteristic function, i.e., equal to 1 on some measurable U, 0 on $I\backslash U$. Then $\xi(\Theta_n T) = \mu(U \cap \Theta_n T)$, and by the μ-mixing property this approaches

$$\mu(T)\mu(U) = \mu(T) \int_I g(t)\, d\mu(t) = \mu(T)\xi(I),$$

so the result is proved for characteristic functions. Hence if g is a simple function (finite linear combination of characteristic functions), it follows as well. Next, let g be an arbitrary member of $L^1(I, \mathcal{C}, \mu)$, and for given ϵ, let h be a simple function such that

$$\int_I |g(t) - h(t)|\, d\mu(t) < \epsilon.$$

Then

$$\begin{aligned}
|\xi(\Theta_n T) - \mu(T)\xi(I)| &= \left| \int_{\Theta_n T} g(t)\, d\mu(t) - \mu(T) \int_I g(t)\, d\mu(t) \right| \\
&\leqq \left| \int_{\Theta_n T} g(t)\, d\mu(t) - \int_{\Theta_n T} h(t)\, d\mu(t) \right| \\
&\quad + \left| \int_{\Theta_n T} h(t)\, d\mu(t) - \mu(T) \int_I h(t)\, d\mu(t) \right| \\
&\quad + \left| \mu(T) \int_I h(t)\, d\mu(t) - \mu(T) \int_I g(t)\, d\mu(t) \right| \\
&\leqq \int_{\Theta_n T} |g(t) - h(t)|\, d\mu(t) \\
&\quad + \left| \int_{\Theta_n T} h(t)\, d\mu(t) - \mu(T) \int_I h(t)\, d\mu(t) \right| \\
&\quad + \mu(T) \int_I |g(t) - h(t)|\, d\mu(t) \\
&< \left| \int_{\Theta_n T} h(t)\, d\mu(t) - \mu(T) \int_I h(t)\, d\mu(t) \right| + 2\epsilon.
\end{aligned}$$

Now, allowing $n \rightarrow \infty$, we find that the first term approaches 0, so the upper limit of $|\xi(\Theta_n T) - \mu(T)\xi(I)|$ must be less than 2ϵ; but since ϵ is arbitrary, this upper limit must vanish, and the lemma is proved.

COROLLARY 15.13. *Let μ be such that $v \ll \mu$, and let $\{\Theta_n\}$ be a μ-mixing sequence. Then*

$$\nu(S \cap \Theta_n T) \to \mu(T)\nu(S)$$

as $n \to \infty$, for all measurable S and T.

Proof. This follows immediately from Lemma 15.12 by setting $\xi(T) = \nu(S \cap T)$.

LEMMA 15.14. *v has a mixing value, which equals ν.*

Proof. Let μ be such that $\nu \ll \mu$. For given $\epsilon > 0$, pick a positive integer m so that $1/m$ is less than the δ provided by Lemma 15.9. Let $\{\Theta_n\}$ be a μ-mixing sequence, and $\mathcal{R}$ a measurable order. From $v \ll \nu \ll \mu$, Proposition 12.8, and Lemma 12.15, we obtain

$$\varphi(v, \Theta_n\mathcal{R})(J(0; \mu, \Theta_n\mathcal{R})) = \nu(J(0; \mu, \Theta_n\mathcal{R})) = 0.$$

Moreover, Lemma 13.5 and the fact that the Θ_n are μ-measure-preserving yields

$$J(\alpha; \mu, \Theta_n\mathcal{R}) = \Theta_n J(\alpha; \mu, \mathcal{R})$$

and

$$(\alpha, \beta]^{\mu}_{\Theta_n\mathcal{R}} = \Theta_n(\alpha, \beta]^{\mu}_{\mathcal{R}}.$$

Hence by Lemma 15.9,

$$\begin{aligned}\varphi(v; \Theta_n\mathcal{R})(S) &= \sum_{j=0}^{m-1} \varphi(v; \Theta_n\mathcal{R}) \left(S \cap \left(\frac{j}{m}, \frac{j+1}{m}\right]^{\mu}_{\Theta_n\mathcal{R}}\right) \\ &= \sum_{j=0}^{m-1} \nu\left(S \cap \Theta_n \left(\frac{j}{m}, \frac{j+1}{m}\right]^{\mu}_{\mathcal{R}}\right) f'\left(\nu\left(\Theta_n J\left(\frac{j}{m}; \mu, \mathcal{R}\right)\right)\right) + \sum_{j=0}^{m-1} \theta_j,\end{aligned}$$

where $|\theta_j| < \epsilon\nu\left(\frac{j}{m}, \frac{j+1}{m}\right]^{\mu}_{\mathcal{R}}$, and hence $|\Sigma_{j=0}^{m-1} \theta_j| < \epsilon\nu(I) = \epsilon$.

Let us for the moment ignore the error term $\Sigma_{j=0}^{m-1} \theta_j$, and allow $n \to \infty$ in the major term on the right side of the above expression for $\varphi(v; \Theta_n\mathcal{R})(S)$. Then by Corollary 15.13 and the continuity of f', this major term approaches the limit

$$\sum_{j=0}^{m-1} \mu\left(\frac{j}{m}, \frac{j+1}{m}\right]^{\mu}_{\mathcal{R}} \nu(S) f'\left(\mu\left(J\left(\frac{j}{m}; \mu, \mathcal{R}\right)\right)\right) = \nu(S) \sum_{j=0}^{m-1} \frac{1}{m} f'\left(\frac{j}{m}\right),$$

which we shall call $\nu(S)\Sigma(m)$. It follows that both the upper and lower limits of $\varphi(v; \Theta_n\mathcal{R})(S)$ as $n \to \infty$ differ from $\nu(S)\Sigma(m)$ by ϵ at most.

Now $\Sigma(m)$ is an approximating Riemann sum to the integral $\int_0^1 f'(x)\,dx$, which equals 1. Letting $m \to \infty$, we find that the upper and lower limits of $\varphi(v;\ \Theta_n\mathfrak{R})(S)$ differ from $\nu(S)$ by less than ϵ. But since ϵ can be chosen arbitrarily small, both limits actually equal $\nu(S)$; so $\lim_{n\to\infty} \varphi(v;\ \Theta_n\mathfrak{R})(S)$ exists and equals $\nu(S)$. If we now set $\mu_v = \nu$, we find that ν is the mixing value of v. This completes the proof of Lemma 15.14.

Proof of Theorem E. Follows from Propositions 15.4 and 15.6 and Lemma 15.14.

Note that we have now supplied an independent proof of the existence of a value on pNA (cf. Theorem B, Section 3). Mixing values can also be used to give an independent proof that the unique value on pNA satisfies (3.1), by showing that the mixing value satisfies (3.1). Indeed, when μ is one-dimensional, (3.1) is equivalent to Lemma 15.14. When μ is multidimensional and f is a polynomial, then (3.1) follows from the one-dimensional case by using Lemma 7.2, the linearity of the mixing value (Proposition 15.6), and the linearity of the right side of (3.1). In general, when f is continuously differentiable on the range of μ, then (3.1) follows from the polynomial case by using the continuity of the mixing value in the variation norm (Proposition 15.2), the continuity of the variation norm in the C^1-norm (7.5), the continuity of the right side of (3.1) in the C^1-norm, and the density of the polynomials in C^1 (Lemma 7.4).

16. An Alternative Proof for Example 9.4

In this section we will prove that the set function of Example 9.4 is not in MIX; this proves, a fortiori, that it is not in pNA, and hence by (8.26) that it is not in $bv'NA$. It also proves that not every set function v in AC is in MIX, and a fortiori[1] that not every v in ORD is in MIX.

It is convenient to let the underlying space I be given by

$$I = \bigtimes_{i=-\infty}^{\infty} J_i,$$

[1] $AC \subset ORD$ (Proposition 12.8).

where J_i is a copy of the 2-point space $\{0, 1\}$, and to let $\mathcal{C}$ be the standard product σ-field. Define λ_i on J_i by

$$\lambda_i(\{0\}) = \lambda_i(\{1\}) = \tfrac{1}{2},$$

and let λ be the product measure of the λ_i on I. Let

$$I^0 = \left(\bigtimes_{i=-\infty}^{-1} J_i\right) \times \{0\} \times \left(\bigtimes_{i=1}^{\infty} J_i\right)$$
$$I^1 = \left(\bigtimes_{i=-\infty}^{-1} J_i\right) \times \{1\} \times \left(\bigtimes_{i=1}^{\infty} J_i\right);$$

then $I^0 \cap I^1 = \emptyset$, $I^0 \cup I^1 = I$. Define λ^0 and λ^1 by

$$\lambda^0(S) = \lambda(S \cap I^0)$$
$$\lambda^1(S) = \lambda(S \cap I^1),$$

let $\mu = 2\lambda^0 - 2\lambda^1$, and let $v(S) = |\mu(S)|$. It is easily verified that this situation is entirely isomorphic to that described in Example 9.4.

Now arrange the integers in the order $(0, 1, -1, 2, -2, \ldots)$. Let $\mathcal{R}$ be the lexicographic order on I w.r.t. the above order on the indices; that is, $x \,\mathcal{R}\, y$ if and only if $x_i = 1$ and $y_i = 0$ when i is the first index—in the above order—for which $x_i \neq y_i$. In the order $\mathcal{R}$, all of I^0 comes "before" all of I^1, and it follows easily that

$$\varphi(v; \mathcal{R}) = 2\lambda^0 - 2\lambda^1 = \mu,$$

since this is so on the initial segments.

Next, let Θ be the "bilateral 1-shift" on I, i.e.

$$(\Theta x)_i = x_{i+1}.$$

In the order $\Theta\mathcal{R}$, we first get half of I^0, then half of I^1, then another half of I^0, and finally another half of I^1. It follows that again

$$\varphi(v; \Theta\mathcal{R}) = 2\lambda^0 - 2\lambda^1 = \mu.$$

In the order $\Theta^n\mathcal{R}$, the space I is divided into 2^{2n} segments of equal λ-measure, in which the odd-numbered segments are contained in I^0 and the even-numbered segments in I^1. Thus we always have

$$\varphi(v; \Theta^n\mathcal{R}) = 2\lambda^0 - 2\lambda^1 = \mu$$

for the initial segments, and hence for all measurable sets. Since the powers of Θ form a λ-mixing sequence (Halmos, 1956) and $v \ll \lambda$, it follows that the mixing value, if it exists, must be μ. But now if we start with the order $\mathcal{R}'$ that is the "reverse" of $\mathcal{R}$—i.e. $x \mathcal{R}' y$ if and only if $y \mathcal{R} x$—and again apply the sequence $\{\Theta, \Theta^2, \ldots\}$, then we find that

$$\varphi(v; \Theta^n \mathcal{R}') = 2\lambda^1 - 2\lambda^0 = -\mu;$$

hence the mixing value, if it exists, must be $-\mu$. Since $\mu \neq 0$, this shows that the mixing value does not exist.

Chapter III. The Asymptotic Approach

17. Introduction and Statement of Results

In this chapter we shall treat an approach, due to Kannai, in which a non-atomic game is regarded as a limit of games with finitely many players. We shall find that the asymptotic value presently to be defined is similar in its properties to the mixing value. This kinship may be attributed to the fact that the value for finite games, though originally defined axiomatically (Shapley, 1953a), is inextricably involved with the notion of random order.

In this section we define the concept of "asymptotic value" and state the basic theorem (Theorem F), which relates it to the value defined in Chapter I; the theorem is proved in Section 18.[1] Section 19 contains a discussion of the asymptotic value in connection with the set function of Example 9.4 and related set functions.

The idea behind the definition of asymptotic value is as follows: One partitions the underlying space I into a large finite number of "small" sets. Each of these sets is a *player* in an auxiliary finite game. The worth of a coalition of such players in the auxiliary game is defined to be the worth of their union in the original non-atomic game. The auxiliary game has a value, which can be considered a measure on the finite ring consisting of unions of the players. As one takes finer and finer partitions, this measure tends to a limit, which is the asymptotic value.

The formal treatment is as follows:

A *partition* Π of the underlying space is a finite family of disjoint subsets whose union is the whole space; a partition Π_2 is a *refinement* of another partition Π_1 if each member of Π_1 is a union of members of Π_2. A partition is called *measurable* if each of its members is measurable. A sequence $(\Pi_1, \Pi_2, \ldots)$ of partitions is said to be *decreasing* if Π_{m+1} is

[1] Sections 17 and 18 cover essentially the same material as Kannai (1966), though it has been reworked and slightly generalized in order to fit into the conceptual framework of this book. In particular, Kannai works only with set functions that are absolutely continuous w.r.t. a given measure, whereas we make no such restriction.

a refinement of Π_m for each m; and *separating* if for all $s, t \in I$ with $s \neq t$, there is an m such that s and t are in different members of Π_m. A decreasing separating sequence of measurable partitions is called an *admissible sequence*. For example, if Π_m is the partition of $[0, 1]$ into the 2^m intervals $[0, 1/2^m]$, $(1/2^m, 2/2^m]$, . . . , $(1 - 1/2^m, 1]$, then for $([0, 1], \mathcal{B})$, $(\Pi_1, \Pi_2, \ldots)$ is an admissible sequence.

If u is a game with finitely many players, we will denote by φu the Shapley value of u (Appendix A), considered as a measure on the set of players of u.

If v is a set function and Π a measurable partition of the underlying space, let v_Π be the finite game whose players are the members of Π, given by

$$v_\Pi(\Xi) = v\left(\bigcup_{j \in \Xi} j\right)$$

for all $\Xi \subset \Pi$. Now let T be a measurable subset of I, and let $\mathcal{P} = (\Pi_1, \Pi_2, \ldots)$ be a decreasing sequence of measurable partitions whose first term Π_1 is the partition $\{T, I \backslash T\}$. For each m let

$$T_m = \{j \in \Pi_m : j \subset T\};$$

T is a coalition in the original infinite game v, and T_m is the corresponding coalition in the corresponding finite game v_{Π_m}. If the numbers $(\varphi v_{\Pi_m})(T_m)$ approach a limit as $m \to \infty$, then this limit will be denoted $(\varphi_{\mathcal{P}} v)(T)$. If the limit exists for all admissible $\mathcal{P}$ starting with $\{T, I \backslash T\}$, and is independent of the choice of such $\mathcal{P}$, then we will denote it by $(\varphi v)(T)$. If that is the case for all measurable T, then we will call the set function φv the *asymptotic value* of T. For given v, the asymptotic value, if it exists, is clearly unique. Note also that if φv exists, it is necessarily finitely additive. The set of all $v \in BV$ having an asymptotic value[2] will be denoted $ASYMP$.

We are now ready for the statement of the main result.

THEOREM F. *$ASYMP$ is a closed symmetric linear subspace of BV which contains pNA, and the operator φ that associates to each v its asymptotic value φv is a value on $ASYMP$.*

[2] There are also set functions outside BV that have asymptotic values, for example the finite but unbounded finitely-additive "measures."

It follows from this that the asymptotic value exists on *pNA* and coincides with the unique value on *pNA* as determined in Chapter I (Propositions 7.6 and 7.11). Our proof will be independent of the proof of Theorem B given in Chapter I, and indeed it will enable us to give yet another proof of Proposition 7.6 (the existence of a value on *pNA*).

The properties of the asymptotic approach discussed up to now are entirely analogous to those of the mixing approach (see Theorem E, Section 14). On the matter of Example 9.4, though, the approaches differ. We have seen (Section 16) that the set function of that example is not in *MIX*; it turns out, however, that it *is* in *ASYMP* (Proposition 19.1). This implies that the maximum of two NA^1 measures is in *ASYMP*, and one might have conjectured that this extends to the maximum of any number of NA^1 measures. We shall show, however, that for $n \geqq 3$, the maximum of n different NA^1 measures is in general *not* in *ASYMP* (Example 19.2).

As for $bv'NA$, it is not known whether this is included in *ASYMP*. Indeed, it is not even known whether the simplest single-jump functions in $bv'NA$ are in *ASYMP*.

18. Proof of Theorem F

Throughout the remainder of this chapter, when $v \in ASYMP$, we will always use φv to denote the asymptotic value of v.

It will be useful to extend the notion of the variation norm to finite games; this is done in the obvious way. Clearly we then have

$$\|v_\Pi\| \leqq \|v\|$$

for any measurable partition Π. If u is a finite game with player space J, note that the value φu, as a measure on J, is itself a finite game. Then from the expression (A.8) for the value of a finite game in terms of random orderings of the players, we obtain

$$\|\varphi u\| \leqq \|u\|.$$

Finally, note that

$$\|\varphi u\| = \sum_{j \in J} |(\varphi u)(\{j\})|,$$

since φu is a measure on J.

PROPOSITION 18.1. $\|\varphi v\| \leqq \|v\|$ *for all* $v \in ASYMP$.

Proof. For a given $\epsilon > 0$, let Ω be a chain

$$\emptyset = S_0 \subset S_1 \subset \cdots \subset S_k = I$$

such that

$$\|\varphi v\|_\Omega \geqq \|\varphi v\| - \epsilon.$$

Let $U_j = S_j \backslash S_{j-1}$, let Π_1 be the partition $(U_1, \ldots, U_k)$, and let $\mathcal{P}$ be an admissible sequence $(\Pi_1, \Pi_2, \ldots)$ starting with this Π_1. For each $T \in \Pi_1$ and each m, let $T_m = \{j \in \Pi_m : j \subset T\}$. Then

$$(18.2) \quad \|\varphi v\| - \epsilon \leqq \|\varphi v\|_\Omega = \sum_{T \in \Pi_1} |(\varphi v)(T)| = \sum_{T \in \Pi_1} \left| \lim_{m \to \infty} (\varphi v_{\Pi_m})(T_m) \right|$$
$$= \sum_{T \in \Pi_1} \lim_{m \to \infty} |(\varphi v_{\Pi_m})(T_m)| = \lim_{m \to \infty} \sum_{T \in \Pi_1} |(\varphi v_{\Pi_m})(T_m)|,$$

since the sum is finite. Moreover, for fixed m, we have

$$\sum_{T \in \Pi_m} |(\varphi v_{\Pi_1})(T_m)| = \sum_{T \in \Pi_1} \left| \sum_{j \in T_m} (\varphi v_{\Pi_m})(\{j\}) \right| \leqq \sum_{T \in \Pi_1} \sum_{j \in T_m} |(\varphi v_{\Pi_m})(\{j\})|$$
$$= \sum_{j \in \Pi_m} |(\varphi v_{\Pi_m})(\{j\})| = \|\varphi v_{\Pi_m}\| \leqq \|v_{\Pi_m}\| \leqq \|v\|.$$

Combining this with (18.2) we get $\|\varphi v\| - \epsilon \leqq \|v\|$, and by letting $\epsilon \to 0$ we complete the proof of Proposition 18.1.

COROLLARY 18.3. *If* $v \in ASYMP$, *then* $\varphi v \in FA$.

Proof. It is necessary only to show that φv is bounded, and this follows from Proposition 18.1.

PROPOSITION 18.4. $ASYMP$ *is a closed linear subspace of* BV.

Proof. That $ASYMP$ is a linear subspace is immediate. Suppose now that $v^k \in ASYMP$ and that $\|v^k - v\| \to 0$. Then by Proposition 18.1, φv^k is a Cauchy sequence, and hence has a limit ν in the variation norm; by Corollary 18.3 and Proposition 4.4, $\nu \in FA$. For given $\epsilon > 0$, choose v^k so that both $\|v^k - v\| < \epsilon$ and $\|\varphi v^k - \nu\| < \epsilon$. Now let $\mathcal{P}$ be an admissible sequence $(\Pi_1, \Pi_2, \ldots)$ starting with $\{T, I \backslash T\}$, and for each m let $T_m = \{j \in \Pi_m : j \subset T\}$. Then

$$|(\varphi v_{\Pi_m})(T_m) - (\varphi v^k_{\Pi_m})(T_m)| \leqq \|\varphi(v_{\Pi_m} - v^k_{\Pi_m})\|$$
$$\leqq \|v_{\Pi_m} - v^k_{\Pi_m}\| \leqq \|v - v^k\| \leqq \epsilon,$$

hence

$$
\begin{aligned}
|(\varphi v_{\Pi_m})(T_m) - \nu(T)| &\leqq |(\varphi v_{\Pi_m})(T_m) - (\varphi v^k_{\Pi_m})(T_m)| \\
&\quad + |(\varphi v^k_{\Pi_m})(T_m) - (\varphi v^k)(T)| + |(\varphi v^k)(T) - \nu(T)| \\
&\leqq \epsilon + |(\varphi v^k_{\Pi_m})(T_m) - (\varphi v^k)(T)| + \|\varphi v^k - \nu\| \\
&\leqq 2\epsilon + |(\varphi v^k_{\Pi_m})(T_m) - (\varphi v^k)(T)|.
\end{aligned}
$$

Letting $m \to \infty$ and using $v^k \in ASYMP$, we deduce that

$$\limsup_{m \to \infty} |(\varphi v_{\Pi_m})(T_m) - \nu(T)| \leqq 2\epsilon.$$

Hence the lim sup vanishes, hence the limit exists and vanishes, and hence ν is the asymptotic value of v. This completes the proof of Proposition 18.4.

PROPOSITION 18.5. *ASYMP is symmetric, and φv is a value on ASYMP.*

Proof. The linearity of φ is obvious. The normalization condition (2.3) follows easily from the corresponding condition for finite games (A.2). Positivity is then a consequence of Proposition 18.1 and Proposition 4.6. The proof that *ASYMP* and φ are symmetric is straightforward; the details will be omitted. This completes the proof.

For $\nu \in NA^+$, let us define a *ν-admissible sequence* to be a decreasing sequence of measurable partitions that is ν-dense, i.e., such that every measurable set in I can be approximated in ν-measure by a member of the field $H(\mathcal{P})$ (*not* the σ-field) generated by the members of all the Π_i.

LEMMA 18.6. *Every admissible sequence is ν-admissible, for all $\nu \in NA^+$.*

Proof. $H(\mathcal{P})$ generates the σ-field $\mathcal{C}$, since $\mathcal{P}$ is separating (Mackey, 1957, p. 139, Theorem 3.3); this implies the ν-denseness (Halmos, 1950, p. 56, Theorem D), completing the proof of Lemma 18.6.

We now establish some notation that will be used throughout the remainder of this chapter. Let us be given an admissible sequence $\mathcal{P} = (\Pi_1, \Pi_2, \ldots)$ of partitions of I; the members of Π_m are the players in the finite game v_{Π_m}. Sometimes we suppress m and write Π for Π_m, n for $n(m) = |\Pi|$. A *random order* $\mathcal{R} = \mathcal{R}_m$ on Π is defined to be a random variable whose values are orders on Π, in which each of the $n!$ such orders have the same probability (namely $1/n!$). Given such a random order, and an h with $1 \leqq h \leqq n$, let $\mathbf{B}_h = \mathbf{B}_h^m$ be the hth member of Π in the order $\mathfrak{R}$, and set

$$\mathbf{S}_h = \mathbf{S}_h^m = \{\mathbf{B}_1, \ldots, \mathbf{B}_h\}$$
$$\mathbf{Q}_h = \mathbf{Q}_h^m = \mathbf{B}_1 \cup \cdots \cup \mathbf{B}_h.$$

LEMMA 18.7. *For each m, let $h(m)$ be an integer satisfying $1 \leqq h(m) \leqq n(m)$, and assume that $h(m)/n(m)$ converges as $m \to \infty$, say to γ. Let ζ be a vector of NA measures. Then $\zeta(\mathbf{Q}_{h(m)}^m)$ converges in probability to $\gamma\zeta(I)$, i.e. for each $\delta > 0$,*

$$\text{Prob}\,\{\|\zeta(\mathbf{Q}_{h(m)}^m) - \gamma\zeta(I)\| \geqq \delta\} \to 0 \textit{ as } m \to \infty.$$

Proof. (See Note 1.) W.l.o.g. we may assume that ζ is a scalar NA^1 measure. Fix m; set $h(m) = h$, $n(m) = n$, $\Pi_m = \Pi$, and $\zeta(\mathbf{Q}_{h(m)}^m) = \mathbf{x}_m = \mathbf{x}$. Denote the members of Π_m by $A_1, \ldots, A_n$, and set $\zeta(A_i) = a_i$. Note that the average of the a_i is $1/n$; hence

$$\text{(18.8)} \qquad \mathrm{E}\mathbf{x} = h/n,$$

where E is the expectation operator. Moreover, it is known[1] that

$$\text{(18.9)} \qquad \text{Var}\,\mathbf{x} = \frac{h}{n-1}\left(1 - \frac{h}{n}\right)\sum_{i=1}^{n}\left(a_i - \frac{1}{n}\right)^2,$$

where Var means "variance." Hence

$$\text{Var}\,\mathbf{x} \leqq \sum_{i=1}^{n}\left(a_i - \frac{1}{n}\right)^2 = \sum_i a_i^2 - \frac{1}{n} \leqq (\max_i a_i)\sum_i a_i$$
$$= \max_i a_i = \max_{A \in \Pi} \zeta(A).$$

Now since $\mathcal{P}$ is admissible, it is ζ-admissible (Lemma 18.6); this is known[2] to imply that $\max_{A \in \Pi} \zeta(A) \to 0$ as $m \to \infty$. Hence Var $\mathbf{x}_m \to 0$ as $m \to \infty$. The proof of Lemma 18.7 is then completed by an appeal to Chebychev's inequality.

The following corollary to Lemma 18.7 will be used in Sections 19 and 43.

[1] See Rosén (1965) p. 385, Theorem 1.1, or Wilks (1962), p. 222.
[2] Halmos (1950), p. 172, Theorem A.

COROLLARY 18.10. *Let ζ be a vector of NA^1 measures on I. Let U be a neighborhood of the "diagonal" in the range of ζ, i.e. the line segment with end points 0 and $\zeta(I)$. Let $\mathcal{P} = (\Pi_1, \Pi_2, \ldots)$ be an admissible sequence, and let $\mathcal{R}_m$ be a random order on Π_m. Then for every ϵ, there is an m_0, such that for $m > m_0$,*

$$\operatorname{Prob}\{\zeta(\mathbf{Q}_h^m) \in U \text{ for all } h \text{ with } 1 \leqq h \leqq n(m)\} \geqq 1 - \epsilon.$$

Remark. The conclusion says that for sufficiently fine partitions, the sequence $\zeta(\dot{\mathbf{Q}}_h)$ will with high probability remain in an arbitrarily small neighborhood of the diagonal.

Proof. Let V be an open neighborhood of the diagonal with $\bar{V} \subset U$. Choose a positive integer l such that

$$\zeta(I)(p-1)/l \leqq y \leqq \zeta(I)p/l \Rightarrow y \in V$$

for all integers p with $1 \leqq p \leqq l$. Then there exist open neighborhoods $V_0, V_1, \ldots V_{l-1}, V_l$ of $0, \zeta(I)/l, \ldots, \zeta(I)(l-1)/l, \zeta(I)$ respectively such that

$$x \in V_l, z \in V_{l+1}, x \leqq y \leqq z \Rightarrow y \in U. \tag{18.11}$$

Let $\epsilon > 0$ be given. For each p, choose a positive integer $h_p = h_p(m)$ so that

$$h_p/n \to p/l \text{ as } m \to \infty.$$

Setting $h(m) = h_p(m)$ in Lemma 18.7, we deduce that if m is chosen sufficiently large, then

$$\operatorname{Prob}\{\zeta(\mathbf{Q}_{h_p}) \in V_p \text{ for all } 0 \leqq p \leqq l\} > 1 - \epsilon.$$

But by (18.11), it follows from $\zeta(\mathbf{Q}_{h_p}) \in V_p$ for all $0 \leqq p \leqq l$ that $\zeta(\mathbf{Q}_h) \in U$ for all $1 \leqq h \leqq n$, and the proof of Corollary 18.10 is complete.

PROPOSITION 18.12. *Let $\nu \in NA^1$, and let k be a positive integer. Then $\nu^k \in ASYMP$.*

Proof. It is just as easy to prove that $f \circ \nu \in ASYMP$, where $f(0) = 0$, $f(1) = 1$, and f is non-decreasing and continuously differentiable on $[0, 1]$; and, notationally, it makes the proof somewhat more transparent. The derivative of f will be denoted f'. Since f' is continuous on a closed interval, it is uniformly continuous there; its modulus of uniform continuity will be denoted by $\eta(\epsilon)$.

Let $v = f \circ \nu$, let $T \in \mathcal{C}$, let $\mathcal{P} = (\Pi_1, \Pi_2, \ldots)$ be an admissible sequence starting with $\{T, I\backslash T\}$, and let $\mathfrak{R}_m$ be a random order on Π_m. For given ϵ, choose l so that $2/l < \eta(\epsilon)$. For each p with $1 \leqq p \leqq l$, choose a positive integer $h_p = h_p(m)$ so that

$$h_p(m)/n(m) \to p/l$$

as $m \to \infty$. Fix m. Let $\mathbf{H}$ be the set of all h between 1 and n such that $\mathbf{B}_h \subset T$, and consider the expression

$$\mathbf{\Delta} = \sum_{h \in \mathbf{H}} [v_\Pi(\mathbf{S}_h) - v_\Pi(\mathbf{S}_{h-1})].$$

The expression $\mathbf{\Delta}$ is the total contribution of the players in T_m (i.e. the players of v_Π who are included in T) to v_Π, when Π is ordered according to $\mathfrak{R}$; thus

(18.13) $$(\varphi v_\Pi)(T_m) = \mathbf{E\Delta}.$$

For $1 \leqq p \leqq l$, let

$$\mathbf{H}_p = \{h \in \mathbf{H}: h_{p-1} < h \leqq h_p\}.$$

Then

$$\begin{aligned} \mathbf{\Delta} &= \sum_{p=1}^{l} \sum_{h \in \mathbf{H}_p} [v(\mathbf{Q}_h) - v(\mathbf{Q}_{h-1})] \\ &= \sum_{p=1}^{l} \sum_{h \in \mathbf{H}_p} [f(\nu(\mathbf{Q}_{h-1}) + \nu(\mathbf{B}_h)) - f(\nu(\mathbf{Q}_{h-1}))] \\ &= \sum_{p=1}^{l} \sum_{h \in \mathbf{H}_p} \nu(\mathbf{B}_h) f'(\mathbf{x}_h), \end{aligned}$$

where

$$\nu(\mathbf{Q}_{h-1}) \leqq \mathbf{x}_h \leqq \nu(\mathbf{Q}_h).$$

If we fix p and look at h in $\mathbf{H}_p$ only, then from Lemma 18.7 it follows that for m sufficiently large,

$$(p-2)/l \leqq \mathbf{x}_h \leqq (p+1)/l$$

with high probability.[3] From this and $2/l < \eta(\epsilon)$ it follows that if we

[3] I.e. with as high a probability as we wish, the size of the probability depending on m.

replace $f'(\mathbf{x}_h)$ by $f'(p/l)$, we will be introducing an error that is with high probability at most ϵ. From this we deduce that with high probability,[4]

$$\mathbf{\Delta} = \sum_{p=1}^{l} f'(p/l) \sum_{h \in \mathbf{H}_p} \nu(\mathbf{B}_h) + \boldsymbol{\theta}$$

where $|\boldsymbol{\theta}| \leqq \nu(I)\epsilon = \epsilon$. Now

$$\sum_{h \in \mathbf{H}_p} \nu(\mathbf{B}_h) = \nu(T \cap \mathbf{Q}_{h_p}) - \nu(T \cap \mathbf{Q}_{h_{p-1}}).$$

Applying Lemma 18.7 to the measure ζ defined by $\zeta(S) = \nu(T \cap S)$, we deduce that

$$\sum_{h \in \mathbf{H}_p} \nu(\mathbf{B}_h) \to \nu(T)/l$$

with high probability. Hence for sufficiently large m,

$$\mathbf{\Delta} = \nu(T)\Sigma(l) + \boldsymbol{\psi},$$

where $|\boldsymbol{\psi}| \leqq 2\epsilon$ with high probability and

$$\Sigma(l) = \sum_{p=1}^{l} f'(p/l)/l.$$

Applying (18.13), we deduce that for sufficiently large m,

$$|(\varphi v_{\Pi_m})(T_m) - \nu(T)\Sigma(l)| \leqq 2\epsilon.$$

It follows that both the upper and lower limits of $(\varphi v_{\Pi_m})(T_m)$ as $m \to \infty$ differ from $\nu(T)\Sigma(l)$ by 2ϵ at most. Now $\Sigma(l)$ is an approximating Riemann sum to the integral $\int_0^1 f'(x)\,dx$, which equals 1. Letting $l \to \infty$ (with[5] fixed ϵ), we find that the upper and lower limits of $(\varphi v_{\Pi_m})(T_m)$ differ from $\nu(T)$ by less than 2ϵ. But since ϵ can be chosen arbitrarily small, both limits actually equal $\nu(T)$; so $\lim_{m\to\infty} (\varphi v_{\Pi_m})(T_m)$ exists and equals $\nu(T)$. This shows that ν is the asymptotic value of v, and completes the proof of Proposition 18.12.

Proof of Theorem F. Follows from Propositions 18.4, 18.5, and 18.12.

[4] We have here a conjunction of l statements each of which occurs with high probability; since ϵ (and therefore l) are for the time being fixed, the conjunction also occurs with high probability.

[5] The only requirement on l was $2/l < \eta(\epsilon)$.

Note that we have again supplied an independent proof of the existence of a value on pNA (cf. Theorem B, Section 3; also the end of Section 15). Asymptotic values can also be used to give an independent proof that the unique value on pNA satisfies (3.1), by showing that the asymptotic value satisfies (3.1) (cf. the end of Section 15).

NOTES

1. This section hinges on Lemma 18.7. Its proof makes use of the principle that "sampling without replacement is better than sampling with replacement," which underlies much of the work on the asymptotic approach, and on values of large, finite games in general (see e.g. Shapley, 1964b). In particular, this implies that the sample mean will have a smaller variance in sampling *without* than in sampling *with* replacement.[6] The reason is clear. In sampling without replacement, if there is a deviation from the population mean in the first $i - 1$ elements of the sample, then the remainder of the population is biased in the *opposite* direction, and this bias extends to the ith element; deviations therefore tend to wipe each other out. In sampling with replacement, the deviations are independent; on the average, they wipe each other out because of independence, but one will not get the deviation-correcting effect that one does in the first case.

In our case, the population consists of the $\zeta(A_i)$, or rather the expressions $n\zeta(A_i)$. The sample consists of the first h in the population, according to a random order, and so the sample mean is $\mathbf{x}$. If one had independence, Lemma 18.7 would follow immediately from the law of large numbers. But the actual variance of $\mathbf{x}$ (see 18.9) is even smaller than the amount $(h/n)\Sigma(a_i - (1/n))^2$ that it would be under independence. So simply by mimicking the standard proof of the law of large numbers, one gets the desired convergence.

19. More on Example 9.4 and Related Set Functions

The set function v of Example 9.4 is defined by $v(S) = |\mu(S)|$, where $I = [-1, 1]$ and $\mu(S) = \int_S \operatorname{sgn} x \, dx$. In Section 9 we showed that $v \notin bv'NA$, and in Section 16 we showed that $v \notin MIX$; thus none of the heretofore developed methods succeeds in assigning a value to this set function. We now show that v *is* in $ASYMP$.

PROPOSITION[1] 19.1. *Let $\mu \in FA$ be such that $\mu(I) = 0$. Define v by $v(S) = |\mu(S)|$. Then $v \in ASYMP$, and $\varphi v \equiv 0$.*

[6] Both sample means are of course distributed around the population mean, i.e. their mean is the population mean.

[1] Proposition 19.1 and its proof were communicated to one of the authors by Y. Kannai, and we are grateful to him for permission to publish it here.

Remark. Note that we do not require $\mu \in NA$, nor even that μ be countably additive.

Proof. Let Π be a partition of I into n sets. To compute φv_Π, divide the $n!$ orderings of Π into pairs, each of which consists of an ordering and its reverse. From $\mu(I) = 0$ it follows that the contribution of a player[2] to v_Π in a given ordering is exactly the negative of his contribution in the reverse ordering. Therefore the contributions cancel, and so φv_Π vanishes identically. Hence φv exists and vanishes identically, and the proposition is proved.

From Proposition 19.1 it follows[3] that the set functions v defined by

$$v(S) = \min(\mu_1(S), \mu_2(S))$$

and by

$$v(S) = \max(\mu_1(S), \mu_2(S)),$$

where $\mu_1, \mu_2 \in FA$ with $\mu_1(I) = \mu_2(I)$, are in *ASYMP*. One might suppose that this result could be extended to the minimum and maximum of more than two measures, but this is not the case. Indeed, we have

Example 19.2. Let $I_1 = [0, \frac{1}{3})$, $I_2 = [\frac{1}{3}, \frac{2}{3})$, $I_3 = [\frac{2}{3}, 1)$, and $I = I_1 \cup I_2 \cup I_3 = [0, 1)$. Define μ_j for $j = 1, 2, 3$ by $\mu_j(S) = 3\lambda(S \cap I_j)$, where λ is Lebesgue measure, and define v by

$$v(S) = \min[\mu_1(S), \mu_2(S), \mu_3(S)].$$

Then v is *not* in *ASYMP*.

The idea of the proof will be to pass to the limit in an unsymmetric way and obtain an impossible condition on the asymptotic value. Specifically, we shall partition I into $3k + 1$ elements, as follows: I_2 and I_3 are each divided into k equal intervals of length $1/3k$ and I_1 is divided into a single interval J_ϵ of length ϵ, and k equal intervals of length $(1 - 3\epsilon)/3k$. Here ϵ is meant to be small, but $1/k$ is much smaller. In a random ordering, we will have approximately equal *numbers* of intervals from each of I_1, I_2, I_3 up to any point. Therefore, up to any point before the appearance of J_ϵ the total *length* of the intervals from I_1 will be approximately $1 - \epsilon$ times the total length from I_2 or I_3. Since v depends only on the minimum of the three measures, it follows

[2] I.e. a member of Π.

[3] Because $\min(x,y) = (x + y - |x - y|)/2$ and $\max(x,y) = (x + y + |x - y|)/2$.

that with high probability most of the increase in v will be attributed to the intervals from I_1, and almost none to those from I_2 or I_3; all this, of course, only before J_ϵ appears. Thus, a given small interval T in I_1 gets $\mu_1(T)$ (with high probability) if it comes before J_ϵ in the random order. Since this event has probability $\frac{1}{2}$, the expected payment to T is approximately $\mu_1(T)/2$, or better. So, even disregarding J_ϵ's contribution, we obtain an expected payment to I_1 of at least $(1 - 3\epsilon)/2$. If there were an asymptotic value, φ, it would follow by letting $\epsilon \to 0$ (much more slowly than $1/k$) that $(\varphi v)(I_1) \geqq \frac{1}{2}$. In similar fashion, it would follow that $(\varphi v)(I_2) \geqq \frac{1}{2}$ and $(\varphi v)(I_3) \geqq \frac{1}{2}$; whence $(\varphi v)(I) \geqq \frac{3}{2}$, contradicting $(\varphi v)(I) = v(I) = 1$.

Proof. Let $0 < \epsilon < \frac{1}{3}$. For each positive integer k, define a partition $\Pi = \Pi_{\epsilon,k}$ of I by dividing I_2 and I_3 each into k intervals of length $1/3k$ and I_1 into $k + 1$ intervals, the first of which has length ϵ and the remaining k length $(1 - 3\epsilon)/3k$.

Denote the interval of length ϵ by J_ϵ, and let $I' = I \backslash J_\epsilon$. By removing J_ϵ from Π, we obtain a partition $\Pi' = \Pi'_{\epsilon,k}$ of I'. For given $\delta > 0$, let $U = U_\delta$ be a δ-neighborhood of the diagonal in the range of $\mu | I'$, where $\mu = (\mu_1, \mu_2, \mu_3)$. Let $\mathfrak{R}'$ be a random order on Π'; with $\mathfrak{R}'$ we may associate random sets $\mathbf{B}'_h$, $\mathbf{S}'_h$ and $\mathbf{Q}'_h$ as in Section 18. From Corollary 18.10 it then follows that if for fixed ϵ, we choose k sufficiently large, then with high probability we have that for all h,

$$\mu(\mathbf{Q}'_h) \in U;$$

i.e. with high probability there is a t such that for all h,

$$\|\mu(\mathbf{Q}'_h) - (t(1 - 3\epsilon), t, t)\| < \delta.$$

If for fixed ϵ, we choose δ sufficiently small and k sufficiently large, then it follows that with high probability, for an arbitrarily high proportion[4] of the h's, we have

$$\mu_1(\mathbf{Q}'_h) < \min\,(\mu_2(\mathbf{Q}'_h), \mu_3(\mathbf{Q}'_h)) \tag{19.3}$$

Each order $\mathfrak{R}$ on Π induces an order $\mathfrak{R}'$ on Π', simply by omitting J_ϵ; similarly to each $\mathfrak{R}'$ there correspond $3k + 1$ different orders $\mathfrak{R}$, obtained by inserting J_ϵ in the $3k + 1$ possible positions. The mapping

[4] All but the relatively small h's.

$\mathcal{R} \to \mathcal{R}'$ induces a (many to one) correspondence from random orders $\boldsymbol{\mathcal{R}}$ on Π to random orders $\boldsymbol{\mathcal{R}}'$ on Π'. Let $\mathbf{h}_\epsilon$ be the serial number of J_ϵ in $\boldsymbol{\mathcal{R}}$, i.e. $\mathbf{B}_{\mathbf{h}_\epsilon} = J_\epsilon$. Then $\boldsymbol{\mathcal{R}}$ coincides with the induced order $\boldsymbol{\mathcal{R}}'$ up to the $\mathbf{h}_\epsilon$th term. From (19.3) it therefore follows that with high probability, for an arbitrary high proportion of the h's that are $<\mathbf{h}_\epsilon$, we have

$$\mu_1(\mathbf{Q}_h) < \min[\mu_2(\mathbf{Q}_h), \mu_3(\mathbf{Q}_h)]. \tag{19.4}$$

Now

$$v_\Pi(\mathbf{S}_h) - v_\Pi(\mathbf{S}_{h-1}) = v(\mathbf{Q}_h) - v(\mathbf{Q}_{h-1}) = \min_{i=1,2,3} \mu_i(\mathbf{Q}_h) - \min_{i=1,2,3} \mu_i(\mathbf{Q}_{h-1}).$$

Hence whenever (19.4) holds, we have

$$v_\Pi(\mathbf{S}_h) - v_\Pi(\mathbf{S}_{h-1}) = \mu_1(\mathbf{Q}_h) - \mu_1(\mathbf{Q}_{h-1}) = \mu_1(\mathbf{B}_h). \tag{19.5}$$

Now let $\mathbf{H}$ be the set of all h between 1 and $3k + 1$ such that $\mathbf{B}_h \subset I_1 \backslash J_\epsilon$. Let $\mathbf{H}_\epsilon$ be the set of all h in $\mathbf{H}$ such that $h < \mathbf{h}_\epsilon$. Let

$$\boldsymbol{\Delta} = \sum_{h \in \mathbf{H}} [v_\Pi(\mathbf{S}_h) - v_\Pi(\mathbf{S}_{h-1})];$$

then $\boldsymbol{\Delta}$ is the total contribution of the players in $I_1 \backslash J_\epsilon$ to v_Π, when Π is ordered according to $\boldsymbol{\mathcal{R}}$. Thus if we denote by $I_{1\Pi}$ the set of members of Π included in I_1, then from the monotonicity of v we obtain

$$(\varphi v_\Pi)(I_{1\Pi}) \geqq \mathrm{E}\boldsymbol{\Delta}.$$

Now $h \in \mathbf{H}_\epsilon$ and h satisfies (19.4), so by (19.5) and the fact that $\mathbf{B}_h \subset I_1$ we have

$$v_\Pi(\mathbf{S}_h) - v_\Pi(\mathbf{S}_{h-1}) = \mu_1(\mathbf{B}_h) = (1 - 3\epsilon)/k.$$

When k is chosen sufficiently large, the probability that (19.4) is satisfied for a large proportion (say $\geqq 1 - \epsilon$) of the h's in $\mathbf{H}_\epsilon$ is high (say $\geqq 1 - \epsilon$); hence

$$\begin{aligned}(\varphi v_\Pi)(I_{1\Pi}) &\geqq \mathrm{E}\boldsymbol{\Delta} \\ &\geqq \mathrm{E} \sum_{h \in \mathbf{H}_\epsilon} [v_\Pi(\mathbf{S}_h) - v_\Pi(\mathbf{S}_{h-1})] \\ &\geqq \frac{1}{k}(1 - \epsilon)^2(1 - 3\epsilon)\mathrm{E}|\mathbf{H}_\epsilon| \\ &= (1 - \epsilon)^2(1 - 3\epsilon)/2.\end{aligned} \tag{19.6}$$

Now we must construct an admissible sequence. For each m, let $\epsilon = \epsilon_m = 1/2^{m+1}$, and let $J_\epsilon = [0, \epsilon]$. Choose $k = k_m$ sufficiently large that (19.6) is satisfied, and in such a way that Π_m is a refinement of Π_{m-1}. If there is an asymptotic value φv, then (19.6) implies that $(\varphi v)(I_1) \geqq \frac{1}{2}$; similarly $(\varphi v)(I_2) \geqq \frac{1}{2}$ and $(\varphi v)(I_3) \geqq \frac{1}{2}$. Hence $(\varphi v)(I) \geqq \frac{3}{2}$, in contradiction to $(\varphi v)(I) = v(I) = 1$. This completes the proof of Example 19.2.

It is not known whether there is a value on the smallest symmetric subspace of BV containing the v of Example 19.2, or on the smallest symmetric subspace containing this set function and, say, pNA. In either case, such a value would be necessarily unique. Indeed, we have

PROPOSITION 19.7 *If the v of Example* 19.2 *is a member of any space on which there is a value φ, then*

$$\varphi v = \tfrac{1}{3}(\mu_1 + \mu_2 + \mu_3) = \lambda.$$

Proof. The automorphism Θ that takes s to $s + \frac{1}{3}$ (mod 1) is a symmetry of v, and hence from the symmetry condition (2.2) it follows that

$$(\varphi v)(I_1) = (\varphi v)(I_2) = (\varphi v)(I_3).$$

Since $(\varphi v)(I) = v(I) = 1$ by the normalization condition (2.3), it follows that $(\varphi v)(I_j) = \frac{1}{3}$ for all j. Reasoning as in the proof of Proposition 6.1, we note that if S_1 and S_2 are subsets of I_1 with $\lambda(S_1) = \lambda(S_2)$, then $(\varphi v)(S_1) = (\varphi v)(S_2)$. Hence $(\varphi v)(S)$ is proportional to $\lambda(S)$ when $S \subset I_1$, and since $(\varphi v)(I_1) = \lambda(I_1)$, it follows that $(\varphi v)(S) = \lambda(S)$ for all $S \subset I_1$. Similarly $(\varphi v)(S) = \lambda(S)$ when $S \subset I_2$ or $S \subset I_3$, and the proof of Proposition 19.7 is complete.

It should be clear that the minimum (or maximum) of more than three measures will behave like Example 19.2. The exceptional behavior of the minimum of *two* measures is not just a consequence of the low dimensionality of the vector measure; a more special relationship is involved. This can be explained in terms of the "dual" of a set function, a concept that is interesting in its own right, though we have not had much occasion to use it in this book (cf. Milnor and Shapley, 1961, and Shapley, 1962ab).

If v is any set function on $(I, \mathcal{C})$ we define the set function v^* given by

$$v^*(S) = v(I) - v(I \backslash S)$$

to be the *dual* of v. Clearly $(v^*)^* = v$ and $\|v^*\| = \|v\|$. Not only BV, but most of the subspaces of BV in this book are closed under the mapping $v \to v^*$, in particular FA, NA, $bv'NA$, pNA, AC, MIX, ORD, and $ASYMP$. Moreover, it is often true that

$$\varphi v = \varphi v^*. \tag{19.8}$$

This is the case when φ is the mixing or asymptotic value, or the value on any space on which there is a unique value. (For finite games (19.8) is easily shown by reversing orders. For the mixing value it again follows by reversing orders; we could just as well have used *final* segments as initial segments, etc., throughout. For the asymptotic value we merely observe that (19.8) holds for finite games. For a space on which there is a unique value φ, we note that the mapping $v \to \varphi v^*$ is also a value.)

Now let v be any set function, and define $\bar{v} = v + v^*$. Let $\{\Pi_m\}$ be an admissible sequence. For each m we have $\varphi v_{\Pi_m} = \varphi(v_{\Pi_m})^*$, by (19.8) for finite games; hence by linearity $\varphi \bar{v}_{\Pi_m} = 2\varphi v_{\Pi_m}$. It follows from this that v is in $ASYMP$ whenever $\bar{v}$ is in $ASYMP$.

In Example 9.4 we have $v^* = -v$ and so $\bar{v} = 0$. (Equivalently, if $v = \min(\mu_1, \mu_2)$, then $v^* = \max(\mu_1, \mu_2)$, and $\bar{v} = \mu_1 + \mu_2$.) Trivially $\bar{v} \in ASYMP$; hence $v \in ASYMP$. The "crease along the diagonal," which causes all the trouble with the axiomatic and mixing values, is exactly canceled out when we add v^* to v. It is only this special occurrence that permits the asymptotic value to exist. Indeed, consider the set function w given by

$$w(S) = \min[\mu_1(S), \mu_2(S), c]$$

where $\mu_1, \mu_2 \in NA^1$ and $0 < c < 1$. Despite the resemblance to the previous case, the "creases" do not cancel, and we have $w \notin ASYMP$. The proof is similar to that for Example 19.2.

Chapter IV. Values and Derivatives

20. Introduction

Let us recall formula (3.1) for the value of a "vector measure game," i.e. a set function of the form $f \circ \mu$, where μ is an n-dimensional vector measure and f is a real function of n real variables. The formula reads

$$\varphi(f \circ \mu)(S) = \int_0^1 f_{\mu(S)}(t\mu(I))\, dt,$$

where $f_{\mu(S)}$ is the derivative of f in the direction $\mu(S)$. This formula is of central importance in the study of values. The purpose of this chapter is to reformulate and generalize it, and thereby also to gain a better insight into what the formula says.

Formula (3.1) may be intuitively understood as follows: Suppose the players could be ordered "at random." Then if T were an initial segment in such a random ordering, $\mu(T)$ would with probability 1 be on the diagonal $[0, \mu(I)]$, i.e. it would be of the form $t\mu(I)$ (compare the discussion at the beginning of Section 14). Now let a coalition S be given. In a random ordering, S would be "evenly spread" over the entire player set; in particular if T is an initial segment with $\mu(T) = t\mu(I)$, then we would have $\mu(S \cap T) = t\mu(S)$ (with probability 1). Suppose now that $T \cup \Delta T$ is another initial segment, where ΔT is a "small" segment disjoint from T; set $\mu(T \cup \Delta T) = (t + \Delta t)\mu(I)$. Then $\mu(S \cap [T \cup \Delta T]) = (t + \Delta t)\mu(S)$, and hence $\mu(S \cap \Delta T) = \mu(S)\, \Delta t$. Now let us appraise the contribution to v—over and above $v(T)$—of that portion of S that is in ΔT. We have

$$(f \circ \mu)(T \cup [S \cap \Delta T]) - (f \circ \mu)(T) = \frac{f(t\mu(I) + \mu(S)\, \Delta t) - f(t\mu(I))}{\Delta t}\, \Delta t.$$

If we think of Δt as an "infinitesimal" segment, then the right side of the above equation becomes, by definition,

$$f_{\mu(S)}(t\mu(I))\, dt.$$

The sum total of all these infinitesimal contributions of S is the total contribution of S in a random ordering, and it is exactly the right side of (3.1).

We wish to apply this reasoning to the more general situation in which v is not of the form $f \circ \mu$. The result will be a reformulation of (3.1) which is valid for a more general class of v's, and which is stated directly in terms of the set function v, rather than in terms of the representation of v in the form $f \circ \mu$.

The above intuitive reasoning is based on the notion of an "evenly spread" measurable set. We know that no such set exists; nevertheless, it is this ideal that must somehow be embodied in the reformulation of (3.1) that we are seeking. Now we can generalize—or idealize—the notion of set by specifying for each point a weight between 0 and 1, which indicates the "degree" to which that point belongs to the set; ordinary sets are then characterized by weights that are 1 or 0, according as the point in question does or does not belong to the set. If we admit this more general—or ideal—kind of set, then we could say that a set that assigns a constant weight to all points of I is "evenly spread" over I; this could form a basis for a generalization and formalization of the above intuitive justification of (3.1).

Formally, let us define a *measurable ideal subset of* $(I, \mathcal{C})$ (or simply *ideal set*) to be a measurable function from $(I, \mathcal{C})$ to $([0, 1], \mathcal{B})$. To an ordinary measurable set—i.e. member S of $\mathcal{C}$—there corresponds naturally an ideal set, namely its characteristic function χ_S. The family of all measurable ideal subsets[1] of $(I, \mathcal{C})$ will be denoted $\mathcal{I}$.

Although formally an ordinary set S is not an ideal set, we will find it convenient in intuitive discussion to identify it with the ideal set χ_S corresponding to it. Under this convention, $\mathcal{C}$ becomes a subset of $\mathcal{I}$. The situation is analogous to that in algebra, where a member of a ring is sometimes identified—especially in intuitive discussion—with the principal ideal that it generates.

Since ideal sets are actually functions, they can, with certain restrictions, be multiplied by constants, added, and subtracted. Thus if

[1] Formally, our ideal sets are similar to the "fuzzy sets" of Zadeh (1965), but intuitively the ideas are somewhat different. In particular, the topology we shall define on $\mathcal{I}$ (the "NA-topology") does not seem to fit in well with the intuitive explanation in that paper.

S is an ideal set, and $0 \leqq \alpha \leqq 1$, then αS is also an ideal set; and if S and T are ideal sets, so is $S \pm T$, as long as the values of $S \pm T$ are in $[0, 1]$. Furthermore, to the set-theoretic notions of union, intersection, and inclusion on ordinary sets, there correspond, respectively, the algebraic notions of max (or sup), min (or inf), and pointwise "less than or equal to" on ideal sets. When identifying ordinary and ideal sets, we may use either the set-theoretic or algebraic terminology on $\mathcal{C}$, whichever is more convenient.

Now let v be a set function. A priori, v is defined on $\mathcal{C}$ only; however, since we are thinking of the ideal sets as being in some sense approximable by ordinary sets, let us suppose for the moment that v has been extended to all of $\mathcal{I}$. Let us return to the intuitive reasoning that we used above in our discussion of formula (3.1). If the players could be ordered at random, then with probability 1 each initial segment would be "evenly spread," i.e. it would be an ideal set of the form tI, where $0 \leqq t \leqq 1$. Now let S be an arbitrary member of $\mathcal{C}$, and let $(t + \Delta t)I$ be an "evenly spread" initial segment that is slightly larger than tI. The contribution to v—over and above $v(tI)$—of that portion of S that is in $(t + \Delta t)I$ but not in tI may be expressed by

$$
\begin{aligned}
v(tI \cup [S \cap (t + \Delta t)I]) - v(tI) &= v(tI + (\Delta t)S) - v(tI) \\
&= \frac{v(tI + (\Delta t)S) - v(tI)}{\Delta t}\,\Delta t.
\end{aligned}
$$

If Δt is "infinitesimal," then

$$\frac{v(tI + (\Delta t)S) - v(tI)}{\Delta t} = \frac{d}{d\tau} v(tI + \tau S);$$

hence the contribution of that portion of S that we are looking at is

$$\frac{d}{d\tau} v(tI + \tau S)\,\Delta t.$$

The value $(\varphi v)(S)$ is the total contribution of S in a "random ordering," and hence is the sum total of all these infinitesimal contributions. Intuitively, therefore, we should have

$$(\varphi v)(S) = \int_0^1 \left[\frac{d}{d\tau} v(tI + \tau S) \right] dt. \tag{20.1}$$

A rigorous statement and proof of formula (20.1), for v in pNA, is the main object of Chapter IV (Theorem H, Section 21). Of central importance in the statement of this result is the notion of *extension* (of v from $\mathcal{C}$ to $\mathcal{I}$), which will be studied in detail in Section 22. Theorem H will be proved in Section 23, and in Section 24 we will discuss the interpretation in terms of Fréchet differentials. In Section 25 we shall relate the mixing and asymptotic approaches studied in Chapters II and III to the notion of extension.

21. Statement of Results

We define a partial order on $\mathcal{I}$ by $f \geqq g$ if $f(s) \geqq g(s)$ for all $s \in I$. A real-valued function w on $\mathcal{I}$ with $w(0) = 0$ is called an *ideal set function* (i.e. a function on ideal sets); it is called *monotonic* if $f \geqq g \Rightarrow w(f) \geqq w(g)$. The characteristic function of a member S of $\mathcal{C}$ is denoted χ_S.

THEOREM G. *There is a unique mapping that associates with each $v \in pNA$ an ideal set function v^*, so that*

$$(\alpha v + \beta w)^* = \alpha v^* + \beta w^* \tag{21.1}$$

$$(vw)^* = v^* w^* \tag{21.2}$$

$$\mu^*(f) = \int_I f \, d\mu \tag{21.3}$$

$$v \text{ monotonic} \Rightarrow v^* \text{ monotonic} \tag{21.4}$$

whenever $v, w \in pNA$, $\alpha, \beta \in E^1$, $\mu \in NA$ and $f \in \mathcal{I}$.

The ideal set function v^* is called the *extension* of v. The extension has other interesting and desirable properties, and in particular the property

$$v^*(\chi_S) = v(S),$$

which, together with the other properties, justifies its name (see (22.12)). Similar extensions can be defined on spaces that are much larger than pNA, some of which are not even contained in BV. These matters will be investigated in Section 22, where also Theorem G will be proved.

We now come to the main object of this chapter, namely a rigorous formulation of the idea embodied in formula (20.1). Denote

$$\partial v^*(t, S) = \frac{d}{d\tau} v^*(t\chi_I + \tau\chi_S),$$

where the derivative on the right is evaluated at $\tau = 0$ (of course no claim is being made about the existence of the derivative, we are merely introducing a notation). To relate this to previously defined notions, we assert[1] that when $v = f \circ \mu$ and f is continuously differentiable on the range of μ, then

$$\partial v^*(t, S) = f_{\mu(S)}(t\mu(I))$$

(compare Theorem B in Section 3); hence when the range of μ has full dimension, we have

$$\partial v^*(t, S) = \sum_i \mu_i(S) f_i(t\mu(I)).$$

THEOREM H. *For each v in pNA and each $S \in \mathcal{C}$, the derivative $\partial v^*(t, S)$ exists for almost all t in* $[0, 1]$, *and is integrable over* $[0, 1]$ *as a function of t; and if φ is the value on pNA, then*

$$(\varphi v)(S) = \int_0^1 \partial v^*(t, S)\, dt.$$

22. Extensions: The Axiomatic Approach

Our first object in this section is to prove Theorem G. Afterwards we will investigate the subject of extensions from a somewhat broader viewpoint.

An ideal set function is said to be of *bounded variation* if it is the difference of two monotonic ideal set functions. The space of all ideal set functions of bounded variation will be denoted IBV. For $v \in IBV$, define

$$\|v\| = \inf(u(\chi_I) + w(\chi_I)),$$

where the inf ranges over all monotonic ideal set functions u and w such that

$$v = u - w.$$

The quantity $\|v\|$ will be called the *variation* of v; it is easily seen that it is a norm.

This definition of the variation of an ideal set function is completely analogous to the definition of variation for an ordinary set function given in Section 3. It also has analogous properties; in particular, IBV

[1] This follows from (22.18) and Proposition 22.16 below.

with the variation norm is a Banach space, in which the set of all monotonic ideal set functions forms a closed cone. Moreover, we may define the notions of chain, link, and subchain analogously to the definitions in Section 4, and the precise analogue of Proposition 4.1 holds. The proof is again entirely analogous.

We next define a topology on $\mathcal{I}$ which will be needed in the proof of the existence part of Theorem G. Each member μ of NA induces a function $\mu^{\#}$ on $\mathcal{I}$ defined by

$$\mu^{\#}f = \int_I f \, d\mu.$$

The *NA-topology* on $\mathcal{I}$ is defined to be the smallest topology for which all these functions are continuous.

The following convention will be useful:

CONVENTION. If the range of integration of an integral is not specified, it shall be taken to be I. An integral with respect to a vector measure is the vector of integrals with respect to its components.

In the study of the NA-topology, we shall use the following lemma:

LEMMA 22.1. *Let μ be a finite-dimensional vector of measures in NA. Let g_1 and g_2 be in $\mathcal{I}$, and $g_2 \geqq g_1$. Then there are T_1 and T_2 in $\mathcal{C}$ with $T_2 \supset T_1$ such that for $i = 1, 2$,*

$$\mu(T_i) = \int g_i \, d\mu.$$

Proof. We use the following result of Dvoretzky, Wald, and Wolfowitz (1951, p. 66, Theorem 4): *Let μ be a finite-dimensional vector of measures in NA, and let m be a positive integer. Then for each m-tuple $f_1, \ldots, f_m$ of non-negative measurable functions on $(I, \mathcal{C})$ such that $f_1 + \cdots + f_m = \chi_I$, there is an ordered partition $(S_1, \ldots, S_m)$ of I into measurable sets such that for all i,*

$$\mu(S_i) = \int f_i \, d\mu. \qquad (22.2)$$

Lemma 22.1 follows from this result if we set

$$m = 3, f_1 = g_1, f_2 = g_2 - g_1, f_3 = \chi_I - g_2, T_1 = S_1, T_2 = S_1 \cup S_2.$$

The following corollary shows that by extending the domain of a non-atomic vector measure to include ideal sets as well as ordinary sets, we do not enlarge its range.

COROLLARY 22.3. *Let μ be a finite-dimensional vector of measures in NA, and let g be in $\mathcal{I}$. Then there is a T in $\mathcal{C}$ with*

$$\mu(T) = \int g \, d\mu.$$

Proof. In Lemma 22.1, set $g_1 = g_2$.

An important application of this corollary is

PROPOSITION 22.4. *In the NA-topology on $\mathcal{I}$, the set of all characteristic functions χ_S is dense in $\mathcal{I}$.*

Proof. If $g \in \mathcal{I}$, then every neighborhood of g contains a neighborhood of the form

$$\{f \in \mathcal{I}: \|\int (f - g) \, d\mu\| < \epsilon\},$$

where μ is a finite dimensional vector of measures in NA, $\epsilon > 0$, and $\| \quad \|$ is the maximum norm. Applying Corollary 22.3, we see that this neighborhood contains a characteristic function, namely χ_T. This completes the proof of Proposition 22.4.

In addition to the NA-topology, and the concept of continuity implied by it, we shall use a concept of "uniform" continuity.[1] A real valued function w on $\mathcal{I}$, or on a subset of $\mathcal{I}$, is said to be *uniformly continuous* if for every $\epsilon > 0$ there is a vector μ of NA measures and a $\delta > 0$ such that for all f and g in the domain of w,

$$\|\int (f - g) \, d\mu\| < \delta \Rightarrow |w(f) - w(g)| < \epsilon.$$

We now proceed to the

Proof of Theorem G. First we prove uniqueness. Let $v \in pNA$. If, in fact, $v \in NA$, then v^* is determined by (21.3). Hence if v is a polynomial in measures, then v^* is determined by (21.1) and (21.2). Since the polynomials in measures are dense in pNA, it remains only to prove that the mapping $v \to v^*$ is a continuous mapping from BV to IBV. First, by an argument similar to that used for (4.16), it may be seen that[2]

(22.5) $\qquad v^*(\chi_I)/v(I)$ is bounded for $v \in (pNA)^+$;

[1] See Note 1.

[2] According to the intuitive interpretation of v^*, we should have $v^*(\chi_I) = v(I)$; but this has not yet been proved. See (22.12) below.

denote the bound by K. Now let $v \in pNA$. By Proposition 7.19, we may find u and w in $(pNA)^+$ such that $v = u - w$ and

$$\|v\| \geqq \tfrac{1}{2}(u(I) + w(I)).$$

Hence by (21.4),

$$\|v^*\| = \|u^* - w^*\| \leqq \|u^*\| + \|w^*\| = u^*(\chi_I) + w^*(\chi_I) \leqq K(u(I) + w(I)) \leqq 2K\|v\|.$$

Thus the mapping $v \to v^*$ is continuous[3], and the proof of uniqueness is complete.

Next, we prove existence. As before we shall identify each S in $\mathcal{C}$ with its characteristic function χ_S. Our first aim is to show that every v in pNA is uniformly continuous on $\mathcal{C}$.

If $v \in NA$, then the uniform continuity of v on $\mathcal{C}$ follows from the definition of uniform continuity. From this and the fact that the sum and the product of uniformly continuous functions are uniformly continuous, it follows that v is uniformly continuous also if it is a polynomial in NA measures. Suppose finally that v is an arbitrary member of pNA. For given $\epsilon > 0$, let v_ϵ be a polynomial in NA measures such that

$$\|v - v_\epsilon\| < \epsilon/4.$$

Then for any two sets S, T, we have

$$(22.6)\quad |(v - v_\epsilon)(S) - (v - v_\epsilon)(T)| \leqq |(v - v_\epsilon)(S)| + |(v - v_\epsilon)(T)| \leqq \|v - v_\epsilon\| + \|v - v_\epsilon\| < \epsilon/2.$$

Now from the uniform continuity of v_ϵ it follows that there is a vector μ of NA measures and a $\delta > 0$ such that

$$\|\mu(S) - \mu(T)\| < \delta \Rightarrow |v_\epsilon(S) - v_\epsilon(T)| < \epsilon/2.$$

From this and (22.6) it follows that

$$\|\mu(S) - \mu(T)\| < \delta \Rightarrow |v(S) - v(T)| < \epsilon,$$

and this establishes the desired uniform continuity.

[3] This can also be proved by means of the Bakhtin-Krasnoselskii-Stetzenko Theorem (see Note 1 to Section 4). In that case, one must first show that every v^* is in IBV; this follows from the internality of pNA (Proposition 7.19) and (21.4).

Having shown that v is uniformly continuous on $\mathcal{C}$, we wish to extend it to a uniformly continuous function v^* on $\mathcal{G}$. For each n, let $\epsilon_n = 1/n$, and let μ^n and δ_n correspond to ϵ_n in accordance with the uniform continuity condition for v on $\mathcal{C}$. Let $g \in \mathcal{G}$; we wish to define $v^*(g)$. Because of the denseness of $\mathcal{C}$ in $\mathcal{G}$ (Proposition 22.4), we may choose for every n a set $S_n = S_n^g$ in $\mathcal{C}$ with

$$\|\int (\chi_{S_n} - g)\, d\mu^n\| < \delta_n/3.$$

Now we may assume w.l.o.g. that δ_n is a decreasing sequence and that for $m \geqq n$, all the coordinates of μ^n are also coordinates of μ^m (thus in general the dimension of μ^m is $\geqq$ the dimension of μ^n). From this it follows that for $m \geqq n$

$$\|\int (\chi_{S_m} - g)\, d\mu^n\| \leqq \|\int (\chi_{S_m} - g)\, d\mu^m\| < \delta_m/3 \leqq \delta_n/3.$$

Hence

$$\|\mu^n(S_m) - \mu^n(S_n)\| = \|\int (\chi_{S_m} - \chi_{S_n})\, d\mu^n\| < 2\delta_n/3 < \delta_n,$$

and hence by the uniform continuity,

$$|v(S_m) - v(S_n)| < 1/n. \tag{22.7}$$

Thus $v(S_m)$ is a Cauchy sequence, and so it has a limit, which we define to be $v^*(g)$. Note that when g is of the form χ_S, we may take $S_n = S$ for all n and deduce that

$$v^*(\chi_S) = v(S).$$

We now wish to show that the function v^* so defined is uniformly continuous on $\mathcal{G}$. Let $\epsilon > 0$ be given. Fix n so that $1/n < \epsilon/3$. Set $\delta = \delta_n/3$, $\mu = \mu^n$. Suppose that

$$\|\int (f - g)\, d\mu\| < \delta;$$

we must show that

$$|v^*(f) - v^*(g)| < \epsilon.$$

If we let $m \to \infty$ in (22.7), we find

$$|v(S_n^g) - v^*(g)| < 1/n < \epsilon/3; \tag{22.8}$$

similarly

$$|v(S_n^f) - v^*(f)| < \epsilon/3. \tag{22.9}$$

On the other hand,

$$\|\mu(S_n^f) - \mu(S_n^g)\| \leqq \|\int(f - \chi_{S_n^f})\, d\mu\| + \|\int(f - g)\, d\mu\| + \|\int(g - \chi_{S_n^g})\, d\mu\| < 3\delta_n/3 = \delta_n.$$

Therefore

$$|v(S_n^f) - v(S_n^g)| < 1/n < \epsilon/3.$$

Combining this with (22.8) and (22.9), we obtain

$$|v^*(f) - v^*(g)| < \epsilon,$$

and so v^* is indeed uniformly continuous, as was to be shown. In particular, it follows that v^* is continuous on $\mathcal{I}$ in the NA-topology.

We claim that v^* satisfies the conditions of Theorem G. Indeed, to establish (21.1) through (21.3), we note that $(\alpha v + \beta w)^* - (\alpha v^* + \beta w^*)$, $(vw)^* - v^*w^*$, and $\mu^* - \mu^\#$ are continuous extensions of 0 and hence must vanish identically. It remains only to establish (21.4).

To this end, let v be monotonic, and let g_1 and g_2 in $\mathcal{I}$ be such that $g_2 \geqq g_1$. Let $\epsilon > 0$ be given, and let μ and δ correspond to ϵ in accordance with the definition of uniform continuity of v^* on $\mathcal{I}$; that is,

$$\|\int(f - g)\, d\mu\| < \delta \Rightarrow |v^*(f) - v^*(g)| < \epsilon. \tag{22.10}$$

By Lemma 22.1, there are sets T_1 and T_2 in $\mathcal{C}$ with $T_2 \supset T_1$ and $\mu(T_i) = \int g_i\, d\mu$. Hence if in (22.10) we set $f = \chi_{T_i}$ and $g = g_i$, we obtain

$$|v(T_i) - v^*(g_i)| = |v^*(\chi_{T_i}) - v^*(g_i)| < \delta.$$

Using $T_2 \supset T_1$ and the monotonicity of v, we deduce

$$v^*(g_2) - v^*(g_1) > -2\epsilon;$$

and therefore, since ϵ may be arbitrarily chosen, $v^*(g_2) - v^*(g_1) \geqq 0$. This completes the proof of (21.4), and with it, the proof of Theorem G.

Theorem G is all we need to know about extensions for the statement and proof of Theorem H, but the subject of the extension of set functions to ideal sets is of some interest in its own right and we would like to investigate it somewhat further. Our remarks fall into two classes: In the remainder of this section we discuss some of the abstract properties of the operator $v \to v^*$, showing in particular how this operator can be uniquely defined on spaces that are much larger than pNA. In Section 25 we discuss some of the more "concrete" properties of v^* for individual v's, showing how $v^*(f)$ can be calculated for given v and f and relating it to procedures previously discussed, such as those leading to the mixing and asymptotic values.

First, let us recall three by-products of the proof of Theorem G that are worth remembering in their own right. The first is that in the NA-topology,

$$(22.11) \qquad v^* \text{ is continuous on } \mathcal{I},$$

and, in fact, uniformly continuous; this was used in the existence proof. The second is

$$(22.12) \qquad v^*(\chi_S) = v(S)$$

for all $S \in \mathcal{C}$; this was also used in the existence proof. The third by-product, which is part of the uniqueness proof, is that the operator $v \to v^*$ is continuous. Together with (22.11), this has a further interesting consequence, namely

$$(22.13) \qquad \|v^*\| = \|v\|.$$

To prove (22.13), let $v \in pNA$ and $\epsilon > 0$; then use Proposition 7.19 to find u and w in $(pNA)^+$ such that $v = u - w$ and $u(I) + w(I) \leqq \|v\| + \epsilon$. By (21.4), both u^* and w^* are monotonic, and hence by (22.12),

$$\|v^*\| = \|u^* - w^*\| \leqq \|u^*\| + \|w^*\| = u^*(\chi_I) + w^*(\chi_I)$$
$$= u(I) + w(I) \leqq \|v\| + \epsilon.$$

Since ϵ may be chosen arbitrarily small, (22.13) follows.

Now (22.13) was established when the variation norm is imposed both on BV and on IBV; but it is true also in another norm, namely the *supremum norm* $\| \quad \|'$. Indeed, define

$$\|v\|' = \sup\{|v(S)| : S \in \mathcal{C}\}$$

for bounded set functions v, and

$$\|w\|' = \sup\{|w(f)| : f \in \mathcal{I}\}$$

for bounded ideal set functions w. Then using (22.11) and Proposition 22.4, we obtain

$$(22.14) \qquad \|v^*\|' = \|v\|',$$

from which we deduce that

(22.15) $v \to v^*$ is continuous in the supremum norm.

This suggests that it might be possible to define the extension operator $v \to v^*$ on a class of set functions much wider than pNA. Let BS be the Banach space of all bounded set functions with the supremum norm, and let pNA' be the subspace of BS spanned by the powers of non-atomic measures. pNA' is much larger than pNA; for example, it contains the set function of Example 9.4 and also set functions of the form $f \circ \mu$, where $\mu \in NA$ and f is a singular continuous function. It even contains set functions outside of BV; for example, the set function of Example 9.3 is in it.

PROPOSITION 22.16. *There is a unique mapping that associates with each $v \in pNA'$ an ideal set function v^* so that* (21.1), (21.2), (21.3) *and* (22.15) *are satisfied. On pNA, this mapping coincides with that of Theorem* G. *Finally, this mapping obeys* (22.11).

The proof of this proposition follows the same lines as that of Theorem G, and the reader will have no difficulty in reconstructing it. In some respects the proof is, in fact, easier than that of Theorem G.

NOTES

1. Uniform continuity is not a topological notion; that is, it cannot be defined in terms of the NA-topology alone. What we are in fact doing is defining $\mathcal{I}$ as a uniform space rather than just a topological space. The uniformity on $\mathcal{I}$ is the smallest uniformity such that all the $\mu^{\#}$ are uniformly continuous; the topology induced by this uniformity is the NA-topology. (For the definition and basic properties of uniform spaces, see, for example, Kelley, 1955, Chapter 6.) Here we do not make formal use of the theory of uniform spaces because it is quicker to prove the relevant facts as we need them.

2. It is possible to ring the changes on Proposition 22.16 in a number of ways. First of all, (22.15) is interchangeable with (22.14); this is readily verified. Next, (21.3) is interchangeable with

(22.17) If $\mu \in NA$, then $\mu^*(\chi_S) = \mu(S)$ for all $S \in \mathcal{C}$ and

$$\mu^*(f + g) = \mu^*(f) + \mu^*(g)$$

whenever f and g are ideal sets with $f + g \leqq \chi_I$.

Condition (22.17) says that μ^* is, in a sense, a measure on ideal sets that extends the given measure; its interchangeability with (21.3) (in the presence of the other conditions, of course) is readily verified. Finally, (21.2) is interchangeable with

(22.18) If f is a continuous real-valued function, then $(f \circ v)^* = f \circ v^*$.

Indeed, to deduce (22.18) from (21.2) we approximate on the range of v to f in the supremum norm by polynomials. The reverse direction is also easy; it uses

$$vw = ((v + w)^2 - (v - w)^2)/4.$$

Still another version of (21.2), as is readily verified, is

$$(v^2)^* = (v^*)^2. \tag{22.19}$$

These interchanges can be made independently of one another. Specifically, we get 12 variations (including the original) on Proposition 22.16, as follows: Let a_1, a_2 and a_3 stand for (21.2), (22.18), and (22.19) respectively; b_1 and b_2 for (21.3) and (22.17) respectively; c_1 and c_2 for (22.14) and (22.15) respectively. Then we have

Remark 22.20. For any choice of i, j and k, Proposition 22.16 remains true if a_i, b_j, and c_k are substituted for (21.2), (21.3), and (22.15), respectively. All the mappings defined in this way coincide with each other.

23. Proof of Theorem H

The subset S of I will be fixed throughout this section; to simplify the notation, therefore, we will write $\partial v^*(t)$ instead of $\partial v^*(t, S)$.

Define

$$|\partial v^*(t)|^+ = \limsup_{\tau \to 0} \left| \frac{v^*(t\chi_I + \tau\chi_S) - v^*(t\chi_I)}{\tau} \right|.$$

LEMMA 23.1. *If v is in pNA, then $|\partial v^*(t)|^+$ is integrable over* [0, 1], *and we have*

$$\int_0^1 |\partial v^*(t)|^+ \, dt \leqq \|v\|.$$

Proof. From (22.11) it follows that $v^*(t\chi_I + \tau\chi_S)$ is continuous in τ. Hence in the definition of $|\partial v^*(t)|^+$, we can let τ vary over the rationals only; therefore $|\partial v^*(t)|^+$ is measurable in t.

To prove the remainder of the lemma, first let v be monotonic. Then v^* is also monotonic, and for $\tau > 0$ we have

$$0 \leqq v^*(t\chi_I + \tau\chi_S) - v^*(t\chi_I) \leqq v^*((t + \tau)\chi_I) - v^*(t\chi_I).$$

For $\tau < 0$ we have

$$0 \geqq v^*(t\chi_I + \tau\chi_S) - v^*(t\chi_I) \geqq v^*((t + \tau)\chi_I) - v^*(t\chi_I).$$

In either case division by τ yields

$$(23.2) \quad 0 \leqq \frac{v^*(t\chi_I + \tau\chi_S) - v^*(t\chi_I)}{\tau} \leqq \frac{v^*((t + \tau)\chi_I) - v^*(t\chi_I)}{\tau}.$$

The function $g(t) = v^*(t\chi_I)$ is monotonic in t, and hence is a.e. differentiable. Hence if we let $\tau \to 0$ in (23.2), then the right side tends a.e. to the limit $g'(t)$. The middle term may not converge, but from (23.2) we obtain

$$(23.3) \qquad |\partial v^*(t)|^+ \leqq g'(t) \qquad \text{a.e.}$$

From the monotonicity of g we obtain

$$(23.4) \qquad \int_0^1 g'(t)\, dt \leqq g(1) - g(0) = v(I) = \|v\|;$$

for example, this follows from decomposing g into an absolutely continuous and a singular part and then using, say, (8.4) (alternatively, see Titchmarsh, 1939, p. 361). The conclusion of Lemma 23.1 follows from (23.3) and (23.4).

In the general case, when v is not necessarily monotonic, let $\epsilon > 0$ be given, and set

$$v = u - w,$$

where u and w in pNA are monotonic and

$$\|v\| + \epsilon \geqq \|u\| + \|w\|;$$

such u and w exist by Proposition 7.19. We then have

$$v^* = u^* - w^*,$$

and hence

$$|\partial v^*(t)|^+ \leqq |\partial u^*(t)|^+ + |\partial w^*(t)|^+.$$

Integrating this inequality, and using the monotonic case of the lemma (which we have already proved), we obtain

$$\int_0^1 |\partial v^*(t)|^+\, dt \leqq \|u\| + \|w\| \leqq \|v\| + \epsilon;$$

and if we let $\epsilon \to 0$, the proof of Lemma 23.1 is complete.

Define

$$(23.5)\quad \Delta_v(t) = \limsup_{\tau \to 0} \frac{v^*(t\chi_I + \tau\chi_S) - v^*(t\chi_I)}{\tau} - \liminf_{\tau \to 0} \frac{v^*(t\chi_I + \tau\chi_S) - v^*(t\chi_I)}{\tau}$$

and

$$\Delta_v = \int_0^1 \Delta_v(t)\, dt.$$

Clearly

$$0 \leqq \Delta_v(t) \leqq 2|\partial v^*(t)|^+,$$

and hence by Lemma 23.1,

$$(23.6)\qquad \Delta_v \leqq 2 \int_0^1 |\partial v^*(t)|^+\, dt \leqq 2\|v\|.$$

Furthermore, since the lim sup is subadditive and the lim inf superadditive, we have

$$\Delta_{v+w}(t) \leqq \Delta_v(t) + \Delta_w(t)$$

and hence

$$(23.7)\qquad \Delta_{v+w} \leqq \Delta_v + \Delta_w.$$

Now it is easily verified that $\Delta_v = 0$ when v is a power of an NA measure, and hence by (23.6), when it is any polynomial in NA measures. If now v is any member of pNA, let $\epsilon > 0$ be given, and let w be a polynomial in measures such that $\|v - w\| < \epsilon$. Then by (23.7) and (23.6),

$$\Delta_v \leqq \Delta_w + \Delta_{v-w} = \Delta_{v-w} \leqq 2\|v - w\| < 2\epsilon.$$

Letting $\epsilon \to 0$, we deduce that $\Delta_v = 0$ for all v in pNA. Hence $\Delta_v(t) = 0$ for almost all t, i.e. $\partial v^*(t)$ exists a.e. for all v. Whenever it exists we have $|\partial v^*(t)| = |\partial v^*(t)|^+$, and hence by Lemma 23.1,

$$(23.8)\qquad \int_0^1 |\partial v^*(t)|\, dt \leqq \|v\|;$$

in particular, this implies the integrability of $\partial v^*(t)$.

Now let

$$\theta v = \int_0^1 \partial v^*(t)\, dt;$$

θv is clearly linear in v, and from (23.8) we obtain

$$|\theta v| \leqq \|v\|.$$

Thus θ is a continuous linear functional on pNA. When v is a power of a measure, it is easily verified that $\theta v - (\varphi v)(S) = 0$. Hence this holds also when v is a polynomial in measures, and hence, since both θv and $\varphi v(S)$ are continuous in v, when v is any member of pNA. This completes the proof of Theorem H.

We close this section with another[1]

Alternative proof for Example 9.4. If $v \in pNA$, then by (22.18),

$$v^*(f) = \left| \int f \, d\mu \right|.$$

We have

$$\int (t\chi_I) \, d\mu = 0.$$

If we set $S = [0, 1]$, then

$$\int (t\chi_I + \tau\chi_S) \, d\mu = \tau.$$

Hence

$$v^*(t\chi_I + \tau\chi_S) - v^*(t\chi_I) = |\tau|.$$

Therefore $\partial v^*(t) = \lim_{\tau \to 0} |\tau|/\tau$ never exists, and so, by Theorem H, v cannot be in pNA. This completes the proof.

Note that though v is not in pNA, it *is* in pNA'.

24. Fréchet Differentials

It is possible to strengthen Theorem H as follows: Not only is it true that for each S, the derivative $\partial v^*(t, S)$ exists for almost all t, but we can reverse the quantifiers and assert that for almost all t, it is true that for each S, $\partial v^*(t, S)$ exists. Furthermore, we will be able to show that for almost all t, $\partial v^*(t, S)$ is a measure in S.

These ideas find their proper expression in terms of Fréchet differentials, and by replacing the ordinary sets S appearing in the derivative $\partial v^*(t, S)$ by ideal sets. First let us define the Fréchet differential; we follow Dunford and Schwartz (1958, p. 92). Let X be a Banach space and let u be a function from an open subset of X to the reals. A

[1] In addition to the original proof in Section 9, a proof was given in Section 16.

Fréchet differential of u at a point x in X is a linear functional $Du(x, \cdot)$ from X to the reals such that

$$u(x + h) = u(x) + Du(x, h) + o(\|h\|)$$

as $\|h\| \to 0$. Of course it does not necessarily exist; but if it does exist, it is necessarily unique, since the difference of two Fréchet differentials is a linear functional in h that is $o(\|h\|)$, and so vanishes. The Fréchet differential is defined similarly when the range space of u is any Banach space rather than the reals; but we do not need this here. Roughly, we may say that u has a Fréchet differential at x if it can be approximated by a linear functional in the neighborhood of x.

In our situation, let $\mathfrak{M}$ be the Banach space of all bounded measurable functions from the underlying space $(I, \mathcal{C})$ to the reals, where the norm is defined by

$$\|f\| = \sup_{t \in I} |f(t)|.$$

The set $\mathcal{I}$ of all ideal sets is a closed subset of $\mathfrak{M}$. We wish to consider the Fréchet differential of v^*; however, since the domain of the function to be differentiated must be open, we will restrict our attention to the restriction of v^* to the interior $\mathcal{I}^\circ$ of $\mathcal{I}$. The set $\mathcal{I}^\circ$ may be characterized as the set of all $f \in \mathcal{I}$ that are bounded away from 0 and from 1 (i.e. such that there exist α and β with $0 < \alpha \leqq f(t) \leqq \beta < 1$ for all $t \in I$).

PROPOSITION 24.1. *Let $v \in pNA$. Then for almost all $t \in (0, 1)$, the following statements hold:*

(i) *The extension v^* has a Fréchet differential $Dv^*(t\chi_I, \cdot)$ at $t\chi_I$.*

(ii) *$\partial v^*(t, S)$ exists for all $S \in \mathcal{C}$.*

(iii) *If we denote $\partial_t v^*(S) = \partial v^*(t, S)$, then $\partial_t v^*$ is a measure in NA, and for all $f \in \mathcal{I}$,*

$$Dv^*(t\chi_I, f) = \int f \, d\partial_t v^*.$$

The integral $\int f \, d\partial_t v^*$ is best viewed as the $\partial_t v^*$-measure of the ideal set f. Thus our theorem states that, for almost all t, the ideal set function v^* behaves like a constant plus a measure in the neighborhood of $t\chi_I$. According to Theorem H, the value is the integral of these measures over t.

Before proceeding with the proof of Proposition 24.1, we state a lemma that we shall use on a number of occasions throughout this proof. Let us call a real function q on a linear space X *subadditive* if $q(x + y) \leqq q(x) + q(y)$ for all x and y in X.

LEMMA 24.2. *Let q be a real non-negative subadditive function on pNA such that for some constant α,*

$$q(v) \leqq \alpha \|v\|$$

for all v, and such that $q(v)$ vanishes whenever v is a polynomial in measures. Then q vanishes identically.

Proof. This was proved in Section 23 in the particular case $q(v) = \Delta_v$ and $\alpha = 2$; the proof is identical here.

Proof of Proposition 24.1. Throughout this proof the letters "a.e." will mean "for almost all t in (0, 1)."

If h is a non-negative function on [0, 1] that is not necessarily measurable, define the *upper integral of h over* [0, 1] by[1]

$$\begin{aligned}\int^{\#} h(t)\,dt &= \int^{\#}_{[0,1]} h(t)\,dt \\ &= \inf\left\{\int_0^1 g(t)\,dt \colon g(t) \geqq h(t) \text{ for all } t, \text{ and } g \text{ is measurable}\right\}.\end{aligned}$$

The properties of the upper integral are in many respects similar to those of the integral; they include the following:

$$\begin{aligned}&\int^{\#} (h + g)(t)\,dt \leqq \int^{\#} h(t)\,dt + \int^{\#} g(t)\,dt;\\ &\int^{\#} h(t)\,dt = 0 \Rightarrow h(t) = 0 \quad \text{a.e.};\\ &h(t) \leqq g(t) \quad \text{a.e.} \Rightarrow \int^{\#} h(t)\,dt \leqq \int^{\#} g(t)\,dt;\\ &h \text{ measurable} \Rightarrow \int^{\#} h(t)\,dt = \int_0^1 h(t)\,dt.\end{aligned}$$

The upper integral is useful because it enables us to bypass awkward measurability proofs. Usually the functions involved are in fact measurable, but it is easier to use the upper integral than to establish this fact formally.

Define

$$|Dv^*(t)|^+ = \limsup_{\|f\|\to 0} \left| \frac{v^*(t\chi_I + f) - v^*(t\chi_I)}{\|f\|} \right|.$$

[1] The upper integral may assume the value $+\infty$, for example when there is no g satisfying the condition in the definition.

When v is monotonic then this lim sup remains the same if we restrict f to be of the form $\tau\chi_I$, since by replacing f by $\|f\|\chi_I$ we cannot decrease the absolute value of the ratio on the right. Hence reasoning as in the proof of Lemma 23.1, we obtain

$$\int^{\#} |Dv^*(t)|^+ \, dt \leqq \|v\|. \tag{24.3}$$

Again reasoning as in the proof of Lemma 23.1, we deduce (24.3) also when v is not necessarily monotonic.

Now let

$$\Gamma_v(t) = \sup_{\|f\| \leqq 1} \left[\limsup_{\tau \to 0} \frac{v^*(t\chi_I + \tau f) - v^*(t\chi_I)}{\tau} - \liminf_{\tau \to 0} \frac{v^*(t\chi_I + \tau f) - v^*(t\chi_I)}{\tau} \right]$$

and

$$\Gamma_v = \int^{\#} \Gamma_v(t) \, dt.$$

Since $\|\tau f\| = |\tau| \, \|f\| \leqq |\tau|$, we have

$$\Gamma_v(t) \leqq 2|Dv^*(t)|^+$$

and hence

$$\Gamma_v \leqq 2 \int^{\#} |Dv^*(t)|^+ \, dt \leqq 2\|v\|.$$

Since Γ_v is non-negative and subadditive, it follows from Lemma 24.2 that it vanishes identically. Hence a.e. we have that for all f with $\|f\| \leqq 1$—and therefore for all $f \in \mathfrak{M}$—the limit

$$\lim_{\tau \to 0} \frac{v^*(t\chi_I + \tau f) - v^*(t\chi_I)}{\tau}$$

exists. This limit will be denoted $Dv^*(t\chi_I, f)$; we are, however, not yet asserting that it is a Fréchet differential. It is easily verified that

$$|Dv^*(t\chi_I, f)| \leqq \|f\| \, |Dv^*(t)|^+ \tag{24.4}$$

and that

$$Dv^*(t\chi_I, \chi_S) = \partial v^*(t, S) \tag{24.5}$$

for all $S \in \mathcal{C}$.

We now claim that a.e., $Dv^*(t\chi_I, f)$ is linear in f. To prove this, define

$$\Phi_v(t) = \sup |Dv^*(t\chi_I, f + g) - Dv^*(t\chi_I, f) - Dv^*(t\chi_I, g)|,$$

where the sup is taken over all f and g with $\|f\| \leqq 1$, $\|g\| \leqq 1$. Define

$$\Phi_v = \int^{\#} \Phi_v(t)\, dt;$$

from (24.4) and (24.3) it follows that Φ_v exists and in fact

$$\Phi_v \leqq 4\|v\|.$$

Next, $\Phi_v(t)$ is clearly non-negative and subadditive in v, and vanishes whenever v is a polynomial in measures (since $Dv^*(t\chi_I, f)$ is linear then, as is easily verified); therefore Φ_v has the same properties. So by Lemma 24.2, Φ_v vanishes for all v. Hence $\Phi_v(t)$ vanishes a.e., i.e. $Dv^*(t\chi_I, f)$ is a.e. additive in f. Since

$$Dv^*(t\chi_I, \alpha f) = \alpha Dv^*(t\chi_I, f)$$

is easily verified, the desired linearity is established. By (24.4), it then follows that $Dv^*(t\chi_I, f)$ is a.e. a continuous linear functional in f.

We now use the fact that the adjoint of $\mathfrak{M}$ is the set FA of bounded finitely additive measures on $(I, \mathcal{C})$ (Dunford and Schwartz, 1958, Theorem IV.5.1, p. 258). In our situation, this yields the fact that a.e. there is a member $\nu_t = \nu_t^v$ of FA such that for all f,

$$Dv^*(t\chi_I, f) = \int f\, d\nu_t.$$

Setting $f = t\chi_S$ and using (24.5), we get $\nu_t(S) = Dv^*(t\chi_I, \chi_S) = \partial v^*(t, S) = \partial_t v^*(S)$; thus $\nu_t = \partial_t v^*$, and we have

$$Dv^*(t\chi_I, f) = \int f\, d\partial_t v^*. \tag{24.6}$$

It remains to prove that a.e. $\partial_t v^*$ is countably additive and non-atomic, and that a.e. $Dv^*(t\chi_I, \cdot)$ is a Fréchet differential.

First we prove that $\partial_t v^*$ is a.e. countably additive. When v is monotonic, then $\partial_t v^*$ is non-negative, and so when $S_1, S_2, \ldots$ is a sequence of disjoint sets in $\mathcal{C}$, we have for all k that

$$\sum_{i=1}^{k} \partial_t v^*(S_i) = \partial_t v^* \Big(\bigcup_{i=1}^{k} S_i\Big) \leqq \partial_t v^* \Big(\bigcup_{i=1}^{\infty} S_i\Big); \tag{24.7}$$

hence $\Sigma_{i=1}^{\infty} \partial_t v^*(S_i)$ exists and is $\leqq \partial_t v^*(\bigcup_{i=1}^{\infty} S_i)$. When v is not necessarily monotonic, we may, for each $\epsilon > 0$, find monotonic u and w in pNA with $v = u - w$ and

$$\|v\| + \epsilon \geqq \|u\| + \|w\| \tag{24.8}$$

(see Proposition 7.19). Then $\partial_t v^* = \partial_t u^* - \partial_t w^*$, and applying (24.7) for u and w separately, we deduce that $\Sigma_{i=1}^{\infty} \partial_t v^*(S_i)$ converges. Using (24.5) and(24.4) we then deduce

$$\begin{aligned} \left| \sum_{i=1}^{\infty} \partial_t v^*(S_i) \right| &\leqq \partial_t u^* \left(\bigcup_{i=1}^{\infty} S_i\right) + \partial_t w^* \left(\bigcup_{i=1}^{\infty} S_i\right) \\ &\leqq |Du^*(t)|^+ + |Dw^*(t)|^+; \end{aligned} \tag{24.9}$$

similarly we have

$$\left| \partial_t v^* \left(\bigcup_{i=1}^{\infty} S_i\right) \right| \leqq |Du^*(t)|^+ + |Dw^*(t)|^+. \tag{24.10}$$

Now let

$$\Psi_v(t) = \sup \left| \partial_t v^* \left(\bigcup_{i=1}^{\infty} S_i\right) - \sum_{i=1}^{\infty} \partial_t w^*(S_i) \right|,$$

where the sup is taken over all sequences of disjoint sets $S_1, S_2, \ldots$ in $\mathcal{C}$, and let

$$\Psi_v = \int^{\#} \Psi_v(t)\, dt.$$

By (24.10) and (24.9) we have

$$\Psi_v(t) \leqq 2(|Du^*(t)|^+ + |Dw^*(t)|^+),$$

and combining this with (24.3) and (24.8), we get

$$\Psi_v \leqq 2(\|u\| + \|w\|) \leqq 2(\|v\| + \epsilon).$$

Since ϵ was arbitrarily chosen, this implies $\Psi_v \leqq 2\|v\|$. Also, $\Psi_v(t)$ is subadditive, non-negative, and vanishes when v is a polynomial in measures; hence Ψ_v has these properties. Thus from Lemma 24.2 we deduce that $\Psi_v(t)$ vanishes a.e., and it follows that $\partial_t v^*$ is a.e. countably additive.

To prove that $\partial_t v^*$ is a.e. non-atomic, set

$$\Xi_v(t) = \sup_{s \in I} |\partial_t v^*(\{s\})|;$$

then, using Lemma 24.2 in a manner that is by now familiar, we deduce the desired non-atomicity.

It remains only to prove that a.e. $Dv^*(t\chi_I, \cdot)$ is indeed a Fréchet differential. This means that a.e. for every $\epsilon > 0$ there is an $\eta > 0$ such that

$$\|f\| < \eta \Rightarrow |v^*(t\chi_I + f) - v^*(t\chi_I) - Dv^*(t\chi_I, f)| \leqq \epsilon \|f\|;$$

or, in other words, that if we set

$$\Theta_v(t) = \limsup_{\|f\| \to 0} \left| \frac{v^*(t\chi_I + f) - v^*(t\chi_I) - Dv^*(t\chi_I, f)}{\|f\|} \right|,$$

then a.e. $\Theta_v(t) = 0$. Now

$$\Theta_v(t) \leqq \limsup_{\|f\| \to 0} \left| \frac{v^*(t\chi_I + f) - v^*(t\chi_I)}{\|f\|} \right| + \limsup_{\|f\| \to 0} \left| \frac{Dv^*(t\chi_I, f)}{\|f\|} \right|;$$

and using the definition of $|Dv^*(t)|^+$ and (24.4) we obtain

$$\Theta_v(t) \leqq 2|Dv^*(t)|^+.$$

Now set $\Theta_v = \int^{\#} \Theta_v(t)\, dt$; then from 24.3 we obtain

$$\Theta_v \leqq 2\|v\|.$$

Since $\Theta_v(t)$ is non-negative, subadditive (in v), and vanishes when v is a polynomial in NA measures, it follows that Θ_v has the same properties; hence, by Lemma 24.2, Θ_v vanishes, and hence $\Theta_v(t)$ vanishes a.e. This completes the proof of Proposition 24.1.

25. Extensions: The Mixing and Asymptotic Approaches

In this section we return to a study of the extension per se and show how it is related to some of the ideas discussed in Chapters II and III.

To give an idea of how the extension is connected with the notion of mixing, let $v \in pNA' \cap AC$. Suppose $\mu \in NA^1$ is such that[1] $v \ll \mu$, let

[1] For $v \in MIX \cap AC$, the measure μ_v appearing in the definition of mixing value can be taken to be any μ for which $v \ll \mu$; see Appendix C.

$\{\Theta_1, \Theta_2, \ldots\}$ be a μ-mixing sequence, and let S be a subset of I. Let T be any subset of I with $\mu(T) = \frac{1}{2}$, say. If ν is any measure with $\nu \ll \mu$, then by Corollary 15.13,

$$(25.1) \qquad \nu(S \cap \Theta_n T) \to \tfrac{1}{2}\nu(S).$$

If (25.1) were true for all ν, rather than just for $\nu \ll \mu$, then we would have $S \cap \Theta_n T \to \frac{1}{2}\chi_S$ in the NA-topology[2], and it would follow from (22.11) that

$$(25.2) \qquad v(S \cap \Theta_n T) \to v^*(\tfrac{1}{2}\chi_S).$$

As it is, although (25.1) is certainly not true for all ν, (25.2) nevertheless does hold. That is because measures that are singular w.r.t. μ have no "relevance" to v^*.

To make this idea more precise, let us define the *NA_μ-topology* to be the smallest topology for which all the functions $\nu^\#$ on $\mathcal{I}$ are continuous, where ν ranges over all measures such that $\nu \ll \mu$. (Recall that $\nu^\# f = \int f \, d\nu$.) Of course the NA_μ-topology is a weaker topology than the NA-topology, i.e. it has fewer open sets. We know that v^* is continuous in the NA-topology; if we could prove the stronger statement that

$$(25.3) \qquad v^* \text{ is continuous in the } NA_\mu\text{-topology},$$

then (25.2) would follow from (25.1).

To demonstrate (25.3), let $\epsilon > 0$ be given. Note that the NA-topology can be defined as the smallest topology such that all the $\nu^\#$ are continuous, where ν ranges over all measures in NA that are either absolutely continuous or singular w.r.t. μ; this follows from the fact that any measure in NA can be decomposed into two such measures. Similar remarks apply to the notion of uniform continuity; therefore there is a finite vector ξ of measures that are $\ll \mu$, a finite vector ζ of measures that are $\perp \mu$, and a $\delta > 0$, such that

$$(25.4) \qquad \left\|\int (f - g)\, d\xi, \int (f - g)\, d\zeta\right\| < \delta \Rightarrow |v^*(f) - v^*(g)| < \frac{\epsilon}{2}.$$

[2] Recall our convention, explained in Section 20, of identifying ideal with ordinary sets in intuitive discussion.

Now let g_0 be a fixed member of $\mathcal{G}$ and let $h \in \mathcal{G}$ be such that

$$(25.5) \qquad \|\int (h - g_0)\, d\xi\| < \delta.$$

Apply Corollary (22.3) to find a $T \in \mathcal{C}$ such that

$$(25.6) \qquad \nu(T) = \int h\, d\nu,$$

where ν is the vector measure (ξ, ζ). Let U be a set that is μ-equivalent[3] to T and such that

$$(25.7) \qquad \zeta(U) = \int g_0\, d\zeta;$$

the existence of such a set U follows from the fact that the components of ζ are $\perp \mu$, together with another application of Corollary 22.3. Since $\xi \ll \mu$, it then follows from (25.6) that

$$\xi(U) = \xi(T) = \int h\, d\xi$$

and so by (25.4), (25.5) and (25.7),

$$|v(U) - v^*(g)| = |v^*(\chi_U) - v^*(g)| < \frac{\epsilon}{2}.$$

But since $v \ll \mu$, we have

$$v(U) = v(T).$$

Finally, from (25.6) and (25.4) applied with $f = \chi_T$ and $g = h$, we get

$$(25.8) \qquad |v(T) - v^*(h)| = |v^*(\chi_T) - v^*(h)| < \frac{\epsilon}{2}.$$

Combining the last three formulas, we get

$$|v^*(h) - v^*(g)| < \epsilon.$$

But h was chosen to be any member of $\mathcal{G}$ satisfying (25.5), and the set of all such h is a neighborhood of g in the NA_μ-topology. This completes the proof of (25.3), and with it the proof of (25.2).

A more general version of (25.2) is the following:

PROPOSITION 25.9. *Let $v \in pNA' \cap AC$, and let $f = \sum_{i=1}^{k} \alpha_i \chi_{S_i} \in \mathcal{G}$, where $\{S_1, \ldots, S_k\}$ is a partition of I. Then for all μ with $v \ll \mu$, all μ-mixing sequences $\{\Theta_1, \Theta_2, \ldots\}$, and all sets $T_1, \ldots, T_k$ with $\mu(T_1) =$*

[3] I.e. differs from T by a set of μ-measure 0.

$\alpha_1, \ldots, \mu(T_k) = \alpha_k$, *we have*

$$v\left(\bigcup_{i=1}^{k} (S_i \cap \Theta_n T_i)\right) \rightarrow v^*(f)$$

as $n \rightarrow \infty$.

The proof uses the same ideas as above. Note that in the statement of Theorem H, we only use[4] ideal sets of the kind appearing in Proposition 25.9—in fact only a very special subclass of these ideal sets, with $k = 2$.

Next we investigate how the extension may be characterized in terms of the ideas leading up to the asymptotic value, in particular the notions of partition and approximation by finite games. Let $v \in pNA'$. Let $(\Pi_1, \Pi_2, \ldots)$ be an admissible sequence of partitions[5] of I all of which have an even number of members. From each partition Π_m choose at random exactly half of the members, in such a way that all subsets of Π_m having half the members of Π_m have equal probability; for example, order them at random and choose the first half. Denote the union of the chosen members by $\mathbf{U}_m$. Then we claim that $v(\mathbf{U}_m) \rightarrow v^*(\frac{1}{2}\chi_I)$ with high probability; or more precisely, for each $\epsilon > 0$,

$$\text{Prob}\{|v(\mathbf{U}_m) - v^*(\tfrac{1}{2}\chi_I)| < \epsilon\} \rightarrow 1 \text{ as } m \rightarrow \infty. \qquad (25.10)$$

To prove this, first let $v \in NA^+$; then (25.10) follows from Lemma 18.7. The case when v is a power of an NA^+ measure follows from the case in which it is a measure, and the case when it is a sum of powers follows from this. Finally, the general case follows by approximation from the case of polynomials in measures, using (22.6). This proves (25.10).

If we wish to characterize $v^*(\frac{1}{2}\chi_S)$, where S is an arbitrary member of $\mathcal{C}$, we must use an admissible sequence $\{\Pi_1, \Pi_2, \ldots\}$ whose first member is $\{S, I\backslash S\}$ and such that, in each Π_m after the first, the number of subsets of S is even. Then if $\mathbf{U}_m$ is the union of one half of the subsets of S in Π_m, randomly chosen, then for all $\epsilon > 0$,

$$\text{Prob}\{|v(\mathbf{U}_m) - v^*(\tfrac{1}{2}\chi_S)| < \epsilon\} \rightarrow 1 \text{ as } m \rightarrow \infty.$$

The proof uses formula (25.10) applied to the underlying space S rather than the underlying space I.

[4] Via the definition of $\partial v^*(t,S)$.

[5] See Section 17.

More generally, we have

PROPOSITION 25.11. *Let $v \in pNA'$, and let $f = \sum_{i=1}^{k} \alpha_i \chi_{S_i} \in \mathcal{I}$, where $\{S_1, \ldots, S_k\}$ is a partition of I. Let $\{\Pi_1, \Pi_2, \ldots\}$ be an admissible sequence whose first member is the partition $\{S_1, \ldots, S_k\}$. Let T_{im} be the set of subsets of S_i belonging to Π_m, and let $|T_{im}|$ be its cardinality. Let p_{im} be positive integers such that*

$$\frac{p_{im}}{|T_{im}|} \to \alpha_i \qquad \text{as } m \to \infty.$$

From each T_{im}, choose p_{im} members at random in such a way that all subsets of T_{im} with exactly p_{im} members have the same probability of being chosen. Let $\mathbf{U}_{im}$ be the union of the p_{im} sets so chosen, and let $\mathbf{U}_m = \bigcup_i \mathbf{U}_{im}$. Then for each $\epsilon > 0$,

$$\text{Prob}\{|v(\mathbf{U}_m) - v^*(f)| < \epsilon\} \to 1 \text{ as } m \to \infty.$$

The proof uses ideas similar to those used above.

NOTES

1. The extension considered in this chapter is closely related to an idea introduced by Owen (1972) for finite-player games (N, v). Owen introduces objects corresponding to our ideal sets, namely points x in the cube $[0, 1]^N$, and defines the "multilinear extension" $\bar{v}$ of v to be $v(S)$ when x is the characteristic vector of S, and to be linear in each x_i separately; these properties characterize $\bar{v}$ uniquely.

Under suitable differentiability conditions, our extensions enjoy an infinite-dimensional analogue of the multilinearity property. For example, let v be a polynomial in NA measures, and let f be an ideal set. Then there are measures μ in NA^+ and ν in NA such that

$$v^*(f + \alpha\chi_S) = v^*(f) + \alpha\nu(S) + o(\nu(S))$$

as $\mu(S) \to 0$, uniformly for all α such that $f + \alpha\chi_S$ is an ideal set. The question as to on what subspaces of set functions the above infinitesimal multilinearity property actually characterizes the extension operator deserves further study.

Owen's multilinear extensions provide an effective tool for investigating the Shapley values of finite games; in fact he shows that

$$(\varphi v)(i) = \int_0^1 (\partial\bar{v}/\partial x_i)(t, \ldots, t)\, dt.$$

From this it follows that if λ is Lebesgue measure and w is the non-atomic game on $[0, n]$ defined by $w = \bar{v} \circ \mu$ where

$$\mu_i(S) = \lambda(S \cap [i - 1, i]),$$

then w is in pNA and

$$(\varphi v)(i) = (\varphi w)([i - 1, i]).$$

Chapter V. The Value and the Core

26. Introduction and Statement of Results

The main object of Chapter V is the proof of

THEOREM I. *Let v be a superadditive set function in pNA that is homogeneous of degree* 1. *Then the core of v has a unique member, which coincides with the value of v.*

Several of the terms used in the statement of Theorem I may not be familiar to the reader. A set function v is *superadditive* if for disjoint S and T,

$$v(S \cup T) \geqq v(S) + v(T).$$

It is *homogeneous of degree* 1 if

$$v^*(\alpha\chi_S) = \alpha v(S)$$

for all α in $[0, 1]$ and all sets $S \in \mathcal{C}$, where v^* is the extension defined by Theorem G (see Secs. 21 and 22). An equivalent[1] formulation of this homogeneity condition is

$$v^*(\alpha f) = \alpha v^*(f)$$

for all α in $[0, 1]$ and all ideal sets f in $\mathcal{I}$. The *core* of v consists of the set of all μ in FA with

$$\mu(S) \geqq v(S)$$

for all S in $\mathcal{C}$, and

$$\mu(I) = v(I).$$

Superadditivity is a very well-known condition in game theory; what it says is that disjoint coalitions do not lose by joining forces.

Homogeneity of degree 1 is a somewhat less known concept. An example of a set function that is homogeneous of degree 1 is any NA measure; the square or cube of such a measure is, however, not homogeneous of degree 1. More generally, let μ be a vector of measures in

[1] This follows from the continuity of v^* in the NA-topology (see (22.11)) and the denseness of $\mathcal{C}$ in $\mathcal{I}$ in that topology (Proposition 22.4).

NA and let f be a real function, differentiable on the range R of μ, that is homogeneous of degree 1, i.e. such that $f(\alpha x) = \alpha f(x)$ for all α in $[0, 1]$ and $x \in R$; then $f \circ \mu$ is in pNA and is homogeneous of degree 1. An example is $\sqrt{\mu^2 + \nu^2}$, where μ and ν are any two measures in NA^+. This, however, is not superadditive. But $-\sqrt{\mu^2 + \nu^2}$ *is* superadditive,[2] and therefore satisfies all the conditions of Theorem I.

Another class of set functions v that are homogeneous of degree 1 is given by

$$(26.1) \qquad v(S) = \max \left\{ \int_S u(\mathbf{x}(s), s)\, d\mu(s) : \int_S \mathbf{x}\, d\mu = \int_S \mathbf{a}\, d\mu \right\}.$$

Here $\mu \in NA^+$; u is a real-valued function of two variables x and s, where x ranges over $[0, \infty)$ and s over I; $\mathbf{a}$ is an integrable function from I to $[0, \infty)$; and the maximum is taken over all integrable functions $\mathbf{x}$ from I to $[0, \infty)$ that satisfy the constraint (i.e. the statement after the colon).

These set-functions can be interpreted as models of productive economies roughly as follows: $u(x, s)\, d\mu(s)$ is the amount of finished good that the producer ds can produce from an amount $x\, d\mu(s)$ of raw material, and $\mathbf{a}(s)\, d\mu(s)$ is the amount of raw material available to ds initially. Hence the total amount of raw material initially available to a coalition S is $\int_S \mathbf{a}(s)\, d\mu(s) = \int_S \mathbf{a}\, d\mu$, and it may reallocate this amount among its members in any way it pleases; that is, if the members of S agree, then any $\mathbf{x}$ that satisfies the constraints can be substituted for $\mathbf{a}$. Therefore, if the maximum in (26.1) exists—i.e. if the supremum is finite and is attained—then the coalition S can reallocate its initial resources in such a way as to produce a total of $v(S)$.

Whether the maximum in (26.1) indeed exists is a nontrivial question; it is treated, in a somewhat broader context, by Aumann and Perles (1965). Here we will content ourselves with stating that if u is non-negative, uniformly bounded, and measurable in both variables simultaneously, and if for each fixed s, $u(\cdot, s)$ is non-decreasing in $\mathbf{x}$ and differentiable over all of $[0, \infty)$, then the maximum exists and, indeed, v is in pNA. However, the same conclusions can be reached under considerably wider conditions. In Chapter VI we shall treat this whole

[2] This follows from the triangle inequality for the euclidean norm.

problem in much greater detail and generality, and also give alternative interpretations for set functions v defined as in (26.1).

To convince ourselves intuitively of the homogeneity of degree 1 of these set functions, let f be an ideal set. If g is any function integrable over I, it seems reasonable to define the "integral of g w.r.t. μ over the ideal set f" by

$$\int_f g\, d\mu = \int_I gf\, d\mu.$$

If in formula (26.1) we substitute the ideal set f for the ordinary set S, this definition of "integral over f" leads us to

$$v^*(f) = \max \left\{ \int_I u(\mathbf{x}(s), s) f(s)\, d\mu(s) : \int_I \mathbf{x} f\, d\mu = \int_I \mathbf{a} f\, d\mu \right\};$$

and $v^*(\alpha f) = \alpha v^*(f)$ would be a trivial consequence of this. Of course it must be proved that v^* is indeed given by the above formula. This question too will be treated in Chapter VI; here we only wanted to illustrate homogenity of degree 1.

The *core* is a basic concept in game theory. In a given game, it is the set of those payoff vectors[3] μ such that no coalition S can assure itself more than it gets under μ, while the all-player set can in fact get μ. For games with player space $(I, \mathcal{C})$, this means that $\mu(S) \geqq v(S)$ for all $S \in \mathcal{C}$ and $\mu(I) = v(I)$, exactly as stipulated at the beginning of this section.

Theorem I is proved in Section 27. We mention also Propositions 27.1, 27.8, and 27.12, and Remark 27.11, which have some independent interest.

NOTES

1. There is a large literature devoted to the core; see, for example, Gillies (1959), Bondareva (1963), Shapley and Shubik (1966, 1969a, 1972), Schmeidler (1972), Shapley (1967 and 1971), Kannai (1969), and Rosenmüller (1971). In particular, Kannai, Rosenmüller, and Schmeidler are concerned with games with infinitely many players. All of these papers are about cores of "side-payment games," i.e. games defined by real-valued set functions such as we have been treating here. Cores of games in the no-side-payment coalition form have also been studied extensively; see Aumann (1961), Burger (1964), Scarf (1967), Billera (1970), and Shapley (1973). The notion of core is especially fruitful in connection with markets and other economic models (of which (26.1)

[3] See Section 2.

is an example); in the side payment case see, for example, the Shapley-Shubik references above. In the no-side-payment case, the literature on the core of market and other economic games is very large indeed, especially when there are infinitely many players. Here we will therefore cite only Aumann (1964), which is one of the relatively early papers on the subject, and Hildenbrand's comprehensive treatise (1974). We should stress that the papers cited in this note were picked rather arbitrarily and are far from constituting a complete bibliography on the core.

27. Proof of Theorem I

PROPOSITION 27.1. *Let v be a superadditive set function in pNA (or, more generally,*[1] *in pNA'). Then v^* is superadditive over the family $\mathcal{I}$ of ideal sets; that is,*

$$v^*(f + g) \geqq v^*(f) + v^*(g)$$

whenever $f, g, f + g \in \mathcal{I}$.

Proof. We will approximate to f and g, in the NA-topology, by *disjoint* ordinary sets; the result will then follow from the continuity of v^* in the NA-topology (see (22.11)) and the superadditivity of v.

Set $g_0 = f$, $g_1 = g$, $g_2 = f + g$. For given $\epsilon > 0$, let μ_0, μ_1, μ_2 be three vector measures, and $\delta_0, \delta_1, \delta_2$ three positive numbers, such that

$$\|\int (h - g_i)\, d\mu_i\| < \delta_i \Rightarrow |v^*(h) - v^*(g_i)| < \epsilon$$

for $i = 0, 1, 2$. Let $\mu = (\mu_0, \mu_1, \mu_2)$. By Lemma 22.1, there are T_1 and T_2 in $\mathcal{C}$, with $T_2 \supset T_1$, such that for $i = 1, 2$,

$$\mu(T_i) = \int g_i\, d\mu;$$

if we set $T_0 = T_2 \backslash T_1$ and note that $g_0 = g_2 - g_1$, then this equation follows for $i = 0$ as well. In particular we have

$$\int \chi_{T_i}\, d\mu_i = \mu_i(T_i) = \int g_i\, d\mu_i;$$

for $i = 0, 1, 2$. Hence

$$\|\int (\chi_{T_i} - g_i)\, d\mu_i\| = 0 < \delta_i,$$

and therefore

$$|v(T_i) - v^*(g_i)| = |v^*(\chi_{T_i}) - v^*(g_i)| < \epsilon.$$

[1] See Section 22 at Proposition 22.16.

But from $T_2 = T_0 \cup T_1$ and $T_0 \cap T_1 = \varnothing$ it follows, by the superadditivity of v that

$$v(T_2) \geqq v(T_0) + v(T_1).$$

Hence

$$\begin{aligned} v^*(f+g) &= v^*(g_2) \geqq v^*(g_0) + v^*(g_1) - 3\epsilon \\ &= v^*(f) + v^*(g) - 3\epsilon, \end{aligned}$$

and letting $\epsilon \to 0$, we obtain the conclusion of Proposition 27.1.

In the remainder of this section, v will be a superadditive set function in pNA that is homogeneous of degree 1, φv will be its value, and v^* will be its extension.

LEMMA 27.2. *Let $S \in \mathcal{C}$. Then $\partial v^*(t, S)$ exists and is the same for all t in* $(0, 1)$.

Proof. By Theorem H, $\partial v^*(t, S)$ exists for almost all t in $(0, 1)$; let t_0 be a value of t for which it exists. Let $0 < t_1 < t_0$; then we have, by homogeneity,

$$\begin{aligned} \frac{v^*(t_1\chi_I + \tau\chi_S) - v^*(t_1\chi_I)}{\tau} &= \frac{\dfrac{t_1}{t_0} v^* \left(t_0\chi_I + \dfrac{t_0}{t_1}\tau\chi_S\right) - \dfrac{t_1}{t_0} v^*(t_0\chi_I)}{\tau} \\ &= \frac{v^*(t_0\chi_I + \tau'\chi_S) - v^*(t_0\chi_I)}{\tau'} \end{aligned}$$

where $\tau' = \dfrac{t_0}{t_1}\tau$. When $\tau \to 0$, so does τ'. Hence when $\tau \to 0$, we have

$$\frac{v^*(t_1\chi_I + \tau\chi_S) - v^*(t_1\chi_I)}{\tau} \to \partial v^*(t_0, S);$$

thus $\partial v^*(t_1, S)$ exists and is equal to $\partial v^*(t_0, S)$. Since t_0 may be chosen arbitrarily close to 1, the proof of Lemma 27.2 is complete.

COROLLARY 27.3. *Let $S \in \mathcal{C}$. Then for all $t \in$* $(0, 1)$,

$$(\varphi v)(S) = \partial v^*(t, S).$$

Proof. Follows from Theorem H and Lemma 27.2.

LEMMA 27.4. *φv is in the core of v.*

Proof. Let $S \in \mathcal{C}$. Fix an arbitrary t in $(0, 1)$. By Proposition 27.1 we have, for all sufficiently small τ in $(0, 1)$,

$$v(S) = v^*(\chi_S) = \frac{v^*(t\chi_I) + v^*(\tau\chi_S) - v^*(t\chi_I)}{\tau}$$

$$\leqq \frac{v^*(t\chi_I + \tau\chi_S) - v^*(t\chi_I)}{\tau}.$$

Letting $\tau \to 0+$ on the right and using Corollary 27.3, we deduce that

$$v(S) \leqq \partial v^*(t, S) = (\varphi v)(S).$$

But $v(I) = (\varphi v)(I)$ is part of the definition of value (2.3); hence the proof of Lemma 27.4 is complete.

LEMMA 27.5. *Let $\mu \in NA$ be in the core of v. Then $\mu = \varphi v$.*

Proof. Fix an arbitrary t in $(0, 1)$. Then from Proposition 27.1 and (21.3) we have

$$v^*(t\chi_I) = tv(I) = t\mu(I) = \mu^*(t\chi_I). \tag{27.6}$$

But since μ is in the core of v, we have $\mu^*(f) \geqq v^*(f)$ for all $f \in \mathcal{I}$; the proof of this is similar to that of Proposition 27.1. In particular, therefore, for $\tau > 0$ sufficiently small, we have, for any $S \in \mathcal{C}$,

$$\mu^*(t\chi_I) + \tau\mu(S) = \mu^*(t\chi_I + \tau\chi_S) \geqq v^*(t\chi_I + \tau\chi_S). \tag{27.7}$$

Combining (27.6) with (27.7), we get

$$\mu(S) = \frac{\mu^*(t\chi_I) + \tau\mu(S) - \mu^*(t\chi_I)}{\tau} \geqq \frac{v^*(t\chi_I + \tau\chi_S) - v^*(t\chi_I)}{\tau}.$$

Letting $\tau \to 0+$, we deduce from Corollary 27.3 that

$$\mu(S) \geqq \partial v^*(t, S) = (\varphi v)(S).$$

Since S was arbitrary, we have also

$$\mu(I \backslash S) \geqq (\varphi v)(I \backslash S).$$

Using $\mu(I) = v(I) = (\varphi v)(I)$ we obtain $\mu(S) = (\varphi v)(S)$, and the proof of Lemma 27.5 is complete.

It remains to prove that the core of v contains only NA measures. In fact, we have the more general

PROPOSITION 27.8. *If $w \in AC$, then every member of the core*[2] *of w is in NA.*

Proof. Related results have been obtained by Schmeidler and by Rosenmüller.[3] Our proof follows Schmeidler's ideas closely. Let $\nu \in NA^+$ be such that $w \ll \nu$, and let μ be in the core of w. Let $T_1, T_2, \ldots$ be a sequence of sets in $\mathcal{C}$; we claim

$$\nu(T_i) \to 0 \text{ implies } \mu(T_i) \to 0. \tag{27.9}$$

Indeed, from $\nu(T_i) \to 0$ and $w \ll \nu$ it follows that

$$w(T_i) \to 0 \text{ and } w(I \backslash T_i) \to w(I). \tag{27.10}$$

Using $\mu(T_i) \geqq w(T_i)$, $\mu(I\backslash T_i) \geqq w(I\backslash T_i)$, and $\mu(I) = w(I)$, we deduce

$$\liminf \mu(T_i) \geqq \liminf w(T_i) = 0$$

and

$$\begin{aligned}\limsup \mu(T_i) &= \mu(I) - \liminf \mu(I\backslash T_i)\\ &\leqq \mu(I) - \liminf w(I\backslash T_i) = \mu(I) - w(I) = 0;\end{aligned}$$

thus $\lim \mu(T_i)$ exists and $=0$, and (27.9) is proved.

Now let $S = \cup_{j=1}^{\infty} S_j$, where the S_j are disjoint. If we set $T_i = S\backslash\cup_{j=1}^{i} S_j$, then the T_i obey the hypotheses of (27.9); hence $\lim \mu(T_i) = 0$, i.e.

$$\mu(S) = \lim \sum_{j=1}^{i} \mu(S_j) = \sum_{j=1}^{\infty} \mu(S_j).$$

Thus μ is countably additive.

To show that μ is non-atomic, let $s \in I$, and in (27.9), let $T_i = \{s\}$ for all i. From $\nu \in NA$ we get $\lim \nu(T_i) = \nu(\{s\}) = 0$. Hence $0 = \lim \mu(T_i) = \mu(\{s\})$, and the proof of Proposition 27.8 is complete.

Theorem I follows immediately from Lemmas 27.4 and 27.5, Proposition 27.8, and the fact that $pNA \subset AC$ (Corollary 5.3).

Remark 27.11. It is perhaps worth noting that Proposition 27.8 continues to hold if we assume $w \in bv'NA$ rather than $w \in AC$. Indeed, for the proof it is sufficient that there exist an NA^+ measure ν such that

[2] Of course, there is no assertion here that the core is non-empty.

[3] See Schmeidler (1972), Theorems 3.2 and 3.10, and Rosenmüller (1971), Theorem 1.2 and Corollary 2.4. Both authors assume $v(S) \geqq 0$ for all S, and this simplifies matters somewhat.

$\nu(T_i) \to 0$ implies (27.10); and for $w \in bv'NA$ there does indeed exist such a measure.

The following proposition will be useful in applications of Theorem I.

PROPOSITION 27.12. *The set of superadditive members of pNA that are homogeneous of degree* 1 *is closed in BV.*

Proof. pNA is closed by definition. The closedness of the set of superadditive set functions follows from the continuity, for each fixed $S \in \mathcal{C}$, of the mapping $v \to v(S)$ from BV to the reals. To show that the space of set functions that are homogeneous of degree 1 is closed, note that $v \to v^*$ is continuous (see (22.13)), and that for each fixed $f \in \mathcal{I}$, $v^* \to v^*(f)$ is continuous. Therefore for each $\alpha \in [0, 1]$, the mapping $v \to v^*(\alpha f) - \alpha v^*(f)$ is continuous, and Proposition 27.12 is proved.

We close this chapter with yet another[4]

Alternative Proof for Example 9.4. If $v \in pNA$, then so is $-v$; then by (22.18) we have

$$-v^*(\alpha f) = -\left|\int \alpha f \, d\mu\right| = -\alpha\left|\int f \, d\mu\right| = \alpha(-v^*(f)),$$

and hence $-v$ is homogeneous of degree 1. Furthermore $-v$ is clearly superadditive. But μ and $-\mu$ are both in the core of $-v$ in contradiction to Theorem I. So $v \notin pNA$, as was to be proved.

[4] Proofs were previously given in Sections 9, 16, and 22.

Chapter VI. An Application to Economic Equilibrium

28. Introduction

In this chapter we will apply the theory developed in the previous chapters to certain economic models. These models may be interpreted either as productive economies similar to—but more general than—the one described in the introduction to Chapter V,[1] or as exchange economies with transferable utility.[2] Our chief result is that under fairly wide conditions, the set function derived from such a model is in pNA, that there is a unique point in its core, and that this unique point coincides with the value. We shall also define the notion of competitive equilibrium for such economies, and show that this too yields a unique payoff, which also coincides with the value, and therefore with the unique core point.

Section 29 is devoted to a careful conceptual discussion of some aspects of economic models with a continuum of economic agents. This is needed for a proper understanding of Section 30, in which we introduce and motivate the particular economic model that is the subject of this chapter. Section 31 contains the statement of the results concerning the relation between the core and the value. In Section 32 we introduce and discuss the competitive equilibrium, and relate it to the previously described concepts. Section 33 is devoted to some examples, Section 34 to a brief discussion of related literature. Sections 35 through 40 are devoted to the proofs. In the last section, Section 41, we discuss some possibilities for extensions of our results.

It is to be stressed that the proof of the main result—i.e. the membership in pNA, the existence of a unique core point, and its coincidence with the value—does not make any use of the notion of competitive equilibrium; rather, it is based directly on Theorem I in Chapter V.

[1] See Formula (26.1) and the subsequent discussion.

[2] I.e. "markets with money" or "markets with side payments"; cf. Shapley and Shubik (1966, 1969a) and Shapley (1964b). These are special cases of the more classical Walrasian exchange economies (cf., e.g., Nikaido, 1956; Debreu and Scarf, 1963; and Aumann, 1964).

29. Conceptual Preliminaries

In this section we would like to clarify some of the ideas used in connection with economic models with a continuum of economic agents, and specifically, the use of integration in connection with such models.

To understand this properly, it is convenient to use an analogy with physics, where continuous models are plentiful and well-understood. Let us recall the problem of computing (or for that matter, defining) the gravitational force exerted by a solid beam I on a given mass point x in space, whose mass is, say, M. One divides I into "small" pieces, calling a typical piece "Δs." Then if ρ is the distance function and s is a point in Δs, all points in Δs have a distance approximately $\rho(x, s)$ from x. Therefore if $\mu(\Delta s)$ denotes the mass of Δs, the gravitational force exerted by Δs on x is approximately

$$\frac{M\mu(\Delta s)(s - x)}{\rho^3(s, x)}$$

(whose magnitude is $M\mu(\Delta s)/\rho^2(s, x)$); and the total gravitational force exerted by I on x is approximately

$$\sum [M(s - x)/\rho^3(s, x)]\mu(\Delta s), \tag{29.1}$$

the sum being taken over all the "small" pieces into which we have divided I. When we say that Δs is "small," what we mean is that its diameter is small; precisely, what is required is that $(s - x)/\rho^3(s, x)$ be almost constant as s ranges over Δs.

The next step is to pass to the limit. As the diameters of the Δs tend to 0, the expression (29.1) tends to

$$\int_I [M(s - x)/\rho^3(s, x)]\mu(ds); \tag{29.2}$$

at the same time the approximations become better and better, and the errors involved tend to 0. Hence we conclude that the total force exerted by I on x is in fact precisely the integral (29.2).

There is also a slightly different way of looking at the integral (29.2). One thinks of I as being divided into "infinitesimal pieces" ds, each with an "infinitesimal mass" $\mu(ds)$. The piece ds has an infinitesimal diameter; if one wishes one can think of it as consisting of a single point, located at s. The force exerted by it on x is

$$[M(s - x)/\rho^3(s, x)]\mu(ds),$$

and the total force is the "sum" of these infinitesimal forces, namely the integral (29.2).

Some readers may be disturbed by the use of terms such as "infinitesimal," which we have not properly defined.[1] Such readers may take the discussion in terms of infinitesimals to be simply an abbreviation for the somewhat more lengthy discussion involving "small pieces" and a limiting process. Imprecise as it may be, though, the discussion in terms of infinitesimals has a certain direct conceptual appeal, which is lacking in the limit discussion. Each infinitesimal piece ds exerts a force that can be calculated exactly—*not* approximately—by a single straightforward application of Newton's formula for the gravitational attraction between two mass points. And the total force is simply the sum of these individual forces. By comparison, the limit approach seems conceptually devious.

In the case of economic models, the "infinitesimal" approach has an additional intuitive advantage. People still think even of very large economies as consisting of individual agents; intuitively, then, such an agent can be associated with an "infinitesimal piece." In the physical analogy, one could think of our beam I as being made up of many individual mass points—as indeed it is, if one considers an atom a point. One replaces this set of mass points by a continuum—both for mathematical convenience and for a better physical understanding of the gravitational field around a beam. But in intuitive discussion of the integral, it may still be convenient to associate an "infinitesimal piece" with one of the individual mass points. It should be stressed, though, that such an association is not necessary, neither in the economic nor in the physical situation. In both situations, the infinitesimal piece can be thought of as a *set* of individuals that has an infinitesimal mass or measure, and all of whose members have the same physical or economic properties (for example the same distance from x in the physical case, the same utility in the economic case).

[1] This is not to say that they cannot be properly defined; cf. Robinson (1966).

In intuitive discussion in the sequel, we shall adopt the convention of associating an infinitesimal with a single individual. This is chiefly because it is easier, e.g., to write "the trader ds" rather than "the set ds of traders" or "one of the traders in ds"; if the reader wishes, he can substitute the alternative interpretation. The reader should be careful to note that we are associating an individual with an infinitesimal *subset* ds of I, not with a point s in I. We will adopt the convention that the point named s is always a member of the set named ds. It will be understood that all functions of s that appear in the analysis are constant on every ds. For example, we shall describe the initial bundle of a trader ds by an expression of the form $\mathbf{a}(s)\mu(ds)$; intuitively, it is to be understood that s is a point in the infinitesimal set ds, and that $\mathbf{a}$ is a function on I that is constant on ds, so that it does not matter which point s in ds is chosen. Readers who prefer to think of the integral in terms of the limiting process may make the necessary re-interpretations, in which ds is replaced by Δs, s is a point in Δs, and Δs is chosen so that $\mathbf{a}$ is "almost constant" on it.

In closing the discussion of this physical analogy, we would like to stress that the whole discussion is concerned exclusively with the passage from the given physical situation to the mathematical model. Once one accepts the integral as properly representing the desired force, the rest of the treatment can be perfectly precise, in the best traditions of modern mathematical analysis. The situation in economics is similar; the mathematical model, once constructed, can be analyzed with the ordinary mathematical tools, with the precision that is characteristic of mathematical analysis. Only in constructing the model, and in relating it to the economic ideas that motivate its construction, is it convenient to make use of words such as "infinitesimal," and of the corresponding ideas.

30. Description of the Model and Economic Interpretation

Let Ω denote the non-negative orthant of a euclidean space E^n, whose dimension n will be fixed throughout. Superscripts will be used to denote coordinates. For x and y in E^n we write $x \geqq y$ if $x^i \geqq y^i$ for all i, $x \geq y$ if $x \geqq y$ but not $x = y$, and $x > y$ if $x^i > y^i$ for all i. A real-valued function f on Ω will be called *non-decreasing* if $x \geqq y$ implies $f(x) \geqq f(y)$, and *increasing* if $x \geq y$ implies $f(x) > f(y)$. The scalar product $\Sigma_{i=1}^{n} x^i y^i$ of two members x and y of E^n will be denoted $x \cdot y$. The symbol 0 will denote both the number zero and the origin of a euclidean space; no confusion will result.

Let $\mu \in NA^1$; μ will be fixed throughout. If g is a μ-integrable function on I and $S \in \mathcal{C}$, we will use the notations $\int_S g$, $\int_S g\, d\mu$, $\int_S g(s)\, d\mu(s)$, and $\int_S g(s)\mu(ds)$ interchangeably. All occur in the literature, and for different purposes one or the other will be more convenient. When the range of integration in an integral is not specified, it will be understood to be I; thus $\int g$ and $\int_I g(s)\mu(ds)$ are the same thing. The phrases "integrable," "almost all," and so on, will be used to mean "μ-integrable," "μ-almost all," and so on.

For each $s \in I$, let $\mathbf{a}(s)$ be in Ω, and let $u(\cdot, s)$ be an increasing non-negative real-valued function on Ω. We will be concerned with the set function v defined by

$$(30.1) \quad v(S) = \max \left\{ \int_S u(\mathbf{x}(s), s)\, d\mu(s) : \mathbf{x}(s) \in \Omega \text{ for all } s \right. \\ \left. \text{and } \int_S \mathbf{x}\, d\mu = \int_S \mathbf{a}\, d\mu \right\},$$

the maximum being taken over all μ-integrable functions $\mathbf{x}$ that satisfy the constraints. Note that the equation in the constraints is a vector equation; thus, when we say that $\mathbf{x}$ is μ-integrable, we mean that all its coordinates are μ-integrable. Naturally, in order that the integrals inside the curly brackets be meaningful, it is necessary to impose certain measurability and integrability conditions on the functions u and $\mathbf{a}$. Furthermore, even if the integrals involved exist, it is by no means clear or even always true that the expression being maximized is bounded; and even if it is bounded, its supremum may not be attained (see Note 1). These matters will be treated in the next section, where sufficient conditions will be imposed on u and $\mathbf{a}$ to ensure that the integrals involved exist and that the maximum exists. In this section we would like to concentrate on the economic interpretations of the set function v.

The model just described will be called a *transferable utility economy*. The reader will recall from the introduction that there are two interpretations, one in terms of exchange economies and one in terms of productive economies. We would like to present the interpretation in terms of productive economies first, since it is simpler. There are n kinds of raw material, and only one kind of finished good. The space I consists of infinitesimal producers ds. Given a bundle (i.e. vector) x in Ω of raw materials, producer ds can produce an amount $u(x, s)\mu(ds)$ of the finished good. Next, $\mathbf{a}(s)\mu(ds)$ is the bundle of raw materials initially available to the producer ds; hence the total bundle of raw materials initially available to a coalition S is $\int_S \mathbf{a}(s)\mu(ds) = \int_S \mathbf{a}\, d\mu$. Now S may reallocate this amount among its members in any way it pleases; that is, if the members of S agree, they may assign to each member ds of S an amount $\mathbf{x}(s)\mu(ds)$ rather than $\mathbf{a}(s)\mu(ds)$, on condition that $\mathbf{x}(s) \in \Omega$ and $\int_S \mathbf{x}\, d\mu = \int_S \mathbf{a}\, d\mu$. Then if the maximum in (30.1) exists, and if S pools and redistributes its resources and then pools the finished goods produced by all the members, then the total amount in the resulting pool of finished goods can be as high as $v(S)$. In short, the coalition S, if it forms, can obtain for its members a total payoff of $v(S)$; in this sense,[1] $v(S)$ is the *worth* of S.

In the interpretation in terms of exchange economies there are $n + 1$ consumer goods, indexed by $0, 1, \ldots, n$. The good indexed by 0 is called *money* and, unlike the others, may appear in negative as well as positive amounts. The space I consists of infinitesimal traders ds, and the amount of any good typically available to ds will also be infinitesimal; a typical bundle will have the form $(x^0, x)\mu(ds)$, where $x^0 \in E^1$ and $x \in \Omega$. Each trader ds has a preference order on the set of all such bundles. Money enters into these preferences in a very special way; specifically, $x^0 + u(x, s)$ is a utility function for the trader ds. In other words, if (x^0, x) and (y^0, y) are in $E^1 \times \Omega$, then ds prefers $(x^0, x)\mu(ds)$ to $(y^0, y)\mu(ds)$ if and only if

$$x^0 + u(x, s) > y^0 + u(y, s).$$

The consumer ds starts out with no money and with the bundle $\mathbf{a}(s)\mu(ds)$ of the other goods. Let S be a coalition (i.e. $S \in \mathcal{C}$) and $\mathbf{x}$ be

[1] Strictly speaking, we do not have "unrestricted side payments" (see Section 2) in this interpretation, since the individual holdings of the finished good must be non-negative. However, since v is monotonic, negative payoffs cannot occur in the value, nor, for that matter, in the core.

such that $\mathbf{x}(s) \in \Omega$ for all s and $\int_S \mathbf{x}\, d\mu = \int_S \mathbf{a}\, d\mu$. This means that the members of S can trade among each other—redistribute their initial resources—in such a way that after the trade, ds will be holding the bundle $\mathbf{x}(s)\mu(ds)$ of goods $1, \ldots, n$, but still no money. The utility to consumer ds of his new bundle will be $u(\mathbf{x}(s), s)\mu(ds)$, and so if we "add" the utilities of all consumers in S we will get a total of $\int_S u(\mathbf{x}(s), s)\mu(ds)$. Let us choose $\mathbf{x}$ so that the maximum in (30.1) is attained; then this total is exactly $v(S)$.

Generally, adding up utilities of different consumers is an economically meaningless procedure. In this case, however, the availability of money lends significance to the total utility of S. Indeed, we claim that *any* distribution of utilities to the consumers in S whose total is $v(S)$ is achievable by the coalition S. This means that if ν is any measure with $\nu(S) = v(S)$, then the coalition S can distribute its total bundle $(0, \int_S \mathbf{a}\, d\mu)$ so that the utility of consumer ds in S will be $\nu(ds)$. To see this, define a measure ξ by

$$\xi(U) = \nu(U) - \int_U u(\mathbf{x}(s), s)\mu(ds)$$

for all $U \in \mathcal{C}$. Then $\xi(S) = 0$, i.e. ξ restricted to S is a feasible redistribution among the traders of S of the initial total—namely 0—of money available to this coalition. If S redistributes its money in this way, then each trader ds will get the bundle $(\xi(ds), \mathbf{x}(s)\mu(ds))$, whose utility is

$$\xi(ds) + u(\mathbf{x}(s), s)\mu(ds) = \nu(ds).$$

Thus this game satisfies the condition of "unrestricted side payments,"[2] and the worth of each coalition S is adequately described by the number $v(S)$.

The literature on transferable utility economies is briefly reviewed in Section 34. Some economists find the exchange interpretation, and in particular the role played in it by "money," somewhat difficult to justify in economic terms. The production version is offered as an alternative interpretation (see Note 2), which when taken as applying to productive situations in which there is only one consumption good (or in which the consumption goods are "aggregated"), may be more

[2] See the end of Section 2.

acceptable. In any event, the exchange interpretation is important as a special case of the more generally accepted Walrasian exchange economies.[3] Indeed, transferable utility economies are mathematically easier to deal with,[4] and thus results obtained for this special class of exchange economies can point the way to similar results for the more general Walrasian exchange economies. Important results relating to general Walrasian economies have on several occasions been first obtained for transferable utility economies; and there is every expectation that this will continue to happen.

NOTES

1. It is quite possible for the sup in (30.1) to exist without the max existing. We have not treated such situations. One reason is that they are conceptually somewhat slippery. It is of course possible to define $v(S)$ by means of the sup, but the idea of the "worth" of a coalition then loses some of its intuitive force. The way we are used to thinking about core and value would presumably also need some revision. If, for example, $\nu(I) = v(I)$, and the sup in the definition of $v(I)$ is not attained, then we cannot really think of ν as a distribution of the amount available to I, since $v(I)$ is not really available to I. A more important reason for insisting that the sup be attained is that the mathematics would otherwise be even more complicated than it now is. For a discussion of where one is led if one replaces "max" by "sup," see Subsection D of Section 33 and Subsection C of Section 41.

2. The relationship between the "exchange" and "production" interpretation of our mathematical model can be seen clearly if we regard consumption as a form of production. That is, we imagine that each trader ds (in the exchange version) inputs the bundle of commodities $\mathbf{x}(s)\mu(ds)$ into a production process that outputs $u(\mathbf{x}(s), s)\mu(ds)$ units of a substance that he actually consumes (or, since utility is transferable, that he transfers to another consumer).

From this comparison, we see that the single consumer good in the production interpretation will enjoy some basic properties both of utility (as the stuff that is ultimately valued) and of money (as a vehicle for the storage and transfer of value).

31. Statement of Main Results

Throughout, the measure μ and the functions u and $\mathbf{a}$ will be as specified at the beginning of Section 30, and the set-function v as defined in (30.1).

[3] See Aumann (1964) and Section 32.

[4] Because they can be modeled as games "with side payments"—i.e. games in coalitional form, with a numerical characteristic function—which Walrasian economies in general cannot. Compare the end of Section 29, also Shapley and Shubik (1966), p. 808.

We shall say that $u(x, s) = o(\|x\|)$ as $\|x\| \to \infty$, *integrably* in s, if for each $\epsilon > 0$ there is an integrable function η on I, such that $|u(x, s)| \leqq \epsilon\|x\|$ whenever $\|x\| \geqq \eta(s)$. If η is bounded, then this is equivalent to saying that $u(x, s) = o(\|x\|)$ as $\|x\| \to \infty$, uniformly in s. But in general, the two concepts are not equivalent; for example, if $n = 1$ and $(I, \mathcal{C}) = ([0, 1], \mathcal{B})$, then $x^{1/2}/s^{1/4} = o(x)$ integrably, but not uniformly. The concept of integrable convergence was introduced by Aumann and Perles (1965) in order to deal with the question of the existence of the maximum in expressions of the form (30.1).

The function u will be called *Borel-measurable* if it is measurable on the product σ-field $\mathcal{B} \times \mathcal{C}$, where $\mathcal{B}$ is the σ-field of Borel subsets of Ω.

THEOREM J. *Assume that* $\mathbf{a}$ *is* μ*-integrable, and that*

(31.1) *u is Borel-measurable;*

(31.2) $u(x, s) = o(\|x\|)$ *as* $\|x\| \to \infty$, *integrably in s;*

(31.3) *for each fixed s, u is continuous on Ω, and for each j, $\partial u(x, s)/\partial x^j$ exists and is continuous at each $x \in \Omega$ for which*[1] $x^j > 0$; *and*

(31.4) $\mathbf{a}(s) > 0$ *for all s.*

Then v (see (30.1)) *is well-defined*[2] *and is in pNA, and the core of v consists of a single payoff vector, which coincides with the value φv.*

Theorem J will be proved in Section 40.

Though it is common enough in economics, condition (31.4)—total positivity of initial resources—has a certain slightly restrictive, unintuitive flavor, and it would be nice if we could dispense with it. Two senses in which this can in fact be done will now be discussed.[3] The first is to demand that there be only a finite number of different utility functions for the members of I. Specifically, let us say that u is *of finite type* if there is a finite set H of functions on Ω such that each of the functions $u(\cdot, s)$ is in H. (Note that this still allows all of the $\mathbf{a}(s)$ to be different.) Then we have

PROPOSITION 31.5. *Theorem J continues to hold if* (31.4) *is replaced by*

(31.6) *u is of finite type.*

[1] I.e., whenever the two-sided partial derivative is defined.

[2] I.e., the maximum is attained for all $S \in \mathcal{C}$.

[3] For a third sense, see Proposition 33.2.

Proposition 31.5 will be proved in Section 39.

The other sense in which (31.4) can be dispensed with is embodied in the following proposition:

PROPOSITION 31.7. *Let u satisfy* (31.1), (31.2), *and* (31.3). *Then v is well defined, the mixing and asymptotic values of v exist and coincide, and the core of v consists of a single payoff vector, which coincides with the mixing and asymptotic value.*

Proposition 31.7 will be proved in Chapter VII (Sections 45 and 46). The proof depends on a basic common property of the mixing and asymptotic approaches called the "diagonal property," which forms the subject of Chapter VII.

Possibilities for other extensions of the results stated here will be discussed in Section 41.

32. The Competitive Equilibrium

Let μ, u, $\mathbf{a}$, and v be as in Section 30. An *allocation* is an integrable function $\mathbf{x}$ from I to Ω such that

$$\int \mathbf{x} = \int \mathbf{a}.$$

A *transferable utility competitive equilibrium* (t.u.c.e.) is a pair $(\mathbf{x}, p)$, where $\mathbf{x}$ is an allocation and $p \in \Omega$, such that for all $s \in I$, $u(x, s) - p\cdot(x - \mathbf{a}(s))$ attains its maximum (over $x \in \Omega$) at $x = \mathbf{x}(s)$. The function on I whose value at s is $u(\mathbf{x}(s), s) - p\cdot(\mathbf{x}(s) - \mathbf{a}(s))$ is called the *competitive payoff density;* its indefinite integral[1] (w.r.t. μ) is called the *competitive payoff distribution;* and p is the vector of *competitive prices.* (All three definitions are, of course, with respect to a given t.u.c.e. $(\mathbf{x}, p)$.)

Intuitively, the vector p is a price vector. Thus, $p\cdot(\mathbf{x}(s) - \mathbf{a}(s))\mu(ds)$ represents the amount that the player[2] ds must pay in order to buy the bundle $\mathbf{x}(s)\mu(ds)$, over and above the amount that he gets by selling his initial bundle $\mathbf{a}(s)\mu(ds)$. This amount must be subtracted from $u(\mathbf{x}(s), s)\mu(ds)$ in order to yield the net "income" of ds, and it is this

[1] If g is an integrable function on I, the *indefinite integral* of g is the measure ν defined by $\nu(S) = \int_S g\, d\mu$.

[2] Producer or trader, according to which interpretation is being used.

income that ds wishes to maximize. If p is such that when all players maximize in this way, the total demand $\int \mathbf{x}$ equals the total supply $\int \mathbf{a}$, then the economy is in equilibrium. Note that in the exchange interpretation, the total excess demand for money at such a point—namely, $\int p \cdot (\mathbf{x} - \mathbf{a})$—also vanishes.

We shall distinguish the concept just defined from the usual concept of competitive equilibrium by calling the latter a *Walrasian competitive equilibrium* (w.c.e.).[3] To relate the two concepts, let us consider the monetary exchange economy interpretation of our game, namely, a market in which there are $n + 1$ goods $0, 1, \ldots, n$, the 0-th good being money. A w.c.e. in such a market takes the form of a pair $((\mathbf{x}^0, \mathbf{x}), (p^0, p))$, and we may assume w.l.o.g. that $p^0 = 1$. It is then easily verified that such a pair is a w.c.e. if and only if $(\mathbf{x}, p)$ is an t.u.c.e. and for all s, $\mathbf{x}^0(s) = p \cdot (\mathbf{a}(s) - \mathbf{x}(s))$. The total utility of the trader ds at this w.c.e. is then seen to be exactly

$$(u(\mathbf{x}(s), s) - p \cdot (\mathbf{x}(s) - \mathbf{a}(s)))\mu(ds).$$

Note that although a w.c.e. remains a w.c.e. when the prices are multiplied by a positive constant, this is not the case for a t.u.c.e.; there the prices have already been normalized, so to speak, by the requirement that the price of money be 1.

PROPOSITION 32.1. *Let u be Borel measurable, and let $\int \mathbf{a} > 0$. Then an integrable $\mathbf{x}$ maximizes $\int u(\mathbf{x}(s), s)\, d\mu(s)$ subject to $\int \mathbf{x} = \int \mathbf{a}$ and $\mathbf{x}(s) \in \Omega$ if and only if there is a p such that $(\mathbf{x}, p)$ is a transferable utility competitive equilibrium.*

This is essentially the content of Theorem 5.1 of Aumann and Perles (1965) (see Section 36 below); it may be considered a form of the Kuhn-Tucker theorem (1951) in an infinite dimensional space. The proposition says that any allocation $\mathbf{x}$ for which $v(I)$ is attained (see (30.1)) is competitive, if the appropriate side payments $p \cdot (\mathbf{x}(s) - \mathbf{a}(s))$ are

[3] This will not be formally defined here; the interested reader is referred to Aumann (1964) (for markets with a continuum of traders) or Debreu (1959) (for finite economies). Some familiarity with the concept of a w.c.e. is needed in certain parts of this section, e.g., in the proof of Proposition 32.5. It is however not needed in most of this section, e.g., for Propositions 32.1, 32.2, or 32.3 or their proofs. Neither is it used in the subsequent sections of this chapter.

made. As for the prices p, when u is differentiable, then

$$p^i = [\partial u/\partial x^i]_{x=\mathbf{x}(s)}$$

for all s such that $\mathbf{x}^i(s) > 0$ (cf. (32.11)). Thus in the production interpretation, p^i is the marginal product of the ith commodity at equilibrium, and in the exchange interpretation, p^i is its marginal utility (assuming, in both cases, that some of the ith commodity is used at equilibrium by the individual in question).

PROPOSITION 32.2. *Assume* (31.1), (31.2), $u(\cdot, s)$ *is continuous on* Ω *for each* s, *and* $\int \mathbf{a} > 0$. *Then there is a t.u.c.e.*

Proof. The main theorem of Aumann and Perles (1965) asserts that under the conditions we have assumed,[4] the maximum in the definition of v is attained (see Section 36 below). The result then follows from Proposition 32.1. This completes the proof of Proposition 32.2.

Without (31.1) and (31.2), there may be no t.u.c.e.; see Section 33.

We now wish to discuss how the competitive equilibrium is related to the core and the value. In the ordinary Walrasian exchange economy with a continuum of traders, it is known that the core coincides with the set of competitive allocations[5] (Aumann, 1964). It is therefore reasonable to conjecture that a similar situation holds for t.u.c.e.'s. This is in fact the case; indeed we have

PROPOSITION 32.3. *Assume* (31.1), (31.2), (31.3), *and* $\int \mathbf{a} > 0$. *Then there is a unique competitive payoff distribution, which coincides with the unique point*[6] *in the core of* v *and so also with the mixing and asymptotic value* φv.

Remark. Note that we are not asserting that the t.u.c.e. is unique. What *is* being asserted is that there is at least one t.u.c.e., and that if $(\mathbf{x}, p)$ is any t.u.c.e., then

$$\int_S (u(\mathbf{x}(s), s) - p\cdot(\mathbf{x}(s) - \mathbf{a}(s)))\, d\mu = (\varphi v)(S)$$

for all $S \in \mathcal{C}$.

[4] And even slightly weaker conditions.

[5] A *competitive allocation* in a Walrasian economy is an allocation $\mathbf{x}$ for which there exists a price vector p such that $(\mathbf{x}, p)$ is a w.c.e.

[6] See Proposition 31.7.

Proof. By Proposition 32.2, there is a t.u.c.e. $(\mathbf{x}, p)$. Let ν be the corresponding competitive payoff distribution. Since $\mathbf{x}$ is an allocation it follows that

$$\nu(I) \leqq v(I). \tag{32.4}$$

Next, if S is any coalition, let $v(S)$ be attained at $\mathbf{y}$, i.e.,

$$v(S) = \int_S u(\mathbf{y}(s), s)\, ds, \int_S \mathbf{y} = \int_S \mathbf{a}, \text{ and } \mathbf{y}(s) \geqq 0 \text{ for all } s.$$

Then, by the definition of t.u.c.e.,

$$u(\mathbf{x}(s), s) - p\cdot(\mathbf{x}(s) - \mathbf{a}(s)) \geqq u(\mathbf{y}(s), s) - p\cdot(\mathbf{y}(s) - \mathbf{a}(s)).$$

Integrating this over S, we obtain

$$\nu(S) \geqq v(S) - p\cdot\int_S (\mathbf{y} - \mathbf{a}) = v(S);$$

together with (32.4), this shows that ν is in the core. But by Proposition 31.7, the core contains a unique point, namely the asymptotic value; so the proof of Proposition 32.3 is complete.

In the above proof, we made use of the fact that there is only one point in the core in order to establish the equivalence between the core and the set of all competitive payoff distributions. The proof of uniqueness for the core, in turn, is intimately bound up with value considerations and with the differentiability of u. But the equivalence between the core and the set of competitive allocations is a much more general phenomenon, which does not depend on differentiability, is not directly connected with value considerations, and in fact continues to hold even when the core has many members. It is therefore of some interest to establish this equivalence under conditions that are more general than those of Proposition 32.3 even though there is no direct connection between this and the value.

Proposition 32.5. *Assume that u is continuous in x for each fixed s and is Borel measurable, that v is well-defined,*[7] *and that* $\int \mathbf{a} > 0$. *Then the core of v coincides with the set of competitive payoff distributions.*

[7] I.e., that for each S, the maximum in the definition of $v(S)$ is attained; (31.2) is a sufficient condition for this (in the presence of the previously mentioned continuity and measurability conditions), but it is not necessary. Incidentally, all that is needed for this proposition is that the max in the definition of $v(I)$ be achieved; if for the other S, $v(S)$ is defined to be the sup rather than the max, the proposition remains true.

Proof. The idea of the proof is to introduce "money" explicitly, as in the monetary exchange interpretation of our economy. We then get an $(n + 1)$-good market whose w.c.e.'s are in $1 - 1$ correspondence with the t.u.c.e.'s of the original economy, and whose core corresponds[8] to the core of v. We now apply the "equivalence theorem" for Walrasian economies (see, e.g., Aumann, 1964), according to which the core of such an economy coincides with the set of all Walrasian competitive allocations (w.c.a.'s)—i.e. allocations associated with some w.c.e. Since the core of the $(n + 1)$-good Walrasian economy corresponds to the core of the original n-good transferable utility economy, we may deduce the equivalence theorem for the original transferable utility economy.

Strictly speaking, we cannot use Aumann's equivalence theorem (1964), for the following reason: In the $(n + 1)$-good Walrasian economy, money is available in negative as well as non-negative quantities, whereas all other goods are available in non-negative quantities only. Therefore the space of all commodity bundles is not the non-negative orthant of E^{n+1}, but rather $E^1 \times \Omega$, where E^1 is the entire real line and Ω is the non-negative orthant of E^n. But though Aumann's theorem is stated only for the case in which the space of commodity bundles is precisely the non-negative orthant, the theorem holds for the case considered here as well; the proof given can be extended without any difficulty. The proof of Proposition 32.5 can now be completed as outlined above.[9]

In the theorems stated in Section 31, the uniqueness—or, better,[10] the *unicity* of the core is established via value considerations, using Theorem I; strong use is thereby made of the "differentiability" of v—i.e. the existence of $\partial v^*(t, S)$—along the diagonal, and this in turn depends on the differentiability of u. Proposition 32.5 gives us an op-

[8] It is in establishing the correspondence between the cores that one uses the assumption that the max in the definition of $v(I)$ is achieved.

[9] An alternative procedure would be to use a more general form of the equivalence theorem due to Hildenbrand (1968), which does not restrict attention to bundles in the non-negative orthant Ω. Hildenbrand's theorem refers to *quasi-competitive* (Debreu, 1962) allocations rather than competitive ones. In this case, though, it turns out that the quasi-competitive allocations are the same as the competitive ones.

[10] It is the point in the core that is unique; the core itself, as a set, is trivially unique, in any game.

portunity to establish the unicity of the core in a different manner, by proving the uniqueness of the competitive payoff distribution. As might be expected, this too depends on the differentiability of u.

In fact, let u satisfy (31.1) and (31.3), and let $\int \mathbf{a} > 0$. Let $(\mathbf{x}, p)$ be a t.u.c.e. From the definition of t.u.c.e. it then follows that

$$u(\mathbf{x}(s), s) - p\cdot\mathbf{x}(s) \geqq u(x, s) - p\cdot x,$$

for all $x \in \Omega$, whence

$$u(x, s) - u(\mathbf{x}(s), s) \leqq p\cdot(x - \mathbf{x}(s)). \tag{32.6}$$

Setting $x = \mathbf{x}(s) + \delta e_j$ for a given j and letting $\delta \to 0+$, we deduce

$$[\partial u/\partial x^j]_{x=\mathbf{x}(s)} \leqq p^j. \tag{32.7}$$

If, moreover, $\mathbf{x}^j(s) > 0$, then we may let $\delta \to 0-$, obtaining the inequality opposite to (32.7); together, they yield

$$[\partial u/\partial x^j]_{x=\mathbf{x}(s)} = p^j \text{ whenever } \mathbf{x}^j(s) > 0. \tag{32.8}$$

Suppose now that $(\mathbf{y}, q)$ is another t.u.c.e. Setting $x = \mathbf{y}(s)$ in (32.6), we obtain

$$u(\mathbf{y}(s), s) - u(\mathbf{x}(s), s) \leqq p\cdot(\mathbf{y}(s) - \mathbf{x}(s)). \tag{32.9}$$

Similarly

$$u(\mathbf{x}(s), s) - u(\mathbf{y}(s), s) \leqq q\cdot(\mathbf{x}(s) - \mathbf{y}(s)). \tag{32.10}$$

Combining these inequalities, we obtain

$$(p - q)\cdot(\mathbf{y}(s) - \mathbf{x}(s)) \geqq 0.$$

Since $\int (p - q)\cdot(\mathbf{y} - \mathbf{x}) = (p - q)\cdot(a - a) = 0$, it follows that a.e.

$$(p - q)\cdot(\mathbf{y}(s) - \mathbf{x}(s)) = 0,$$

and hence a.e. the inequalities (32.9) and (32.10) are equalities. Combining the first of these equalities with (32.6), we obtain a.e.

$$u(x, s) - u(\mathbf{y}(s), s) \leqq p\cdot(x - \mathbf{y}(s)).$$

Proceeding as in the derivation of (32.8), we deduce that a.e. when $\mathbf{y}^j(s) > 0$,

$$[\partial u/\partial x^j]_{x=\mathbf{y}(s)} = p^j. \tag{32.11}$$

But from (32.8) itself, applied to $(\mathbf{y}, q)$ rather than $(\mathbf{x}, p)$, we obtain

$$(32.12) \qquad [\partial u/\partial x^j]_{x=\mathbf{y}(s)} = q^j \text{ whenever } \mathbf{y}^j(s) > 0.$$

Since $\int \mathbf{y}^j = \int \mathbf{a}^j > 0$, it follows that there are s with $\mathbf{y}^j(s) > 0$ obeying (32.11); together with (32.12), this then yields

$$p^j = [\partial u/\partial x^j]_{x=\mathbf{y}(s)} = q^j.$$

Hence $p = q$, and so the competitive prices are uniquely determined. But then there can be at most one competitive payoff density, namely

$$(32.13) \qquad \max(u(x, s) - p\cdot(x - \mathbf{a}(s))),$$

and so at most one competitive payoff distribution.

If, moreover, u also satisfies (31.2), then, by Proposition 32.2, there is a t.u.c.e., i.e. the max in (32.13) is attained. Thus we have provided an alternative proof, which does not depend on the value concept, of all but the last clause of Proposition 32.3.

Theorem I provides a direct connection between the value and the core, and what we have just said provides the corresponding connection between the t.u.c.e. and the core. To complete the triangle, we now demonstrate directly how the value is connected with the t.u.c.e., without considering the core. Unlike the previous demonstrations, though, this demonstration will have a heuristic rather than a strictly rigorous nature.[11]

If $\mathbf{x}$ is a measurable function from I to Ω, we will find it convenient slightly to abuse our notation by writing $u(\mathbf{x})$ for the function on I whose value at s is $u(\mathbf{x}(s), s)$.

Assume that it has been established that $v \in pNA$. If f is an ideal set (see Chapter IV), it then seems reasonable to suppose that

$$(32.14) \quad v^*(f) = \max\left\{\int u(\mathbf{x})f : \int \mathbf{x}f = \int \mathbf{a}f \text{ and } \mathbf{x}(s) \geqq 0 \text{ for all } s\right\}.$$

Assuming (32.14), let us, for given $S \in \mathcal{C}$ and $t \in (0, 1)$, calculate the expression

$$\partial v^*(t, S) = \lim_{\tau \to 0} \frac{v^*(t\chi_I + \tau\chi_S) - v^*(t\chi_I)}{\tau}.$$

[11] Though there are some gaps in the previous demonstrations, they are relatively easy to fill in. The gaps in the present argument are more serious.

From (32.14) it follows that

$$v^*(t\chi_I) = tv(I) = \int tu(\mathbf{x}),$$

where $\mathbf{x}$ is the allocation at which $v(I)$ is achieved. Now let $v^*(t\chi_I + \tau\chi_S)$ be achieved at $\mathbf{y}$. Then for sufficiently small τ, we have

$$\begin{aligned}(32.15)\quad v^*(t\chi_I + \tau\chi_S) - v^*(t\chi_I)&\\ &= \int (u(\mathbf{y}) - u(\mathbf{x}))(t\chi_I + \tau\chi_S) + \tau \int_S u(\mathbf{x})\\ &\leqq \int [p\cdot(\mathbf{y} - \mathbf{x})](t\chi_I + \tau\chi_S) + \tau \int_S u(\mathbf{x})\\ &= p\cdot\int (\mathbf{a} - \mathbf{x})(t\chi_I + \tau\chi_S) + \tau \int_S u(\mathbf{x})\\ &= tp\cdot\int (\mathbf{a} - \mathbf{x}) + \tau p\cdot\int_S (\mathbf{a} - \mathbf{x}) + \tau \int_S u(\mathbf{x})\\ &= 0 + \tau \int_S (u(\mathbf{x}) - p\cdot(\mathbf{x} - \mathbf{a})),\end{aligned}$$

and hence

$$(32.16)\qquad \partial v^*(t, S) \leqq \int_S (u(\mathbf{x}) - p\cdot(\mathbf{x} - \mathbf{a})).$$

We can, however, say more, namely that equality holds in (32.16). To show this, it is only necessary to point to a $\mathbf{y}$ satisfying

$$(32.17)\qquad \int (t\chi_I + \tau\chi_S)\mathbf{y} = \int (t\chi_I + \tau\chi_S)\mathbf{a} \quad\text{and}\quad \mathbf{y}(s) \geqq 0 \text{ for all } s$$

for which the inequality in (32.15) becomes an equality up to a term that is $o(\tau)$. Now it is always possible to find a $\mathbf{y}$ satisfying (32.17) that will have the property that $\mathbf{y}^j(s) = 0$ whenever $\mathbf{x}^j(s) = 0$; and moreover, such that $\mathbf{y}^j(s) = \mathbf{x}^j(s) + K^j(s)\tau$ for all s, where $K^j(s)$ is a constant that depends on j and on whether s is or is not in S, but otherwise does not depend on s or on τ. From this and (32.8) it follows that equality holds in (32.15), up to a term that is $o(\tau)$. Hence equality holds in (32.16). But then it follows from Theorem H that the value φv coincides with the competitive payoff distribution.

The two major gaps in this argument are the unproved assumptions that $v \in pNA$ and that v^* is given by (32.17). Given $v \in pNA$, (32.17) is probably not too hard to prove, e.g. by the theorems of Section 25. But to prove $v \in pNA$, say from the assumptions of Theorem J, is a serious bit of work. Indeed, it is precisely this that constitutes the most difficult part of the proof of Theorem J, and it will require every bit of Sections 35 through 40 before it is done.

We repeat, though, that our proof of Theorem J will not depend on the above argument (nor will it depend explicitly on the t.u.c.e. at all); rather, it will use Theorem I, i.e. it will depend on core considerations only. The above arguments were given only to shed light on the relations between core, value, and competitive equilibrium, from several different viewpoints. From the point of view of this book, the t.u.c.e. is, strictly speaking, not needed at all; and if it *is* introduced, it is most directly related to the core and the value via the first proof of Proposition 32.3 given above.

33. Examples

In all the numbered examples of this section,[1] I will be the unit interval $[0, 1]$, $\mathcal{C}$ the Borel σ-field $\mathcal{B}$, and μ Lebesgue measure λ.

A. THE CASE $n = 1$

In the production interpretation, $n = 1$ means that the finished good is produced from only one kind of raw material, though the efficiency of production of the various traders ds—i.e. the production functions $u(\cdot, s)\mu(ds)$—may be different.[2] In the exchange interpretation, we are dealing with a market in which only one kind of good is being bought and sold (for money), the demand for (and supply of) this one good being created by the different utility functions $u(\cdot, s)\mu(ds)$ that the traders ds have for the good. Conceptually and computationally, this case is somewhat easier to deal with than the one of general n. Yet it is far from trivial, and its analysis involves most of the basic ideas that are met with in the general case.

Example 33.1. Let $n = 1$, and for all s, let

$$u(x, s) = \sqrt{x + s} - \sqrt{s}$$

(see Fig. 8) and

$$\mathbf{a}(s) = \tfrac{1}{32}.$$

[1] Examples 33.1, 33.3, 33.6, 33.9, 33.11, 33.12, 33.13.

[2] It may even happen that one trader produces more efficiently than another at a certain level, whereas the other produces more efficiently than the first at a different level.

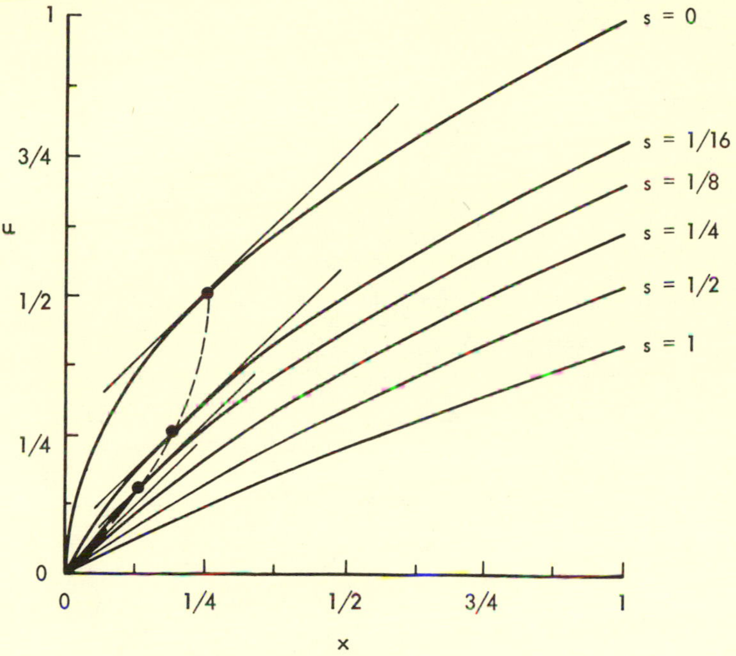

Figure 8. The function $u(x, s)$ for Example 33.1

This market satisfies (31.1), (31.2), (31.3), and (31.4). Therefore, from Theorem J and Proposition 32.3, it follows that $v \in pNA$, the core of v and the t.u.c.e. are unique, and both coincide with the value. It is easiest to compute the t.u.c.e., making use of (32.10) and (32.11). The idea of the computation is that the higher the price p is, the smaller will be the total demand for the good x; we must find a price at which the total demand exactly matches the total supply $\int \mathbf{a} = \frac{1}{32}$. Suppose then that the price is p; let $\mathbf{x}(s)\mu(ds)$ be the demand of ds. Then by (32.10), if $\mathbf{x}(s) > 0$, we have

$$p = [\partial u/\partial x]_{x=\mathbf{x}(s)} = \frac{1}{2\sqrt{\mathbf{x}(s) + s}};$$

hence

$$\mathbf{x}(s) = \frac{1}{4p^2} - s \quad \text{and} \quad u(\mathbf{x}(s), s) = \frac{1}{2p} - \sqrt{s}.$$

By (32.11), if $\mathbf{x}(s) = 0$, then

$$\frac{1}{2\sqrt{s}} = [\partial u/\partial x]_{x=0} \leqq p;$$

hence,

$$\frac{1}{4p^2} - s \leqq 0.$$

Thus, we conclude that in any case

$$\mathbf{x}(s) = \max\left(0, \frac{1}{4p^2} - s\right).$$

Hence,

$$\tfrac{1}{32} = \int \mathbf{a} = \int \mathbf{x} = \int_0^{\frac{1}{4p^2}} \left(\frac{1}{4p^2} - s\right) ds$$

$$= \int_0^{\frac{1}{4p^2}} s\, ds = \left[\frac{s^2}{2}\right]_0^{\frac{1}{4p^2}} = \frac{1}{32p^4}.$$

Hence $p = 1$, and it follows that

$$v(I) = \int_0^1 u(\mathbf{x}(s), s)\, ds = \int_0^{\frac{1}{4}} (\tfrac{1}{2} - \sqrt{s})\, ds = \tfrac{1}{2} \cdot \tfrac{1}{4} - \tfrac{2}{3}(\tfrac{1}{4})^{3/2} = \tfrac{1}{24}.$$

The competitive payoff density is

$$[u(\mathbf{x}(s), s) - p\cdot\mathbf{x}(s)] + p\cdot\mathbf{a}(s);$$

when $s \geqq \frac{1}{4}$ this consists simply of $p\cdot\mathbf{a}(s) = \frac{1}{32}$. When $s \leqq \frac{1}{4}$, we have in addition to this amount, the amount

$$\tfrac{1}{2} - \sqrt{s} - (\tfrac{1}{4} - s) = \tfrac{1}{4} + s - \sqrt{s} = (\tfrac{1}{2} - \sqrt{s})^2,$$

which ranges from 0 at $s = \frac{1}{4}$ to $\frac{1}{4}$ at $s = 0$. The situation is pictured in Figure 9 (solid lines). The slopes of the u-curves at $x = \mathbf{x}(s)$ (dashed line in Fig. 8) are all equal to the competitive price of 1 when $\mathbf{x}(s) > 0$, but when $\mathbf{x}(s) = 0$ the tangent at 0 may have a slope smaller than 1.

Note that the competitive payoff density may be thought of as consisting of two parts, namely $p\cdot\mathbf{a}(s)$ and $u(\mathbf{x}(s), s) - p\cdot\mathbf{x}(s)$. In the production interpretation these two parts may be thought of as follows: the first part is compensation to ds in his role as supplier, and is *always*

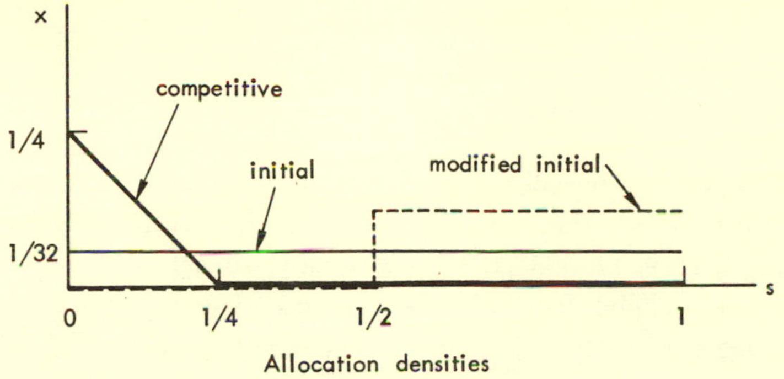

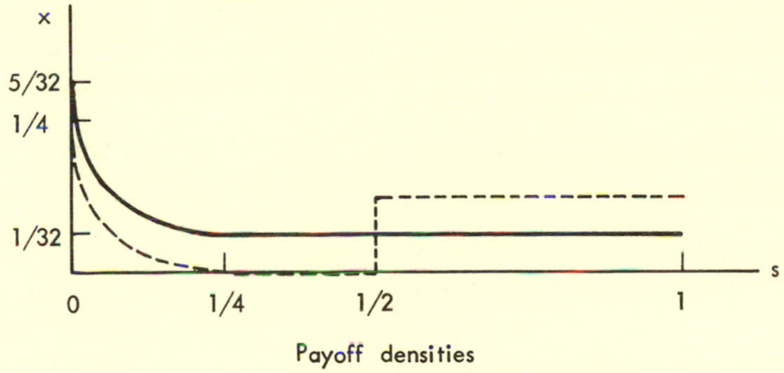

Figure 9. Competitive solution for Example 33.1

divided among the players in proportion[3] *to* **a**(s). The second part is attributable to his role as producer, i.e. to his u-function $u(\cdot, s)$, and *does not depend in any way on his initial bundle*[4] **a**(s).

[3] This needs no interpretation when $n = 1$. When $n > 1$, it means in proportion to $p \cdot \mathbf{a}(s)$. However, even when $n > 1$, if one trader's initial bundle density is exactly twice that of another—in the vectorial sense—then that part of his payoff density attributable to his role as a supplier will also be twice that of the other trader.

[4] It is interesting to remark that this division into two parts with the above properties is far from trivial if one looks at the payoff from the value or core point of view.

If, for example, we redefine the initial bundle distribution here by

$$\mathbf{a}(s) = \begin{cases} 0, & 0 \leqq s \leqq \frac{1}{2} \\ \frac{1}{16}, & \frac{1}{2} < s \leqq 1, \end{cases}$$

then our calculation will remain essentially unchanged, the only difference occurring in the payoff density. This is illustrated by the dashed lines in Figure 9. The players ds for $s \in [0, \frac{1}{4}]$ will act as producers, and will obtain a payoff density of $\frac{1}{4} + s - \sqrt{s}$; they will obtain nothing as suppliers, since they have no initial bundles. For $s \in [\frac{1}{4}, \frac{1}{2}]$, the players have no initial supplies, neither are they sufficiently efficient to produce; therefore, their payoff is 0. The remaining players (those between $\frac{1}{2}$ and 1) get a payoff density of $\frac{1}{16}$, in proportion to their initial holdings; but they are not sufficiently efficient as producers to act in this capacity. One might say that they sell their initial holdings to the efficient producers between 0 and $\frac{1}{4}$ in return for a promise of manufactured goods.

This second version of the example does not satisfy (31.4), and therefore we cannot deduce from Theorem J that $v \in pNA$. However, we have

PROPOSITION 33.2. *In Theorem J,* (31.4) *may be replaced by the assumption that* $n = 1$.

This proposition will be proved in Section 40, together[5] with Theorem J. Readers who have followed the preceding example will realize that even when $n = 1$, the assumption $\mathbf{a}(s) > 0$ is by no means of a trivial nature;[6] players for whom $\mathbf{a}(s) = 0$ may still have considerable significance as producers (in the production interpretation), even though they supply none of the initial good. Thus Proposition 33.2 is by no means an easy consequence of Theorem J.

B. THE FINITE TYPE CASE

Suppose that u satisfies (31.1), (31.2), and (31.3), and moreover that it is of finite type; that is, there are finitely many functions $f_1, \ldots, f_k$

[5] More precisely, a common generalization (Proposition 40.26) of Theorem J and Proposition 33.2 will be proved.

[6] As it would be (because of $n = 1$) in most discussions of a Walrasian economy. When money is introduced explicitly, a transferable utility economy with $n = 1$ becomes a Walrasian economy with $n = 2$ (cf. the proof of Proposition 32.5).

on Ω such that each of the functions $u(\cdot, s)$ is one of the f_i. Define a k-dimensional vector η of measures on I by

$$\eta_i(S) = \mu\{s: u(\cdot, s) = f_i\},$$

and an n-dimensional vector ζ of measures on I by

$$\zeta(S) = \int_S \mathbf{a}.$$

Let $\nu = (\eta, \zeta)$. Then v is a function[7] of the $n + k$ dimensional vector ν, say $v = g \circ \nu$. Let us calculate g in a specific example.

Example 33.3. Let $n = 1$ and for all s, let

$$u(x, s) = \sqrt{x + 1} - 1$$

and

$$\mathbf{a}(s) = 8s.$$

In this case, both η (which $= \lambda$) and ζ are one-dimensional, so $\nu = (\eta, \zeta)$ is two-dimensional. The range R of ν is depicted in Figure 10; note that it is not symmetric around the diagonal, but rather (as always) around the center of the diagonal. It may be seen that $v = g \circ \nu$, where

$$g(y, z) = y\left[\sqrt{\frac{z}{y} + 1} - 1\right] = \sqrt{y(y + z)} - y.$$

We would now like to apply Theorem B to deduce[8] that $v \in pNA$ and to obtain the value φv. Unfortunately, this is impossible, because the conditions of Theorem B fail; g is not continuously differentiable on the range R. Indeed, we have

$$\partial g/\partial z = \tfrac{1}{2}\sqrt{y/(y + z)}. \tag{33.4}$$

If we let $(y, z) \to 0$ along the diagonal, then $\partial g/\partial z \to \frac{1}{2}\sqrt{\frac{1}{2}}$, whereas if we let $(y, z) \to 0$ along the bottom boundary of R, then $\partial g/\partial z \to \frac{1}{2}$. Hence, $\partial g/\partial z$ cannot be extended to all of R so that it will be continuous at 0.

Though Theorem B is not applicable, Proposition 10.17 is, and we apply it to deduce that $v \in pNA$. To calculate the value φv, we would

[7] For a detailed discussion, see Section 39, in particular formulas (39.7) and (39.18).

[8] Of course we know from Proposition 31.5 that $v \in pNA$; what we are investigating here is whether a simple proof can be obtained for this very simple special case.

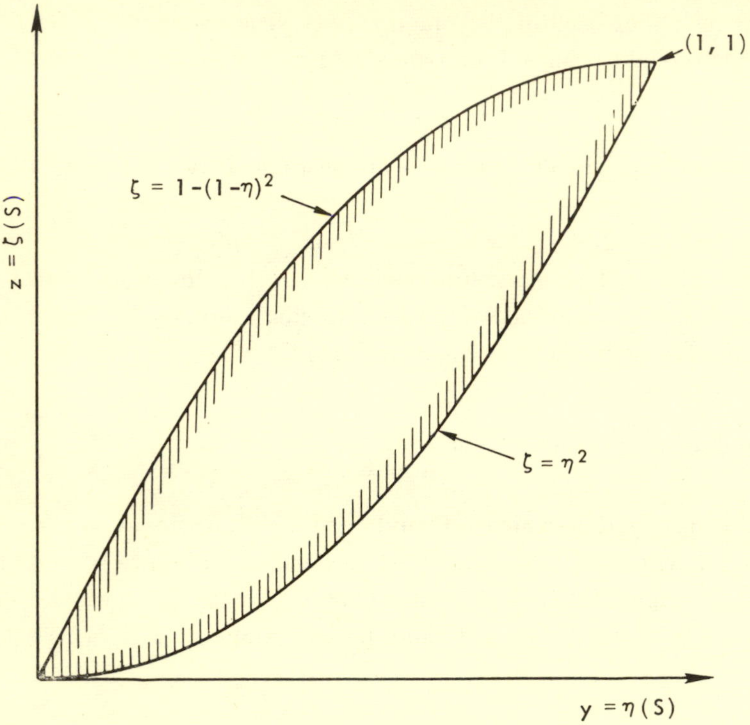

Figure 10. The range of $\nu = (\eta, \zeta)$ in Example 33.3

like to use the "diagonal formula" (3.2). Though we have not heretofore proved this under the conditions of Proposition 10.17, it does in fact hold under those conditions.[9] Using (33.4) and

$$\frac{\partial g}{\partial y} = \frac{2y + z}{2\sqrt{y(y+z)}} - 1,$$

we thus obtain

$$(33.5)\qquad (\varphi v)(S) = \eta(S)\int_0^1 \frac{\partial g}{\partial y}(t, t)\,dt + \zeta(S)\int_0^1 \frac{\partial g}{\partial z}(t, t)\,dt$$

$$= \left(\frac{3}{2\sqrt{2}} - 1\right)\eta(S) + \frac{1}{2\sqrt{2}}\zeta(S).$$

[9] Probably the easiest way to establish this at this stage of the game is to use Theorem H. If one wants to restrict oneself to more elementary methods, it is not difficult to devise a proof using Proposition 10.7 and the fact that $f^\delta \circ \mu$ satisfies the conditions of Theorem B.

In the production interpretation, the first and second terms may be considered compensation to the members of S in their roles as producers and suppliers, respectively (cf. the discussion of Example 33.1). As we shall see in Section 39, $g \circ \nu$ satisfies the conditions of Proposition 10.17 whenever u is of finite type, and so the value formula applies. This implies a decomposition of φv into a term involving η only (production) and a term involving ζ only (supply), so that for the finite type case we have a derivation of this phenomenon from value considerations as well.

In this example, $v(S)$ is always achieved by $\mathbf{x}(s) \equiv \zeta(S)$; in particular, $v(I)$ is achieved by $\mathbf{x}(s) \equiv 1$. Hence in the t.u.c.e., we have

$$p = [\partial u/\partial x]_{x=1} = \left[\frac{1}{2\sqrt{x+1}}\right]_{x=1} = \frac{1}{2\sqrt{2}}.$$

Hence, for all s, the competitive payoff density is given by

$$\begin{aligned} u(\mathbf{x}(s), s) - p \cdot (\mathbf{x}(s) - \mathbf{a}(s)) &= \sqrt{\mathbf{x}(s) + 1} - 1 - \frac{1}{2\sqrt{2}}(\mathbf{x}(s) - s) \\ &= \sqrt{2} - 1 - \frac{1}{2\sqrt{2}} + \frac{1}{2\sqrt{2}} s \\ &= \frac{3}{2\sqrt{2}} - 1 + \frac{1}{2\sqrt{2}} s; \end{aligned}$$

hence the competitive payoff distribution is given by (33.5), which is as it should be. In the general finite type case as well, it may be seen by direct computations that the diagonal formula for the value yields the competitive payoff distributions (cf. Section 39, especially the material following (39.7)).

The main point of Example 33.3 was to show that Theorem B is not sufficient to deal even with the simplest[10] finite type cases and that Proposition 10.17 is needed. In Section 39 we shall see that Proposition 10.17 is indeed sufficient to cover the general finite type case.

[10] We used $\sqrt{x+1} - 1$ rather than simply $\sqrt{x}$ in order to show that even when u is differentiable on the n-dimensional nonnegative orthant, g may not be differentiable on the $n + k$-dimensional orthant, and Theorem B may not be applicable.

C. DIFFERENTIABILITY

To show that condition (31.3) cannot be dispensed with, consider the following market:

Example 33.6. Let $n = 1$, and for all s, let

$$u(x, s) = \begin{cases} x, & \text{for } x \leqq 1 \\ \sqrt{x}, & \text{for } x \geqq 1, \end{cases}$$

and

$$\mathbf{a}(s) = \begin{cases} \frac{3}{2}, & \text{for } s \leqq \frac{1}{2}, \\ \frac{1}{2}, & \text{for } s > \frac{1}{2}. \end{cases}$$

The function $u(x, s)$ is shown in Figure 11. It is not differentiable at $x = 1$; the left derivative is 1, and the right derivative is $\frac{1}{2}$. Thus if we define $\mathbf{x}(s) \equiv 1$, then $(\mathbf{x}, p)$ is a t.u.c.e. whenever $\frac{1}{2} \leqq p \leqq 1$, because the line through (1, 1) with slope p supports the graph of u for those values of p. The competitive payoff density corresponding to a given value of p will therefore be

$$u(\mathbf{x}(s), s) - p\cdot(\mathbf{x}(s) - \mathbf{a}(s)) = \begin{cases} 1 + \frac{1}{2}p, & \text{for } s \leqq \frac{1}{2} \\ 1 - \frac{1}{2}p, & \text{for } s > \frac{1}{2}. \end{cases}$$

The competitive payoff distribution is therefore given by

$$(33.7) \quad \xi_p(S) = (1 + \tfrac{1}{2}p)\lambda(S \cap [0, \tfrac{1}{2}]) + (1 - \tfrac{1}{2}p)\lambda(S \cap [\tfrac{1}{2}, 1]).$$

It follows from Proposition 32.5 that the core of the market is the set of all ξ_p, where p ranges from $\frac{1}{2}$ to 1. In particular, it consists of more than

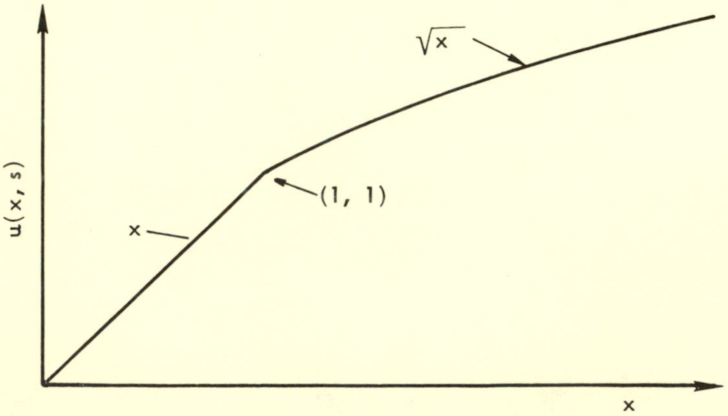

Figure 11. The function $u(x, s)$ for Example 33.6

one point, so that the conclusion of Theorem J (and also that of Proposition 31.7) fails.

We can also calculate the core directly. Let $\zeta(S) = \int_S \mathbf{a}$; the range of (λ, ζ) is depicted in Figure 12. It may be verified that

$$v = \min(\zeta, \sqrt{\zeta\lambda}). \tag{33.8}$$

Since $\frac{1}{2}(\zeta + \lambda) \geqq \sqrt{\zeta\lambda}$ and $\zeta(I) = \lambda(I) = 1$, it follows that the core of v contains all convex combinations of the form $t\zeta + (1 - t)\lambda$, with $\frac{1}{2} \leqq t \leqq 1$; these are precisely the ξ_p of (33.7).

The reader will note the similarity between formulas (33.8) and (3.4); in neither case does v belong to pNA. Indeed, the proof that the v of (33.8) does not belong to pNA can be carried out along the same lines as the proof of Example 9.4 appearing at the end of Section 27. Rather than doing this in detail, though, we will present another non-differentiable market more directly related to (3.4) and Example 9.4.

Example 33.9. Define a function f on the non-negative half line by

$$f(x) = \begin{cases} x, & \text{for } 0 \leqq x \leqq 1 \\ 2 - \dfrac{1}{x}, & \text{for } 1 \leqq x \end{cases}$$

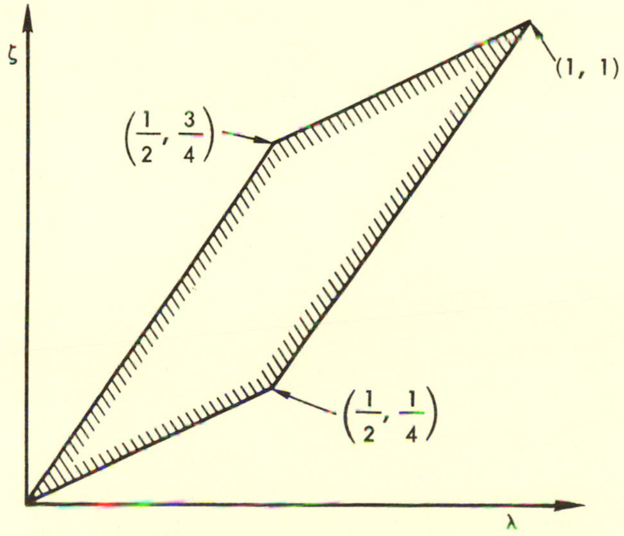

Figure 12. The range of (λ, ζ) in Example 33.6

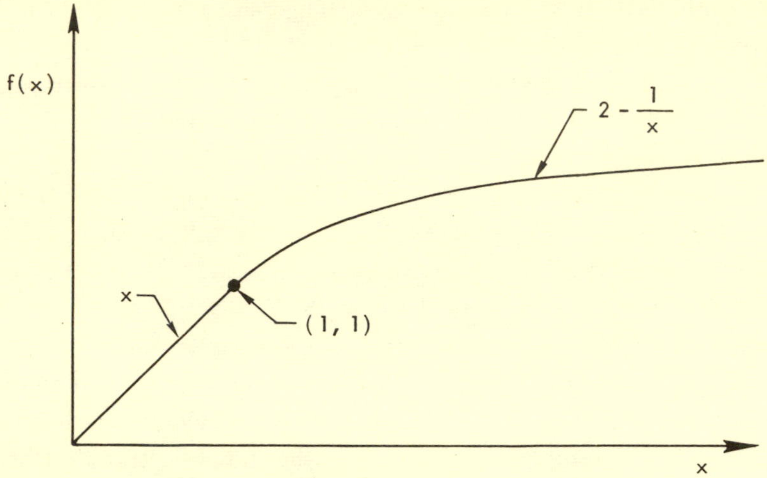

Figure 13. The function $f(x)$ in Example 33.9

(see Fig. 13). Let $n = 2$, and for all s, let

$$u(x, s) = f(\min(x^1, x^2) + \tfrac{1}{10}(x^1 + x^2))$$

and

$$\mathbf{a}(s) = \begin{cases} (\frac{3}{2}, \frac{1}{2}) & \text{for } 0 \leqq s \leqq \frac{1}{2} \\ (\frac{1}{2}, \frac{3}{2}) & \text{for } \frac{1}{2} < s \leqq 1. \end{cases}$$

This is often called the "glove market"; it has the following (exchange) interpretation:[11] The commodities 1 and 2 are left and right gloves, respectively. Individual gloves are next to useless, being usable only for the material in them (this accounts for the term $(x^1 + x^2)/10$, which is needed here so that u be strictly increasing). Pairs of gloves, however, can be used as gloves. The utility for pairs of gloves and for material is bounded, being governed by the function f (this is needed here to ensure that (31.2) is obeyed).

It is easy to see that $v(I)$ is achieved for $\mathbf{x}(s) \equiv (1, 1)$. The prices p must satisfy

$$p^1 \geqq 0.1, \quad p^2 \geqq 0.1, \quad p^1 + p^2 = 1.2;$$

otherwise, however, they are arbitrary. In other words, p may be any convex combination of (1.1, 0.1) and (0.1, 1.1). The set of all competi-

[11] Cf. Shapley and Shubik (1969b), pp. 342–347, and Owen (1972), p. P-74.

tive payoff distributions—i.e. the core—may be easily calculated from this. Alternatively we may proceed as follows: Define $\zeta(S) = \int_S \mathbf{a}$. Then

$$v = \min(\zeta^1, \zeta^2) + \tfrac{1}{10}(\zeta^1 + \zeta^2), \tag{33.10}$$

and it is easily verified from this that any convex combination of $\zeta^1 + \frac{1}{10}(\zeta^1 + \zeta^2)$ and $\zeta^2 + \frac{1}{10}(\zeta^1 + \zeta^2)$ is in the core of v. Next, if we set

$$\lambda_1(S) = \lambda(S \cap [0, \tfrac{1}{2}])$$
$$\lambda_2(S) = \lambda(S \cap [\tfrac{1}{2}, 1]),$$

then from (33.10) it follows that

$$v = \tfrac{6}{10}(\zeta^1 + \zeta^2) + \tfrac{1}{2}|\zeta^1 - \zeta^2| = \tfrac{6}{5}\lambda + \tfrac{1}{2}|\lambda_1 - \lambda_2|,$$

and hence it follows immediately from Example 9.4 that $v \notin pNA$. From Section 16 it follows that v is not in MIX either. Thus the conclusions of Theorem J and Proposition 31.7 fail in two ways: The core contains more than one point, and v is not in pNA and has no mixing value.

As for the asymptotic value, it follows from Proposition 19.1 that v *does* have an asymptotic value, namely $\frac{6}{10}(\zeta^1 + \zeta^2)$. But by increasing the dimension of the example by 1, the asymptotic value is eliminated as well. Indeed, let $n = 3$ and define

$$u(x, s) = f(\min(x^1, x^2, x^3) + (x^1 + x^2 + x^3)/10)$$

and

$$\mathbf{a}(s) = \begin{cases} (\tfrac{3}{2}, \tfrac{3}{4}, \tfrac{3}{4}) & \text{for } 0 \leqq s \leqq \tfrac{1}{3} \\ (\tfrac{3}{4}, \tfrac{3}{2}, \tfrac{3}{2}) & \text{for } \tfrac{1}{3} < s \leqq \tfrac{2}{3} \\ (\tfrac{3}{4}, \tfrac{3}{4}, \tfrac{3}{2}) & \text{for } \tfrac{1}{3} < s \leqq 1. \end{cases}$$

This may be thought of as a "3-handed" glove market, in the same way that Example 33.9 is a "2-handed" glove market. One can calculate v, getting an expression similar to (33.10), in which there appears a minimum of 3 rather than of 2 measures. The core turns out to be large, as before; in fact it is 2-dimensional (a triangle), as compared to the 1-dimensional core (an interval) in Example 33.9. Again, v has no mixing value, as can be seen from arguments similar to those used in Section 16; a fortiori it is not in pNA. This time, however, v has no asymptotic value either; this follows from Example 19.2. Thus one sees that without differentiability, both Theorem J and Proposition 31.7 fail in the strongest way imaginable.

D. ACHIEVEMENT OF THE MAX IN THE DEFINITION OF v

If condition (31.2) is not obeyed, the max in the definition of $v(S)$ may not be achieved, even though the sup may be finite. The following example of this is from Aumann and Perles (1965):

Example 33.11. Let $n = 1$, and for all s, let

$$u(x, s) = xs \quad \text{and} \quad \mathbf{a}(s) = 1.$$

In this case, the integral appearing in the definition of $v(I)$ is $\int s\mathbf{x}(s)\, ds$, and this must be maximized subject to $\int \mathbf{x} = 1$; the supremum in this case is 1, but it is not achieved.

Since $v(I)$ is not achieved, it follows from Proposition 32.1 that this economy has no t.u.c.e. Therefore, one cannot hope to extend Proposition 32.3, according to which the core, value, and competitive payoff distributions all coincide, to this situation. But possibly an extension of Theorem J could be proved, i.e. maybe we could show that if in the definition (31.1) of v we replace max by sup, then the core would consist of a single point, v would be in pNA, and the value φv would coincide with the single point in the core.

Under this new definition of v, the v for Example 33.11 is given by[12]

$$v(S) = \lambda(S)\ (\text{ess. sup. } S).$$

The core of this v consists of a single point, namely λ. Indeed, it is easy to see that λ is in the core. Suppose that the core also contains another point, say ν. Since $\nu \neq \lambda$, there is a set S such that $\nu(S) < \lambda(S)$; let k be sufficiently large so that $1/k < \lambda(S) - \nu(S)$. Divide I into k disjoint sets, each of which has essential supremum 1. For at least one of these sets—let us call it T—we must have

$$\nu(T) \leqq \nu(I)/k = 1/k.$$

Since ν is non-negative (because $\nu(S) \geqq v(S)$), it follows that

$$\nu(S \cup T) \leqq \nu(S) + \nu(T) < \lambda(S);$$

[12] ess. sup. S is the *essential supremum* of S, i.e. the smallest number α with the property that $\lambda(S \cap [\alpha, 1]) = 0$.

on the other hand

$$\nu(S \cup T) \geqq v(S \cup T) = \lambda(S \cup T)(\text{ess. sup. } (S \cup T)) \geqq \lambda(S)(\text{ess. sup. } T) = \lambda(S).$$

This contradiction proves that the core indeed contains only the point λ.

Unfortunately, v is not in pNA; in fact, it is not even in AC. To see this, let us define an arbitrary set function v to be *continuous at* S, where $S \in \mathcal{C}$, if for all non-decreasing sequences $\{S_i\}$ such that $\bigcup S_i = S$, and all non-increasing sequences $\{S_i\}$ such that $\bigcap S_i = S$, we have

$$\lim v(S_i) = v(S).$$

It is easily verified that every member of AC is continuous at every $S \in \mathcal{C}$. But the v of Example 33.11 is almost never[13] continuous. For example, if $S = [0, \frac{1}{2}]$ and $S_i = [0, \frac{1}{2}] \cup [1 - (1/i), 1)$, then $\{S_i\}$ is a monotone non-increasing sequence and $\bigcap S_i = S$; but

$$\lim v(S_i) = \tfrac{1}{2} > \tfrac{1}{4} = v(S).$$

Therefore, $v \notin AC$ and a fortiori $v \notin pNA$, and so Theorem J cannot be generalized to this situation.

Neither is Proposition 31.7 generalizable to this situation; indeed the v of Example 33.11 is not[14] in ORD, and a fortiori not in MIX. This v does however have an asymptotic value, which coincides with the unique point in the core. We have already seen that the core consists of the unique point λ. To see that the asymptotic value exists and equals λ, consider a partition of I into a large number of small sets[15] $S_1, \ldots, S_k$. In a random ordering of the S_i, there will with high probability be an S_i near the beginning of the ordering whose essential supremum is close to 1. This means that with high probability most of the S_i will be contributing approximately $\lambda(S_i)$ to $v(S)$, and our assertion about the asymptotic value follows from this. Of course the argument as given

[13] It is continuous only at those S for which ess. sup. $S = 0$ (i.e. $\lambda(S) = 0$) or ess. sup. $S = 1$.

[14] Define $\mathcal{R}$ on $[0, 1]$ as follows: first comes $[0, \frac{1}{2})$ in the natural order; then comes $[\frac{1}{2}, 1)$ in the reverse of the natural order; then $\{1\}$. (I.e. $\mathcal{R}$ is the same as $>$ except that within $[\frac{1}{2}, 1)$ the order is reversed.) Then $\varphi(v; \mathcal{R})$ does not exist.

[15] For definiteness one can think of intervals of equal length, but the argument goes through perfectly well without any such assumption.

here is heuristic, but the reader may convince himself that it is easily made precise.

Thus there is still some hope that a weaker form of Proposition 31.7 might be salvaged, in which one refers only to the asymptotic value. But this is not the case either; unfortunately, Proposition 31.7 fails in a much more decisive way: we now present an example of an economy satisfying all our assumptions except (31.2), in which the core of v contains many points.[16]

Example 33.12. Let $n = 2$, let

$$u(x, s) = \begin{cases} (x^1 + x^2) - ((x^1)^{1/s} + (x^2)^{1/s})^s & \text{when } s \in (0, 1), \\ x^1 + x^2 & \text{when } s = 0 \text{ or } s = 1, \end{cases}$$

and let

$$\mathbf{a}(s) = \begin{cases} (\tfrac{1}{2}, \tfrac{3}{2}) & \text{when } s \in [0, \tfrac{1}{2}], \\ (\tfrac{3}{2}, \tfrac{1}{2}) & \text{when } s \in (1, \tfrac{1}{2}]. \end{cases}$$

Here, the exact form of u is of no importance; what is needed is only that for fixed s, u be increasing, differentiable, and homogeneous of degree 1 in x, that for fixed x, u be decreasing in s for $s \in (0, 1)$, and that

$$\lim_{s \to 0} u(x, s) = \min(x^1, x^2).$$

The form of u at $s = 0$ and at $s = 1$ is of no importance.

Define a vector measure ζ by

$$\zeta(S) = \int_S \mathbf{a}.$$

Then,[17]

$$v(S) = \begin{cases} \min(\zeta^1(S), \zeta^2(S)), & \text{when ess. inf. } S = 0 \\ u(\zeta(S), \text{ess. inf. } S), & \text{when ess. inf. } S > 0. \end{cases}$$

From this it follows that ζ^1, ζ^2, and any convex combination of ζ^1 and ζ^2 are in the core of v, and so the core contains more than one point. Therefore without (31.2) or at least some condition that guarantees that $v(I)$ is attained, there is no hope for generalizing Proposition 31.7 either.

[16] When the max in (30.1) is replaced by sup.

[17] ess. inf. S is the *essential infimum* of S; it is defined to be the largest α such that $\lambda(S \cap [0, \alpha]) = 0$.

Proposition 31.5 also cannot be extended. Rather than describing the counterexample in detail, we will indicate it by means of a figure.

Example 33.13. Let $n = 2$, and let

$$f(x) = \min(x^1, x^2) + x^1 + x^2.$$

For all s, $u(x) = u(x, s)$ is defined to be $f(x)$ when x is *not* in the interior of the central region C in Figure 14, while in the interior of C it is defined so that it is non-negative, differentiable, increasing, and $\leqq f(x)$. The initial bundle $\mathbf{a}$ is defined by

$$\mathbf{a}(s) = \begin{cases} (\frac{1}{2}, \frac{3}{2}), & \text{when } s \in [0, \frac{1}{2}], \\ (\frac{3}{2}, \frac{1}{2}), & \text{when } s \in (\frac{1}{2}, 1]. \end{cases}$$

The smallest concave function that is $\geqq u(x)$ is $f(x)$. Readers familiar with the methods of Aumann and Perles (1965) (compare also Proposition 39.3) will be able to deduce without difficulty that $v = f \circ \zeta$, where ζ is given by $\zeta(S) = \int_S \mathbf{a}$; this may also be seen directly via Lyapunov's

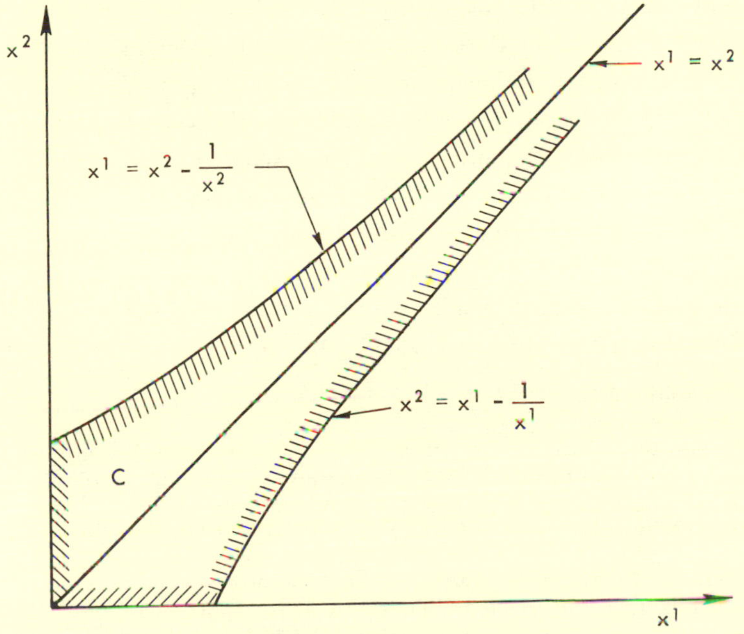

Figure 14. The region C in Example 33.13

theorem.[18] In any case we see that the core of v contains $2\zeta^1 + \zeta^2$, $\zeta^1 + 2\zeta^2$, and all convex combinations of these two measures.

What happens if we require (instead of (31.2)) that $u(\cdot, s)$ be concave for each fixed s? If u is not required to be of finite type, then this does not help; indeed, the u of Example 33.12 has this property.[19] If u is of finite type, then we are in a situation where all the $v(S)$ are attained in spite of the fact that (31.2) does not hold. This situation will be discussed further in Section 41.

34. Discussion of the Literature

Transferable utility economies with finitely many agents[1] made their debut on the modern scene in a paper by Shubik (1959), though their ancestry can be traced considerably further back (see Auspitz and Lieben, 1889; Böhm-Bawerk, 1891; and von Neumann and Morgenstern, 1944). They were subsequently studied intensively by Shapley (1964ab) and Shapley and Shubik (1966, 1969ab, 1972).

The fact that in a Walrasian economy with a continuum of traders, the core coincides with the set of all competitive allocations has been discussed extensively in the literature.[2] The same principle, in a different form, is embodied in the theorem of Debreu and Scarf (1963); there it is proved that under appropriate conditions, the core of an n-person Walrasian economy "tends," in a certain sense, to the set of all competitive allocations as $n \to \infty$. To distinguish these two kinds of theorems, let us call the former a *continuous* theorem, the latter an *asymptotic*[3] theorem.

The problem studied in this chapter from a "continuous" viewpoint was investigated by Shapley (1964b) from an "asymptotic" viewpoint.

[18] Proposition 1.3.

[19] It is even strictly concave.

[1] We use this term to mean either "trader" or "producer," according to the interpretation. Transferable utility economies with finitely many agents are defined in a manner entirely analogous to the continuous economies defined in this paper.

[2] See, for example, Aumann (1964), Vind (1964), Hildenbrand (1968, 1974), and Cornwall (1969).

[3] The first of these "asymptotic" theorems, on the convergence of the core in a special class of markets, is due to Edgeworth (1881). Recent generalizations of the Debreu-Scarf result may be found in Kannai (1970), Hildenbrand (1970b), and Arrow and Hahn (1971).

Specifically, consider a transferable utility economy with a fixed finite number m of types of agents, where unlike here, the *type* of an agent is determined both by his initial bundle and his utility[4] function. Let there be k agents of each type. Assume that the utility function of each agent is concave and differentiable (in the sense of (31.3)). Then as $k \to \infty$, the value of each trader tends, uniformly, to what he would get under a t.u.c.e.

This theorem can be compared and contrasted with our present results in a number of respects; the comparison turns out to be typical of similar comparisons between continuous and asymptotic theorems in other cases.

First of all, continuous theorems are usually "cleaner" in their statement: they assert equality, whereas asymptotic theorems assert only that a certain limiting relation holds. This is exactly the situation here. To some extent, of course, the difference is illusory; in the continuous result, the limit notion is often built into the definition of the objects about which one is asserting equalities. Let us, for example, compare the main theorem of Shapley (1964b) with Proposition 31.7: in the former, one considers the limit of values of a certain sequence of finite economies; in the latter, one considers immediately a continuous economy, but defines its value via a sequence of values of finite games. One may avoid such a process by using the "axiomatic" value, as in Theorem J; nevertheless, the axiomatic value appearing there does, in fact, equal the asymptotic value. The core notion is less directly related to finite games; but even this could easily be defined asymptotically. One must remember also that the whole notion of a game with a continuum of players is intuitively appealing only in the sense that it somehow approximates a large finite game; in a sense, therefore, the asymptotic approach is more direct. Thus we may sum up by saying that here, as usual in such situations, the continuous approach yields cleaner results but is somewhat more sophisticated, conceptually as well as in its use of mathematical tools.

A more important difference, perhaps, is that asymptotic theorems usually require far stronger assumptions than continuous theorems.

[4] We shall use this term to mean utility function in the exchange interpretation, production function in the production interpretation.

Outstanding among such assumptions—here as in other cases[5]—is that the number of types of traders is a fixed finite number. Concavity (or quasi-concavity) of preferences is also often required in asymptotic theorems, but not in continuous ones, and this is the case here as well. There is one respect in which the asymptotic result of Shapley (1964b) assumes less than we do here, and that is in the behaviour of u; we assume here that u is increasing and that $u(x) = o(\|x\|)$ as $\|x\| \to \infty$, whereas neither assumption is needed there. However if we were to assume concavity and finite type, then we might be able to dispense with these two assumptions[6] on u.

Finally, it may be remarked that the asymptotic results imply a framework within which the manner and rate of convergence can be discussed. The continuous formulation, by its nature, precludes such considerations.

35. The Space $\mathfrak{U}_1$

The proof of Theorem J will proceed as follows: First we shall prove Proposition 31.5, which assumes that u is of finite type. To go from this to Theorem J, we shall approximate to general u's by u's of finite type. Now each u may be viewed as a family of functions $u(\cdot, s)$ on Ω. We shall say that two u's are close, if roughly speaking, they are close for all but a small set of s's. But for this, one needs a metric on the space of functions on Ω of which the functions $u(\cdot, s)$ are typical. In this section we shall define such a metric, use it to define precisely the above-mentioned notion of closeness between two u's, and finally, prove that any u can then be approximated by a u of finite type.

We shall assume w.l.o.g. that $\mu(I) = 1$. For $x \in \Omega$, we shall write Σx to mean $\Sigma_{i=1}^n x^i$.

Let $\mathfrak{F}_0$ denote the set of all real-valued functions f on Ω that are continuous, are non-decreasing, vanish at 0, and satisfy

$$(35.1) \qquad f(x) = o(\Sigma x) \text{ as } \Sigma x \to \infty .$$

[5] But some of the more modern results on the convergence of the core avoid this assumption, at the cost of considerable complexity in the statements of the results. See Hildenbrand (1974).

[6] Cf. the end of Section 41.

Let $\mathfrak{F}_1$ denote the set of all f in $\mathfrak{F}_0$ that are increasing (rather than just non-decreasing) and

(35.2) have continuous partial derivatives $f^j(x) = \partial f/\partial x^j$ at each $x \in \Omega$ for which $x^j > 0$.

Note that (35.1) is equivalent to the condition

$$f(x) = o(\|x\|) \text{ as } \|x\| \to \infty, \tag{35.3}$$

since we have

$$\frac{1}{n} \Sigma x \leqq \|x\| \leqq \Sigma x \tag{35.4}$$

for all $x \in \Omega$.

Let $\mathfrak{F}$ be $\mathfrak{F}_0$ or $\mathfrak{F}_1$. If $\mathcal{E}$ is the linear span of $\mathfrak{F}$ (i.e. the set of all finite linear combinations of members of $\mathfrak{F}$), then we impose a norm on $\mathcal{E}$ by

$$\|g\| = \sup_{x \in \Omega} |g(x)|/(1 + \Sigma x);$$

that this norm is finite follows from (35.1). This norm induces a metric and hence a topology on $\mathcal{E}$, and hence on $\mathfrak{F}$.

PROPOSITION 35.5. *$\mathfrak{F}_1$ has a denumerable dense subset.*

Proof. Let Ω' denote the one-point compactification of Ω, which is obtained from Ω by adding a point that we shall call ∞. If $f \in \mathfrak{F}_1$, then the function $f(x)/(1 + \Sigma x)$ can be extended in a natural way from Ω to Ω' by defining it to be 0 at ∞; it will then be continuous on all of Ω'. Let $\mathfrak{F}_1'$ be the set of all functions f' on Ω', such that $f'(\infty) = 0$ and for some $f \in \mathfrak{F}_1$, we have

$$f'(x) = f(x)/(1 + \Sigma x)$$

for all $x \in \Omega$. $\mathfrak{F}_1$ is in 1–1 correspondence with $\mathfrak{F}_1'$ under the correspondence

$$f'(x) \leftrightarrow f(x)/(1 + \Sigma x).$$

Now $\mathfrak{F}_1'$ is a subspace of the space $C(\Omega')$ of all continuous functions on Ω'; if we impose the uniform convergence metric on $C(\Omega')$, it then follows from the compactness of Ω' and the Stone-Weierstrass theorem that $C(\Omega')$ has a denumerable dense subset. But since it is metric, it follows that it has a denumerable basis; hence also $\mathfrak{F}_1'$ has a denumerable basis, in the uniform convergence topology. Now the above 1–1 correspondence takes the uniform convergence topology on $\mathfrak{F}_1'$ onto the norm topology on $\mathfrak{F}_1$ that we have defined previously; hence, $\mathfrak{F}_1$ has a denumerable basis in that topology, and so also a denumerable dense subset. This completes the proof of Proposition 35.5.

Let $\mathfrak{F}$ be $\mathfrak{F}_0$ or $\mathfrak{F}_1$, $\mathcal{E}$ its linear span. We remark that from the compactness of Ω' and from the fact that $g(x) = o(\Sigma x)$ for all $g \in \mathcal{E}$, it follows that the sup in the definition of $\|g\|$ is attained, so that we may write

$$\|g\| = \max_{x \in \Omega} |g(x)|/(1 + \Sigma x).$$

Let $\mathfrak{U}_0$ be the space of all functions u on $\Omega \times I$ that satisfy (31.1) and (31.2), and such that $u(\cdot, s)$ is non-decreasing and continuous on Ω and vanishes at 0. Let $\mathfrak{U}_1$ be the space of all $u \in \mathfrak{U}_0$ satisfying (31.3), and such that $u(\cdot, s)$ is increasing for each fixed s. Note that if $u \in \mathfrak{U}_i$ then $u(\cdot, s) \in \mathfrak{F}_i$, where $i = 0$ or 1.

Let $u \in \mathfrak{U}_0$. For $s \in I$, we write u_s for the function on Ω whose value at x is $u(x, s)$. For $\delta > 0$, a *δ-approximation* to u is defined to be a member $\hat{u}$ of $\mathfrak{U}_0$ such that $\|\hat{u}_s - u_s\| \leqq \delta$ for all s except possibly a set of μ-measure $\leqq \delta$, in which $\hat{u}_s(x) = \sqrt{\Sigma x}$.

PROPOSITION 35.6. *For every $\delta > 0$ and $u \in \mathfrak{U}_1$, there is a $\hat{u} \in \mathfrak{U}_1$ that is a δ-approximation to u and is of finite type.*[1]

Proof. Let $\{f_1, f_2, \ldots\}$ be a denumerable dense subset of $\mathfrak{F}_1$ (Proposition 35.5). Let $\delta > 0$ be given. For each s in I, $u_s \in \mathfrak{F}_1$; let $\mathbf{i}(s)$ be the first i such that

$$\|u_s - f_{\mathbf{i}(s)}\| \leqq \delta.$$

It may be seen that $\mathbf{i}$ is Borel measurable. For each k, define $u^k \in \mathfrak{U}_1$ by

$$u_s^k(x) = \begin{cases} f_{\mathbf{i}(s)}(x), & \text{if } \mathbf{i}(s) \leqq k, \\ \sqrt{\Sigma x}, & \text{otherwise.} \end{cases}$$

Let $S_i = \{s \in I : \mathbf{i}(s) = i\}$. Clearly, $\cup_{i=1}^{\infty} S_i = I$, and hence for k sufficiently large,

$$\mu\left(I \backslash \bigcup_{i=1}^{k} S_i\right) \leqq \delta;$$

[1] See Section 31.

for such k, u^k is a δ-approximation to u of finite type. This completes the proof of Proposition 35.6.

36. Further Preparations

In this section we shall introduce some notation and quote some results from Aumann and Perles (1965)[1] that will be used throughout the sequel. Let us denote that paper, in this section, by [A-P]. The most important of these results give sufficient conditions for the attainability of the max in expressions of the form (30.1), and necessary and sufficient conditions for a specific measurable $\mathbf{x}$ actually to attain this max.

Let u be a Borel-measurable function on $\Omega \times I$. If $\mathbf{x}$ is a μ-integrable function from I to Ω, we will as before use the notation $u(\mathbf{x})$ for the function on I whose value at s is $u(\mathbf{x}(s), s)$. For all $a \in \Omega$ and $S \in \mathcal{C}$, write

$$u_S(a) = \max \left\{ \int_S u(\mathbf{x}) : \mathbf{x}(s) \in \Omega \text{ for all } s \text{ and } \int_S \mathbf{x} = a \right\}.$$

We shall say that *$u_S(a)$ is attained at* $\mathbf{x}$ if $\mathbf{x}$ is an integrable function from I to Ω such that $\int_S \mathbf{x} = a$ and $\int_S u(\mathbf{x}) = u_S(a)$.

Proposition 36.1. *Let $u \in \mathfrak{U}_0$. Then for all S and a, $u_S(a)$ exists, i.e. the max is attained and is finite.*

This is essentially the main theorem of [A-P]. Note that

$$v(S) = u_S(\textstyle\int_S \mathbf{a}).$$

Next, we explain the concept of concavification. Let f be a non-negative real-valued function on Ω and let

$$F = \{(\nu, x) \in E^1 \times \Omega : 0 \leqq \nu \leqq f(x)\}.$$

Let F^* be the convex hull of F. If there is a function f^* on Ω such that

$$F^* = \{(\nu, x) \in E^1 \times \Omega : 0 \leqq \nu \leqq f^*(x)\},$$

then f is said to be *spannable*, and f^* is called the *concavification*[2] of f;

[1] In citing these results, the more general hypothesis in [A-P] of upper-semicontinuity of u has been replaced by the present assumption of continuity.

[2] In this book the word "concavification" applies only to spannable functions. This coincides with the usage of [A-P] but differs slightly from that of Shapley and Shubik (1966), where it is defined in terms of the *closure* of the convex hull.

clearly f^* is unique and concave. If f is concave, then f is spannable and $f^* = f$.

PROPOSITION 36.2. *If $f \in \mathfrak{F}_0$, then f is spannable and f^* is non-decreasing and continuous.*

This is essentially Proposition 3.1 of [A-P].

If $u \in \mathfrak{U}_0$, then $u_s \in \mathfrak{F}_0$ for all s, and so $u_s{}^*$ is defined. Define a function u^* on $\Omega \times I$ by

$$u^*(x, s) = u_s{}^*(x).$$

Then $u^*{}_s = u_s{}^*$, and so we will henceforth write u_s^* for their joint value.

PROPOSITION 36.3. *Let $u \in \mathfrak{U}_0$. Then $u^* \in \mathfrak{U}_0$, and for all S and a*

$$u^*{}_S(a) = u_S(a).$$

In particular, it follows that u_S is concave on Ω.

This is an immediate consequence of Lemmas 3.3 and 3.5 and Propositions 3.1 and 4.1 of [A-P]. From the concavity of u_S and $u^*{}_S = u_S$ it follows that $u^*{}_S = u_S{}^*$; so we will henceforth write u_S^* for their joint value.

PROPOSITION 36.4. *Let $u \in \mathfrak{U}_0$, let $a \in \Omega$ be >0, let $S \in \mathfrak{C}$, and let $\mathbf{x}$ be a measurable function from I to Ω. Then a necessary and sufficient condition for $u_S(a)$ to be attained at $\mathbf{x}$, i.e. for*

$$\int_S u(\mathbf{x}) = u_S(a) \text{ and } \int_S \mathbf{x} = a,$$

is that there be a $p \in \Omega$ such that

$$u(x, s) - u(\mathbf{x}(s), s) \leqq p \cdot (x - \mathbf{x}(s))$$

for all $x \in \Omega$ and almost all $s \in S$. If u is increasing for each fixed s, then $p \in \Omega$ (i.e. $p \geqq 0$) may be replaced by $p > 0$. If $u \in \mathfrak{U}_1$, then for $i = 1, \ldots, n$,

$$p^i = [\partial u / \partial x^i]_{x=\mathbf{x}(s)}$$

for almost all s such that $\mathbf{x}^i(s) > 0$.

This is essentially Proposition 5.1 of [A-P].

We close this section with a statement of the Measurable Choice Theorem.

PROPOSITION 36.5. *Let* $(X, \mathfrak{X})$ *be a standard measurable space, i.e. one that is isomorphic to* $([0, 1], \mathfrak{B})$. *Let G be a subset of* $I \times X$ *that is measurable in the product σ-field* $\mathfrak{C} \times \mathfrak{X}$, *and whose projection on I is I. Then there is a measurable function* $g: I \to X$ *such that for almost all s,* $(g(s), s) \in G$.

This theorem is due to von Neumann (1949, p. 448, Lemma 5). Von Neumann's proof uses Assumption 2.1, namely that $(I, \mathfrak{C})$ is also standard, but the theorem remains true without this.[3]

37. Basic Properties of δ-Approximations

The main goal of this section is Lemma 37.8, in which it is shown that for given u, there is a fixed integrable function η that bounds all coordinates of all functions $\mathbf{y}$ that maximize $\hat{u}_S(b)$, whenever S is not too small, b is bounded away from 0 and ∞, and $\hat{u}$ is a sufficiently good δ-approximation to u. The existence of such a fixed η is important in, among other things, compactness arguments in many places in the sequel. One example of such a use is in Proposition 37.13, in which the continuity of u_S on Ω is established (the difficulty, of course, occurs on the boundary of Ω).

LEMMA 37.1. *Let* $u \in \mathfrak{U}_0$. *For each* $\delta > 0$ *let* η_δ *be an integrable function with* $\eta_\delta(s) \geqq 1$ *for each s, such that*

$$u(x, s) < \delta \Sigma x$$

and

$$\sqrt{\Sigma x} < \delta \Sigma x$$

whenever $\Sigma x \geqq \eta_\delta(s)$. *Then if* $\hat{u}$ *is a δ-approximation to u, then*

$$\hat{u}(x, s) < 3\delta \Sigma x$$

whenever $\Sigma x \geqq \eta_\delta(s)$.

Proof. We have $\hat{u}(x, s) = \sqrt{\Sigma x}$ or

$$|\hat{u}(x, s) - u(x, s)| \leqq \delta(1 + \Sigma x).$$

[3] See Aumann (1969).

In the first case there is nothing to prove, and in the second case, if $\Sigma x \geqq \eta_\delta(s)$, then by using $\eta_\delta(s) \geqq 1$ we obtain

$$\hat{u}(x, s) \leqq \delta(1 + \Sigma x) + u(x, s) < 2\delta\Sigma x + \delta\Sigma x = 3\delta\Sigma x.$$

This completes the proof of Lemma 37.1.

For each $f \in \mathfrak{F}_0$ and $x \in \Omega$, let

$$P(x;f) = \{p \in \Omega : f(y) - f(x) \leqq p\cdot(y - x) \text{ for all } y \in \Omega\}.$$

Let $\xi(x;f)$ be the infimum[1] value that any coordinate of any point in $P(x;f)$ can achieve; more precisely,

$$\xi(x;f) = \min_i \inf \{p_i : p \in P(x;f)\}.$$

If x is an interior point of Ω and f is concave and differentiable, then $P = P(x;f)$ contains precisely one point, namely the gradient $f'(x)$; in that case $\xi = \xi(x;f)$ is simply the smallest partial derivative $f^i(x)$. If f is differentiable but not concave, then P can contain at most a single point (namely the gradient), but may also be empty. If it is concave but not necessarily differentiable, then P is non-empty, and consists of the points p in Ω such that $(p, -1)$ is the normal vector to a hyperplane that supports the subgraph[2] of f at the point $(x, f(x))$. This, in fact, is the general characterization of P, also when f need be neither differentiable nor concave, and when x may be on the boundary of Ω.

LEMMA 37.2. *Let $u \in \mathfrak{U}_0$ be increasing for each fixed s. Then for each $\epsilon > 0$ and each real α there is a $\delta > 0$ such that if $\hat{u} \in \mathfrak{U}_0$ is a δ-approximation to u, $S \in \mathfrak{C}$ is such that $\mu(S) \geqq \epsilon$, and $\mathbf{x}$ is an integrable function from I to Ω such that*

$$\xi(\mathbf{x}(s); \hat{u}_s) < \delta$$

for all $s \in S$, then

$$\Sigma \int_S \mathbf{x} > \alpha.$$

Proof. First we prove:

(37.3) If C is a compact subset of Ω, then for each s there is a $\delta = \delta(C, s) > 0$ such that $\xi(x;f) \geqq \delta$ for all $x \in C$ and all $f \in \mathfrak{F}_0$ such that $\|f - u_s\| \leqq \delta$.

[1] As usual, the infimum of the empty set is taken to be $+\infty$; thus if $P(x;f) = \emptyset$ then $\xi(x;f) = +\infty$.

[2] The set of all points underneath or on the graph.

Indeed, if not, let $\{x_1, x_2, \ldots\}$ be a sequence in C, and $\{f_1, f_2, \ldots\}$ a sequence in $\mathfrak{F}_1$ such that $\|f_k - u_s\| \to 0$ and $\xi(x_k; f_k) \to 0$. Let x_0 be a limit point of $\{x_k\}$, w.l.o.g. a limit. Further, assume w.l.o.g. that $\xi(x_k; f_k)$ is "assumed at p^1" for all j, i.e. that

$$\inf\{p^1: p \in P(x_k; f_k)\} = \xi(x_k; f_k).$$

It follows that for $k = 1, 2, \ldots$, there is a $p_k \in P(x_k; f_k)$ such that $p_k^1 < \xi(x_k; f_k) + \frac{1}{k}$; then $p_k^1 \to 0$, and

$$f_k(y) - f_k(x_k) \leqq p_k \cdot (y - x_k)$$

for all $y \in \Omega$. Now for $k = 0, 1, 2, \ldots$ set $y_k = x_k + (1, 0, \ldots, 0)$. Since $y_k \in \Omega$, we have

$$f_k(y_k) - f_k(x_k) \leqq p_k \cdot (y_k - x_k) = p_k^1 \to 0. \tag{37.4}$$

Now

$$u(y_k, s) - f_k(y_k) \leqq (1 + \Sigma y_k)\|f_k - u_s\|$$

and

$$f_k(x_k) - u(x_k, s) \leqq (1 + \Sigma x_k)\|f_k - u_s\|.$$

Hence

$$u(y_k, s) - u(x_k, s) \leqq [f_k(y_k) - f_k(x_k)] + [(1 + \Sigma y_k) + (1 + \Sigma x_k)]\|f_k - u_s\|.$$

Since C is bounded and $x_k \in C$, it follows that $1 + \Sigma x_k$ is bounded; hence also $1 + \Sigma y_k = 2 + \Sigma x_k$ is bounded. Since $\|f_k - u_s\| \to 0$ by assumption, the second term of the right side of this inequality approaches 0. The first term is non-negative because f is non-decreasing, and so by (37.4), it approaches 0 as well. Hence the left side tends to 0, and from the continuity of u it follows that $u(y_0, s) - u(x_0, s) \leqq 0$, contradicting the fact that u_s is increasing. This contradiction establishes (37.3).

Let us now define $\gamma = \gamma(s, \delta)$ by

$$\gamma(s, \delta) = \inf\{\Sigma x: (\exists f \in \mathfrak{F}_0)(\|f - u_s\| \leqq \delta \text{ and } \xi(x; f) < \delta)\}. \tag{37.5}$$

Clearly γ is non-decreasing as δ decreases. Suppose γ is bounded as $\delta \to 0$, say $\gamma < \gamma_0\ (=\gamma_0(s))$. Then in the compact set

$$C = \{x \in \Omega: \Sigma x \leqq \gamma_0\},$$

$\xi(x; f)$ comes arbitrarily close to 0 for f arbitrarily close in norm to u_s, contradicting (37.3). Hence for each s,

$$\gamma(s, \delta) \to \infty \text{ as } \delta \to 0. \tag{37.6}$$

If the lemma is false, then for each k there is a set S_k of measure $\geqq \epsilon$, a $\frac{1}{k}$-approximation $\hat{u}$ to u, and an integrable function $\mathbf{x}_k$ such that

$$\xi(\mathbf{x}_k(s); \hat{u}_s) < \frac{1}{k}$$

for $s \in S_k$ and

$$\Sigma \int_{S_k} \mathbf{x}_k \leqq \alpha. \tag{37.7}$$

Then $\|\hat{u}_s - u_s\| \leqq 1/k$ for all s except for s in a set V_k, where $\mu(V_k) \leqq 1/k$. From (37.5) it then follows that for $s \in S_k \backslash V_k$, we have $\Sigma \mathbf{x}_k(s) \geqq \gamma(s, 1/k)$, and so from (37.6) we deduce that for such s,

$$\Sigma \mathbf{x}_k(s) \to \infty \quad \text{as} \quad k \to \infty.$$

Now define $\mathbf{g}_k \colon I \to E^1$ by

$$\mathbf{g}_k(s) = \begin{cases} \Sigma \mathbf{x}_k(s) & \text{if } s \in S_k \backslash V_k \\ k & \text{otherwise.} \end{cases}$$

Then for each $s \in I$, $\mathbf{g}_k(s) \to \infty$ as $k \to \infty$. Hence from Egoroff's theorem it follows that $\mathbf{g}_k(s) \to \infty$ as $k \to \infty$ *uniformly* for s in a subset U of I of measure $1 - \frac{1}{2}\epsilon$; thus for $s \in U$ we have $\mathbf{g}_k(s) \geqq \gamma_k \to \infty$, say. In particular, it follows that for $s \in (S_k \cap U) \backslash V_k$, we have

$$\Sigma \mathbf{x}_k(s) \geqq \gamma_k \to \infty.$$

From this and (37.7) we set

$$\alpha \geqq \Sigma \int_{S_k} \mathbf{x}_k = \int_{S_k} \Sigma\, \mathbf{x}_k \geqq \int_{(S_k \cap U) \backslash V_k} \Sigma\, \mathbf{x}_k \geqq \left(\frac{\epsilon}{2} - \frac{1}{k}\right) \gamma_k \to \infty,$$

an absurdity. This completes the proof of Lemma 37.2.

LEMMA 37.8. *Let $u \in \mathfrak{U}_0$ be increasing for each fixed s. Then for each $\epsilon > 0$ and each $\alpha \geqq 0$ there is a $\delta > 0$ and an integrable function $\boldsymbol{\eta}$ such* *that if* $S \in \mathfrak{C}$ *is such that* $\mu(S) \geqq \epsilon$, *b in* Ω *satisfies* $\Sigma b < \alpha$, $\hat{u} \in \mathfrak{U}_0$ *is a* δ*-approximation to u, and* $\hat{u}_S(b)$ *is attained at* $\hat{\mathbf{y}}$, *then*

$$\hat{\mathbf{y}}(s) \leqq \boldsymbol{\eta}(s)e$$

for almost all $s \in S$.

Proof. Lemma 37.2 yields a δ—which we call δ_1 to distinguish it from the δ of this lemma—that obeys the conclusions of that lemma. Let $\delta = \delta_1/3n$. Because of Lemma 37.1, there is an integrable function $\boldsymbol{\eta}$ such that $\hat{u}(z, s) < 3\delta\Sigma z$ whenever $\Sigma z \geqq \boldsymbol{\eta}(s)$ and $\hat{u}$ is a δ-approximation to u. We will prove that this δ and $\boldsymbol{\eta}$ satisfy Lemma 37.8.

Suppose that they do not. Then there is an S with $\mu(S) \geqq \epsilon$, a δ-approximation $\hat{u}$ to u, a j with $1 \leqq j \leqq n$, a b in Ω with $\Sigma b < \alpha$, and a subset U of S of positive measure such that

$$\hat{\mathbf{y}}^j(s) > \boldsymbol{\eta}(s)$$

for all $s \in U$. W.l.o.g. we may assume that $\hat{\mathbf{y}}^j(s) = \max_i \hat{\mathbf{y}}^i(s)$ for all s in U. Now for fixed but arbitrary s_0 in U, let x be the vector whose jth coordinate vanishes and all of whose other coordinates are equal to the corresponding coordinates of $\hat{\mathbf{y}}(s_0)$. Let p be the price vector corresponding to $\hat{\mathbf{y}}$ and S in accordance with Proposition 36.4. Then

$$\begin{aligned} -\hat{u}(\hat{\mathbf{y}}(s_0), s_0) \leqq \hat{u}(x, s_0) - \hat{u}(\hat{\mathbf{y}}(s_0), s_0) &\leqq p\cdot(x - \hat{\mathbf{y}}(s_0)) \\ &= p^j(x^j - \hat{\mathbf{y}}^j(s_0)) = -p^j\hat{\mathbf{y}}^j(s_0). \end{aligned}$$

But since $\Sigma\hat{\mathbf{y}}(s_0) \geqq \hat{\mathbf{y}}^j(s_0) > \boldsymbol{\eta}(s_0)$, it follows that

$$\hat{u}(\hat{\mathbf{y}}(s_0), s_0) < 3\delta\Sigma\hat{\mathbf{y}}(s_0) \leqq 3n\delta \max_i \hat{\mathbf{y}}^i(s_0) = \delta_1\hat{\mathbf{y}}^j(s_0);$$

hence $p^j\hat{\mathbf{y}}^j(s_0) < \delta_1\hat{\mathbf{y}}^j(s_0)$, and therefore $p^j < \delta_1$. But we have chosen p so that $p \in P(\mathbf{y}(s); \hat{u}_s)$ for all $s \in S$. Hence for $s \in S$, we have

$$\xi(\hat{\mathbf{y}}(s); \hat{u}_s) \leqq p^j < \delta_1.$$

Furthermore, since $\delta = \delta_1/3n < \delta_1$ and $\hat{u}$ is a δ-approximation to u, it is a fortiori a δ_1-approximation. Since $\mu(S) \geqq \epsilon$, it follows from Lemma 37.2 that $\Sigma\int_S \hat{\mathbf{y}} > \alpha > \Sigma b$, contradicting $\int_S \hat{\mathbf{y}} = b$. This proves Lemma 37.8.

LEMMA 37.9. *Let $u \in \mathcal{U}_0$. Then for any ϵ, there is a μ-integrable real function ζ, such that for all s in I and all x in Ω,*

$$u(x, s) \leqq \epsilon(\zeta(s) + \Sigma x).$$

Proof. Let ζ be an integrable real function such that $u(y, s) \leqq \epsilon\Sigma y$ whenever $\Sigma y \geqq \zeta(s)$; such a ζ exists because (35.1) must hold integrably in s. Then because u_s is non-decreasing, we have

$$u(x, s) \leqq u\left(x + \frac{\zeta(s)}{n} e, s\right) \leqq \epsilon(\zeta(s) + \Sigma x),$$

as was to be proved.

COROLLARY 37.10. *Let $u \in \mathcal{U}_0$. Then if $\mathbf{x}$ is integrable, so is $u(\mathbf{x})$.*

PROPOSITION 37.11. *Let $u \in \mathcal{U}_0$. Then for each $\epsilon > 0$, there is a $\delta > 0$, such that if $\hat{u} \in \mathcal{U}_0$ is a δ-approximation to u, then for all $S \in \mathcal{C}$ and all $b \in \Omega$ we have*

$$|u_S(b) - \hat{u}_S(b)| < \epsilon(1 + \Sigma b).$$

Proof. Let $u_1(x, s) = \sqrt{\Sigma x}$. Apply Lemma 37.9 using $\epsilon/3$ instead of ϵ, both to u and to u_1, obtaining functions ζ and ζ_1 with

$$u(x, s) \leqq \frac{\epsilon}{3}(\zeta(s) + \Sigma x)$$

$$u_1(x, s) \leqq \frac{\epsilon}{3}(\zeta_1(s) + \Sigma x).$$

Next, choose δ sufficiently small so that $\int_U(\zeta + \zeta_1) \leqq 2$ whenever $\mu(U) \leqq \delta$, and also so that $\delta \leqq \epsilon/3$. Letting U be the exceptional set in the definition of δ-approximation, we obtain for any $\mathbf{x}$,

$$\begin{aligned}
\int_S |u(\mathbf{x}) - \hat{u}(\mathbf{x})| &\leqq \int_{S\setminus U} \frac{\epsilon}{3}(1 + \Sigma\mathbf{x}) + \int_{U\cap S}(u(\mathbf{x}) + u_1(\mathbf{x})) \\
&\leqq \frac{\epsilon}{3}\int(1 + \Sigma\mathbf{x}) + \int_U \frac{\epsilon}{3}(\zeta + \zeta_1) + \int_U \frac{2\epsilon}{3}\Sigma\mathbf{x} \\
&\leqq \frac{\epsilon}{3}\left(1 + \Sigma\int\mathbf{x}\right) + \frac{2\epsilon}{3}\left(1 + \Sigma\int\mathbf{x}\right) \\
&= \epsilon\left(1 + \Sigma\int\mathbf{x}\right).
\end{aligned}$$

Now let $u_S(b)$ and $\hat{u}_S(b)$ be achieved at $\mathbf{y}$ and $\hat{\mathbf{y}}$, respectively. Then $b = \int_S \mathbf{y} = \int_S \hat{\mathbf{y}}$, and we have

$$\begin{aligned} u_S(b) &= \int_S u(\mathbf{y}) \geqq \int_S u(\hat{\mathbf{y}}) = \int_S \hat{u}(\hat{\mathbf{y}}) + \int_S (u(\hat{\mathbf{y}}) - \hat{u}(\hat{\mathbf{y}})) \\ &\geqq \int_S \hat{u}(\hat{\mathbf{y}}) - \int_S |u(\hat{\mathbf{y}}) - \hat{u}(\hat{\mathbf{y}})| \geqq \int_S \hat{u}(\hat{\mathbf{y}}) - \epsilon\left(1 + \Sigma \int_S \hat{\mathbf{y}}\right) \\ &= \hat{u}_S(b) - \epsilon(1 + \Sigma b). \end{aligned}$$

Hence $\hat{u}_S(b) - u_S(b) \leqq \epsilon(1 + \Sigma b)$. Similarly

$$\begin{aligned} \hat{u}_S(b) &= \int_S \hat{u}(\hat{\mathbf{y}}) \geqq \int_S \hat{u}(\mathbf{y}) = \int_S u(\mathbf{y}) + \int_S (\hat{u}(\mathbf{y}) - u(\mathbf{y})) \\ &\geqq \int_S u(\mathbf{y}) - \epsilon\left(1 + \Sigma \int_S \mathbf{y}\right) = u_S(b) - \epsilon(1 + \Sigma b), \end{aligned}$$

and so $u_S(b) - \hat{u}_S(b) \leqq \epsilon(1 + \Sigma b)$. This completes the proof of Proposition 37.11.

We close this section with a proposition (Proposition 37.13) that is a consequence of Lemma 37.8, though not directly connected with the concept of δ-approximation. First we require another lemma.

LEMMA 37.12. *Let $f \in \mathfrak{F}_0$ be increasing. Then the concavification f^* of f is also increasing.*

Proof. Let $x \in \Omega$. Since f is spannable (Proposition 36.2), there exist points $x_1, \ldots, x_k$ in Ω, and positive numbers $\alpha_1, \ldots, \alpha_k$ summing to 1, such that

$$\sum_{i=1}^{k} \alpha_i x_i = x$$

and

$$\sum_{i=1}^{k} \alpha_i f(x_i) = f^*(x).$$

If $y \geq x$ and $z = y - x$, then $z \geq 0$, and we have

$$\sum_{i=1}^{k} \alpha_i (x_i + z) = x + z = y$$

and so by the concavity of f^* and the fact that f is increasing, we get

$$f^*(y) \geqq \sum_{i=1}^{k} \alpha_i f^*(x_i + z) \geqq \sum_{i=1}^{k} \alpha_i f(x_i + z) > \sum_{i=1}^{k} \alpha_i f(x_i) = f^*(x).$$

This completes the proof of Lemma 37.12.

PROPOSITION 37.13. *Let $u \in \mathcal{U}_0$ be increasing for each s. Then for each $S \in \mathcal{C}$, u_S is continuous on Ω.*

Proof. Let $b \in \Omega$; we wish to prove that u_S is continuous at b. Let

$$L = \{i: b^i > 0\}, \quad M = \{i: b^i = 0\},$$

and let

$$\Omega^L = \{a \in \Omega: a^i = 0 \quad \text{for all } i \in M\},$$
$$\Omega^M = \{a \in \Omega: a^i = 0 \quad \text{for all } i \in L\};$$

thus $b \in \Omega^L$. Our proof will proceed in two stages: First, we show that

(37.14) $\quad u_S|\Omega^L$ is continuous at b.

Second, setting $\Omega_b^L = \{a \in \Omega^L: b/2 \leqq a \leqq 2b\}$, we shall show that

(37.15) for every ϵ there is a δ such that if $c \in \Omega^M$, $\|c\| \leqq \delta$, and $a \in \Omega_b^L$, then $u_S(a + c) - u_S(a) < \epsilon$.

Together, (37.14) and (37.15) prove the desired continuity of u_S at b.

In proving (37.14), we will never "leave" the space Ω^L; therefore we may assume w.l.o.g. that $\Omega^L = \Omega$, i.e. that $b > 0$. Then (37.14) turns into the assertion that u_S is continuous at b. By Proposition 36.3, u_S is concave on Ω, and since $b > 0$, it follows that b is in the interior of Ω. Since every concave function is continuous in the interior of its domain of definition, it follows that u_S is continuous at b, and so (37.14) is proved.

Next, we prove (37.15). By Proposition 36.3 and Lemma 37.12, we may assume w.l.o.g. that u is concave for each fixed s (otherwise, replace it by its concavification u^*). Suppose now that (37.15) is false. Then we can find an $\epsilon > 0$, a sequence $\delta_j \to 0$, and sequences $\{c_j\}$ and $\{a_j\}$ such that $c_j \in \Omega^M$, $\|c_j\| \leqq \delta_j$, $a_j \in \Omega_b^L$, and

(37.16) $$u_S(a_j + c_j) - u_S(a_j) \geqq \epsilon.$$

Since a_j is in Ω_b^L, which is compact, it follows that $\{a_j\}$ has a limit point a in Ω_b^L; w.l.o.g. let it be the limit. Note that since $a \in \Omega_b^L$, we have $a^i > 0$ for all $i \in L$; hence applying (37.14) to a instead of b, we get that

(37.17) $$u_S(a_j) \to u_S(a)$$

as $j \to \infty$.

Now let $u_S(a_j + c_j)$ be attained at $\mathbf{y}_j$. From Lemma 37.8 it follows that there is an integrable function η such that $\mathbf{y}_j(s) \leqq \eta(s)e$ for all j and almost all s. The space of all integrable functions on S can be considered as $L^1(S \times \{1, \ldots, n\})$. Since the set of all $\mathbf{x}$ in this space such that $0 \leqq \mathbf{x}(s) \leqq \eta(s)e$ a.e. is weakly sequentially compact,[3] it follows that the sequence $\{\mathbf{y}_n\}$ has a subsequence that is weakly convergent, say to $\mathbf{y}$. Then there is a sequence of functions converging strongly (i.e. in the L^1-norm) to $\mathbf{y}$, each one of which is a (finite) convex combination of $\mathbf{y}_1, \mathbf{y}_2, \ldots$.[4] Now every strongly convergent sequence in L^1 has a subsequence that converges a.e. to the same limit; so there is a sequence $\{\mathbf{z}_j\}$ of convex combinations of $\mathbf{y}_1, \mathbf{y}_2, \ldots$ that converges a.e. to $\mathbf{y}$. Since $\mathbf{y}_j(s) \leqq \eta(s)e$ a.e., it follows also that $\mathbf{z}_j(s) \leqq \eta(s)e$ a.e. Hence $u(\mathbf{z}_j)$ is pointwise $\leqq u(\eta e)$, which is integrable by Corollary 37.10. Moreover, from the continuity of $u(\cdot, s)$ for each fixed s, it follows that $u(\mathbf{z}_j(s), s) \to u(\mathbf{y}(s), s)$ a.e. as $j \to \infty$. Hence the Lebesgue dominated convergence theorem applies, and we deduce

$$\int_S u(\mathbf{z}_j) \to \int_S u(\mathbf{y}). \tag{37.18}$$

Now from the concavity of u for each fixed s and the fact that the $\mathbf{z}_j$ are convex combinations of the $\mathbf{y}_j$, it follows that

$$\int_S u(\mathbf{z}_j) \geqq \min_j \int_S u(\mathbf{y}_j) = \min_j u_S(a_j + c_j).$$

Hence by (37.18), it follows that

$$\int_S u(\mathbf{y}) \geqq \min_j u_S(a_j + c_j).$$

But if we had chopped off any finite number of terms from the originally given sequences $\{a_j\}$ and $\{c_j\}$, this would not have changed $\mathbf{y}$ nor any of the foregoing considerations. Hence for all k we have

$$\int_S u(\mathbf{y}) \geqq \min_{j \geqq k} u_S(a_j + c_j),$$

and letting $k \to \infty$, we deduce

$$\int_S u(\mathbf{y}) \geqq \liminf_{j \to \infty} u_S(a_j + c_j).$$

[3] Dunford and Schwartz (1958), p. 292, Theorem IV.8.9.
[4] *Ibid*, p. 422, Corollary V.3.14.

Applying (37.16) and (37.17), we then deduce

$$\int_S u(\mathbf{y}) \geqq u_S(a) + \epsilon. \tag{37.19}$$

On the other hand, since $\mathbf{y}_j \to \mathbf{y}$ weakly, we have

$$\int_S \mathbf{y} = \lim_{j\to\infty} \int_S \mathbf{y}_j = \lim_{j\to\infty} (a_j + c_j) = a + \lim_{j\to\infty} c_j = a.$$

Thus by definition we must have $\int_S u(\mathbf{y}) \leqq u_S(a)$, in contradiction to (37.19). This completes the proof of Proposition 37.13.

38. The Derivatives of the Function u_S

In this section we shall establish the existence and some continuity properties of the derivatives of the function u_S.

PROPOSITION 38.1. *Let $u \in \mathfrak{U}_1$ and $S \in \mathfrak{C}$. Then for each j such that $1 \leqq j \leqq n$, the partial derivative $u_S^j = \partial u_S/\partial x^j$ exists at each point $b \in \Omega$ such that $b^j > 0$.*

Proof. Without loss of generality let $j = 1$. Because of the concavity of u_S (Proposition 36.3),

$$\alpha = \lim_{\delta\to 0+} (u_S(b + \delta e_1) - u_S(b))/\delta$$

and

$$\beta = \lim_{\delta\to 0-} (u_S(b + \delta e_1) - u_S(b))/\delta$$

both exist, though they may a priori be different; in any case we have $\alpha \leqq \beta$. If $\alpha = \beta$ our theorem is proved, so let us assume $\alpha < \beta$.

We now show that[1]

$$u_S(b + \gamma e_1) - u_S(b) \leqq \min(\alpha\gamma, \beta\gamma) \tag{38.2}$$

for all $\gamma > -b^1$. Indeed, suppose that

$$u_S(b + \gamma e_1) - u_S(b) > \alpha\gamma$$

for some $\gamma > 0$ or

$$u_S(b + \gamma e_1) - u_S(b) > \beta\gamma$$

[1] The right side of (38.2) is $\alpha\gamma$ when $\gamma > 0$ and $\beta\gamma$ when $\gamma < 0$.

for some $\gamma < 0$. In the first case we will have

$$u_S(b + \gamma e_1) - u_S(b) = \alpha'\gamma \tag{38.3}$$

for some $\alpha' > \alpha$ and some $\gamma > 0$. Now the left side of (38.3) is a concave function of γ that vanishes for $\gamma = 0$, and hence

$$u_S(b + \delta e_1) - u_S(b) \geqq \alpha'\delta$$

for all δ such that $0 \leqq \delta \leqq \gamma$. Hence the right-hand partial derivative of u_S at b is $\geqq \alpha' > \alpha$, contradicting the fact that it equals α. In the second case a contradiction is similarly obtained, and (38.2) is proved.

Suppose now that $u_S(b)$ is attained at $\mathbf{y}$. Then we claim that for almost all $s \in S$ and all $\gamma > -\mathbf{y}^1(s)$,

$$u(\mathbf{y}(s) + \gamma e_1, s) - u(\mathbf{y}(s), s) \leqq \min(\alpha\gamma, \beta\gamma). \tag{38.4}$$

Indeed, if this is not so, then for each s in a subset U of positive measure, there is a $\boldsymbol{\gamma}(s)$ such that

$$u(\mathbf{y}(s) + \boldsymbol{\gamma}(s)e_1, s) - u(\mathbf{y}(s), s) > \min(\alpha\boldsymbol{\gamma}(s), \beta\boldsymbol{\gamma}(s)).$$

By the measurable choice theorem (Proposition 36.5), we may assume that $\boldsymbol{\gamma}(s)$ is measurable, and clearly it may be chosen integrable. Furthermore, either $\boldsymbol{\gamma}(s) > 0$ in a set of positive measure, or $\boldsymbol{\gamma}(s) < 0$ in a set of positive measure. In the first case, let V be that set, define $\mathbf{z}(s) = \mathbf{y}(s) + \boldsymbol{\gamma}(s)e_1$ for $s \in V$, $\mathbf{z}(s) = \mathbf{y}(s)$ otherwise. Setting $c = \int_S \mathbf{z}$ and $\gamma = \int_S \boldsymbol{\gamma}$, note that $\int_S u(\mathbf{z}) \leqq u_S(c)$ and that $c = b + \gamma e_1$; hence

$$\begin{aligned} u_S(b + \gamma e_1) - u_S(b) &= u_S(c) - u_S(b) \geqq \int_S (u(\mathbf{z}) - u(\mathbf{y})) \\ &= \int_V (u(\mathbf{z}) - u(\mathbf{y})) > \alpha\gamma, \end{aligned}$$

contradicting (38.2). In the second case (when $\boldsymbol{\gamma}(s) < 0$ in a set of positive measure), a contradiction is similarly obtained, using β instead of α. This establishes (38.4).

Since $\int_S \mathbf{y}^1 = b^1 > 0$, there must be some set of s of positive measure in which $\mathbf{y}^1(s) > 0$. Now from (38.4), it follows that for almost all $s \in S$ with $\mathbf{y}^1(s) > 0$, the right-hand derivative of u_s w.r.t. x^1 at $x = \mathbf{y}(s)$ is $\leqq \alpha$, and the left-hand derivative is $\geqq \beta$. Since $\beta > \alpha$, these two derivatives are unequal. So u_s is not differentiable at $\mathbf{y}(s)$ w.r.t. x^1, contrary to $u \in \mathfrak{U}_1$. This proves Proposition 38.1.

If f is a function differentiable at a point of E^n, we will denote by f' the vector $(f^1, \ldots, f^n)$ of its partial derivatives, i.e. its gradient. In particular, we have

$$u'_S = (u^1_S, \ldots, u^n_S).$$

PROPOSITION 38.5. *Let $u \in \mathfrak{U}_1$, let $b \in \Omega$ be >0, let $S \in \mathcal{C}$, and let $u_S(b)$ be attained at* $\mathbf{y}$. *Then for all $j = 1, \ldots, n$, and almost all $s \in S$ we have $u^j_S(b) = u^j(\mathbf{y}(s), s)$ when $\mathbf{y}^j(s) > 0$. Furthermore, for all $x \in \Omega$ and almost all $s \in S$ we have*

$$u(x, s) - u(\mathbf{y}(s), s) \leqq u'_S(b) \cdot (x - \mathbf{y}(s)).$$

Proof. By Proposition 36.4, there is a vector $p > 0$ such that

$$u(x, s) - u(\mathbf{y}(s), s) \leqq p \cdot (x - \mathbf{y}(s)) \tag{38.6}$$

for all $x \in \Omega$ and almost all s in S; furthermore, $p^j = u^j_s(\mathbf{y}(s))$ for almost all s for which $\mathbf{y}^j(s) > 0$.

Now for an arbitrary $\gamma \geqq -b^j$, let $u_S(b + \gamma e_j)$ be attained at $\mathbf{z}$. Then by (38.6), a.e.

$$u(\mathbf{z}(s), s) - u(\mathbf{y}(s), s) \leqq p \cdot (\mathbf{z}(s) - \mathbf{y}(s)).$$

Integrating this inequality over S, we obtain

$$u_S(b + \gamma e_j) - u_S(b) \leqq p \cdot (b + \gamma e_j - b) = \gamma p^j.$$

By Proposition 38.1, the partial derivative u^j_S exists. Letting $\gamma \to 0+$, we deduce $u^j_S(b) \leqq p^j$; letting $\gamma \to 0-$, we deduce $u^j_S(b) \geqq p^j$. Hence $u^j_S(b) = p^j$ and Proposition 38.5 is proved.

The next proposition asserts that the gradient $u'_S(b)$ is continuous in b, and that this continuity has certain uniformity properties, both in b and in S.

For $\epsilon > 0$ and $\alpha > 0$, we denote

$$A(\epsilon, \alpha) = \{x \in \Omega : x \geqq \epsilon e \text{ and } \Sigma x \leqq \alpha\}.$$

PROPOSITION 38.7. *Let $u \in \mathfrak{U}_1$. Then for every $\epsilon > 0$ and every $\alpha > 0$ there is a $\delta > 0$ such that for all S with $\mu(S) \geqq \epsilon$ and all b and c in $A(\epsilon, \alpha)$ with $\|b - c\| \leqq \delta$, we have*

$$\|u'_S(b) - u'_S(c)\| < \epsilon.$$

Outline of Proof. It is not difficult to prove that a function that is concave and possesses all its partial derivatives at every point in the interior of Ω is necessarily continuously differentiable there (cf. Proposition 39.1). The function u_S satisfies these conditions (Propositions 36.3 and 38.1), and so it is continuously differentiable in the interior of Ω; since $A(\epsilon, \alpha)$ is compact, the continuity must be uniform w.r.t. b in $A(\epsilon, \alpha)$. Unfortunately, this line of argument will not yield the uniformity of the continuity w.r.t. S, which is essential for the applications in Section 40. We must therefore use a different attack.

Let b and c in $A(\epsilon, \alpha)$ be close to each other and let $u_S(b)$ and $u_S(c)$ be attained at $\mathbf{y}$ and $\mathbf{z}$ respectively.[2] Let

$$\mathbf{\Psi}(s) = (u'_S(c) - u'_S(b))\cdot(\mathbf{y}(s) - \mathbf{z}(s));$$

from Proposition 38.5 it follows that $\mathbf{\Psi}$ is non-negative. Since

$$\int \mathbf{\Psi} = (u'_S(c) - u'_S(b))\cdot(b - c),$$

it follows that $\int \mathbf{\Psi}$ is small. But since $\mathbf{\Psi}$ is non-negative, it follows that $\mathbf{\Psi}(s)$ itself is usually[3] small.

Suppose now that the conclusion of the proposition is false, i.e. that for some j, $u^j_S(b)$ and $u^j_S(c)$ are not close; say $u^j_S(b)$ is considerably larger than $u^j_S(c)$. Since $\mathbf{\Psi}(s)$ is usually small, it follows that $\mathbf{y}^j(s)$ is usually close to $\mathbf{z}^j(s)$. Furthermore, from Lemma 37.8 we know that $\mathbf{y}^j$ and $\mathbf{z}^j$ are bounded by some η, so they cannot usually vanish; it follows that there must be some s for which $\mathbf{z}^j(s)$ is not close to 0, and moreover, $\mathbf{y}^j(s)$ and $\mathbf{z}^j(s)$ are close. If we now proceed in the positive x^j-direction from $\mathbf{z}^j(s)$, then near $\mathbf{z}(s)$, u will be rising at a rate given approximately[4] by

$$\left[\frac{\partial}{\partial x^j} u(x, s)\right]_{x=\mathbf{z}(s)};$$

by Proposition 38.5, this is equal to $u^j_S(c)$. On the other hand, if k is a coordinate for which $\mathbf{y}^k(s)$ differs considerably[5] from $\mathbf{z}^k(s)$, then since

[2] If it could be shown that for some s, $\mathbf{y}(s)$ and $\mathbf{z}(s)$ are close to each other and neither almost vanishes, then our result would follow from the continuous differentiability of $u(\cdot, s)$ and Proposition 38.5. But this is not necessarily true.

[3] I.e. for all s except for a set of small measure.

[4] Because $u(\cdot, s)$ is continuously differentiable.

[5] If there is such a coordinate. If not, then $\mathbf{y}(s)$ is close to $\mathbf{z}(s)$, and so in the argument below, H_b automatically passes close to the graph of $u(\cdot, s)$ at $\mathbf{z}(s)$.

$\Psi(s)$ is small, $u_S^k(b)$ must be close to $u_S^k(c)$. Therefore along the line connecting $\mathbf{z}(s)$ with $\mathbf{y}(s)$, the hyperplanes H_b and H_c in the $(n+1)$-dimensional (x, u)-space given respectively by

$$u = u(\mathbf{y}(s), s) + u'_S(b)\cdot(x - \mathbf{y}(s))$$

and

$$u = u(\mathbf{z}(s), s) + u'_S(c)\cdot(x - \mathbf{z}(s))$$

must be almost parallel. But these hyperplanes support the graph of $u(\cdot, s)$ (Proposition 38.5) and pass through it at the points corresponding to $\mathbf{y}(s)$ and $\mathbf{z}(s)$, respectively. Therefore H_b and H_c almost coincide along the line from $\mathbf{y}(s)$ to $\mathbf{z}(s)$, and in particular, H_b passes close to the graph of $u(\cdot, s)$ at $\mathbf{z}(s)$. But since H_b supports the graph of $u(\cdot, s)$, it then follows that the rate of rise of u in the positive x^j-direction from $\mathbf{z}(s)$ cannot be much greater than $u_S^j(b)$, at least if we average over a large enough x^j-interval. But this is in contradiction to the fact that this rate is approximately $u_S^j(c)$, as shown above.

This line of proof yields the required uniformity w.r.t. S, because the estimates involved depend on S only via the function η, which by Lemma 37.8 can be chosen to depend on u, ϵ, and α only.

Proof of Proposition 38.7. Let η correspond to ϵ and α in accordance with Lemma 37.8, let b and c be in $A(\epsilon, \alpha)$, and let $u_S(b)$ and $u_S(c)$ be attained at $\mathbf{y}$ and $\mathbf{z}$ respectively. We first wish to prove that there is a number β, depending on u, ϵ, and α only (and not on the choices of S, b or c), such that

$$\text{(38.8)} \qquad \Sigma u'_S(b) \leqq \beta \qquad \text{and} \qquad \Sigma u'_S(c) \leqq \beta.$$

Indeed, setting $x = 0$ in Proposition 38.5, we obtain

$$u'_S(b)\cdot\mathbf{y}(s) \leqq u(\mathbf{y}(s), s) \leqq u(\eta(s)e, s).$$

Integrating over S, we obtain

$$u'_S(b)\cdot\int_S \mathbf{y} \leqq \int_S u(\eta e) \leqq \int u(\eta e).$$

Since $b \in A(\epsilon, \alpha)$, we have $b \geqq \epsilon e$; therefore

$$\epsilon\Sigma u'_S(b) = u'_S(b)\cdot\epsilon e \leqq u'_S(b)\cdot b = u'_S(b)\cdot\int_S \mathbf{y} \leqq \int u(\eta e).$$

Thus $\Sigma u'_S(b) \leqq \int u(\eta e)/\epsilon$, and similarly $\Sigma u'_S(c) \leqq \int u(\eta e)/\epsilon$. Setting $\beta = \int u(\eta e)/\epsilon$, we deduce (38.8).

In the remainder of the proof, let j be a fixed index. We next claim that there is a number $\delta_1 > 0$ (depending on u, ϵ, α, and j), such that

$$\text{(38.9)} \qquad \begin{cases} \mu\{s \in S: \mathbf{y}^j(s) > \delta_1\} > \delta_1, & \text{and} \\ \mu\{s \in S: \mathbf{z}^j(s) > \delta_1\} > \delta_1. \end{cases}$$

Indeed, for a fixed δ_1, let $U = \{s \in S: \mathbf{y}^j(s) > \delta_1\}$. If $\mu(U) \leqq \delta_1$, then

$$\int_S \mathbf{y}^j \leqq \int_U \mathbf{y}^j + \int_{S\setminus U} \mathbf{y}^j \leqq \int_U \eta + \delta_1\mu(S\setminus U).$$

Since η is integrable, it follows that the right side of this inequality will be $<\epsilon$ if δ_1 is chosen sufficiently small; but this contradicts $\epsilon \leqq b^j = \int_S \mathbf{y}^j$. The same reasoning applies to $\mathbf{z}$. This proves (38.9).

Because $u^j(x, s)$ is continuous in x for each fixed s whenever $x^j > 0$, it is, for each fixed s, uniformly continuous in the set

$$\boldsymbol{C}(s) = \{x \in \Omega: x^j \geqq \delta_1,\ x \leqq \eta(s)e\}.$$

So for each fixed s we may find a number $\boldsymbol{\delta}_2(s) > 0$ such that

$$|u^j(y, s) - u^j(z, s)| < \epsilon/3$$

whenever $\|y - z\| \leqq \boldsymbol{\delta}_2(s)$ and y, $z \in \boldsymbol{C}(s)$. Furthermore, it may be shown that $\boldsymbol{\delta}_2$ may be chosen measurable, and we may assume w.l.o.g. that

$$\boldsymbol{\delta}_2(s) \leqq 1$$

for all s. Since $\boldsymbol{\delta}_2$ is measurable and bounded, it is integrable. Since it is always positive, it follows that if we define

$$\delta_3 = \inf\left\{\int_U \boldsymbol{\delta}_2 : \mu(U) \geqq \delta_1/2\right\},$$

then $\delta_3 > 0$. Finally, choose δ so that

$$\delta < \epsilon\delta_3/3n\beta.$$

Let

$$\boldsymbol{\Psi}(s) = (u'_S(c) - u'_S(b))\cdot(\mathbf{y}(s) - \mathbf{z}(s)).$$

From Proposition 38.5 it follows that for $s \in S$,

$$u'_S(b)\cdot(\mathbf{y}(s) - \mathbf{z}(s)) \leqq u(\mathbf{y}(s), s) - u(\mathbf{z}(s), s) \leqq u'_S(c)\cdot(\mathbf{y}(s) - \mathbf{z}(s));$$

hence

$$\mathbf{\Psi}(s) \geqq 0. \tag{38.10}$$

Hence by the definition of β,

$$\begin{aligned} 0 &\leqq \int_S \mathbf{\Psi} = (u'_S(c) - u'_S(b))\cdot \int (\mathbf{y} - \mathbf{z}) \\ &= (u'_S(c) - u'_S(b))\cdot(b - c) \leqq \|u'_S(c) - u'_S(b)\| n \|b - c\| \\ &\leqq \beta n\delta < \epsilon\delta_3/3 < \epsilon\delta_3/2. \end{aligned} \tag{38.11}$$

Setting

$$W = \{s \in S: \mathbf{\Psi}(s) \geqq \epsilon\boldsymbol{\delta}_2(s)/2\},$$

we obtain

$$\mu(W) < \delta_1/2; \tag{38.12}$$

for if $\mu(W) \geqq \delta_1/2$, then because $\mathbf{\Psi}$ is non-negative ((38.10)),

$$\int_S \mathbf{\Psi} \geqq \int_W \mathbf{\Psi} \geqq \frac{\epsilon}{2}\int_W \boldsymbol{\delta}_2 \geqq \epsilon\delta_3/2,$$

contradicting (38.11).

Suppose now that the proposition is false, i.e. that

$$|u^j_S(b) - u^j_S(c)| > \epsilon$$

for some j, w.l.o.g. the one we have fixed. Assume first that $u^j_S(c) - u^j_S(b) > \epsilon$. From (38.12) and (38.9) it follows that there must be an s in S such that $\mathbf{z}^j(s) > \delta_1$ and $s \notin W$. Choose such an s. Since $s \notin W$, we deduce that $\mathbf{\Psi}(s) < \epsilon\boldsymbol{\delta}_2(s)/2$. Set $\Psi = \mathbf{\Psi}(s)$, $\delta_2 = \boldsymbol{\delta}_2(s)$, $y = \mathbf{y}(s)$, $z = \mathbf{z}(s)$, $w = z + \frac{1}{2}\delta_2 e_j$, $w^\# = z + \delta_2 e_j$. Then

$$(u'_S(c) - u'_S(b))\cdot(y - w) = \Psi - (u^j_S(c) - u^j_S(b))\frac{\delta_2}{2} < \frac{\epsilon\delta_2}{2} - \frac{\epsilon\delta_2}{2} = 0.$$

Hence

$$u'_S(b)\cdot(w - y) < u'_S(c)\cdot(w - y).$$

Hence by Proposition 38.5,

$$\begin{aligned} u(w^\#, s) - u(y, s) &\leqq u'_S(b)\cdot(w^\# - y) = u'_S(b)\left(w + \frac{\delta_2}{2}e_j - y\right) \\ &= u'_S(b)\cdot(w - y) + u^j_S(b)\frac{\delta_2}{2} < u'_S(c)\cdot(w - y) + u^j_S(b)\frac{\delta_2}{2}. \end{aligned}$$

On the other hand, for an appropriate $\theta \in [0, 1]$,

$$\begin{aligned} u(w^{\#}, s) - u(y, s) &= (u(w^{\#}, s) - u(z, s)) - (u(y, s) - u(z, s)) \\ &= u^j(z + \theta\delta_2 e_j, s)\delta_2 - (u(y, s) - u(z, s)) \\ &\geqq \left(u^j(z, s) - \frac{\epsilon}{3}\right)\delta_2 - u'_S(c)\cdot(y - z), \end{aligned}$$

because of the definition of $\delta_2 = \boldsymbol{\delta}_2(s)$ and Proposition 38.5. Again using Proposition 38.5 and $z^j = \mathbf{z}^j(s) > 0$, we deduce that

$$\begin{aligned} u(w^{\#}, s) - u(y, s) &\geqq u^j_S(c)\delta_2 - \frac{\epsilon\delta_2}{3} - u'_S(c)\cdot(y - z) \\ &= u^j_S(c)\,\frac{\delta_2}{2} - \frac{\epsilon\delta_2}{3} + u'_S(c)\cdot(w - y). \end{aligned}$$

Combining the first and third inequalities for $u(w^{\#}, s) - u(y, s)$, we deduce that

$$u^j_S(b)\,\frac{\delta_2}{2} > u^j_S(c)\,\frac{\delta_2}{2} - \epsilon\,\frac{\delta_2}{3}.$$

Hence

$$\epsilon\,\frac{\delta_2}{3} > (u^j_S(c) - u^j_S(b))\,\frac{\delta_2}{2} > \epsilon\,\frac{\delta_2}{2}.$$

Hence $2 > 3$, an absurdity. The case $u^j_S(b) - u^j_S(c) > \epsilon$ is handled similarly. This completes the proof of Proposition 38.7.

Corollary 38.13. *Let $u \in \mathfrak{U}_1$. Then for all $j = 1, \ldots, n$ and all $S \in \mathcal{C}$, u'_S is continuous at each b in Ω with $b > 0$.*

Proposition 38.14. *Let $u \in \mathfrak{U}_1$. Then for every $\epsilon > 0$ and every $\alpha > 0$ there is a $\delta > 0$ such that for all S with $\mu(S) \geqq \epsilon$, all b in $A(\epsilon, \alpha)$, and all δ-approximations $\hat{u}$ to u, we have*

$$\|\hat{u}'_S(b) - u'_S(b)\| < \epsilon.$$

Proof. Fix j. Let $\epsilon_1 = \frac{1}{2}\epsilon$, and let δ_1 correspond to ϵ_1 and $\alpha + \epsilon_1$ in accordance with Proposition 38.7; furthermore, choose $\delta_1 < \epsilon_1$. Let δ correspond to $\delta_1\epsilon/(4(1 + \delta_1 + \alpha))$ in accordance with Proposition 37.11. Then since $\hat{u}$ is a δ-approximation to u, we have

$$|\hat{u}_S(b) - u_S(b)| < \delta_1\epsilon/4$$

and

$$|\hat{u}_S(b + \delta_1 e_j) - u_S(b + \delta_1 e_j)| < \delta_1 \epsilon/4.$$

Hence

$$\left| \frac{\hat{u}_S(b + \delta_1 e_j) - \hat{u}_S(b)}{\delta_1} - \frac{u_S(b + \delta_1 e_j) - u_S(b)}{\delta_1} \right| < \frac{\epsilon}{2}.$$

Similarly

$$\left| \frac{\hat{u}_S(b) - \hat{u}_S(b - \delta_1 e_j)}{\delta_1} - \frac{u_S(b) - u_S(b - \delta_1 e_j)}{\delta_1} \right| < \frac{\epsilon}{2}.$$

But because of the concavity of $\hat{u}_S$ (Proposition 36.3), we have

$$\frac{\hat{u}_S(b) - \hat{u}_S(b - \delta_1 e_j)}{\delta_1} \geqq \hat{u}_S^j(b) \geqq \frac{\hat{u}_S(b + \delta_1 e_j) - \hat{u}_S(b)}{\delta_1}.$$

Thus, again because of the concavity,

$$\begin{aligned} \hat{u}_S^j(b) &\geqq \frac{u_S(b + \delta_1 e_j) - u_S(b)}{\delta_1} - \frac{\epsilon}{2} \\ &\geqq u_S^j(b + \delta_1 e_j) - \frac{\epsilon}{2} \\ &\geqq u_S^j(b) - \frac{\epsilon}{2} - \frac{\epsilon}{2} = u_S^j(b) - \epsilon. \end{aligned}$$

Similarly $\hat{u}_S^j(b) \leqq u_S^j(b) + \epsilon$. This completes the proof of Proposition 38.14.

39. The Finite Type Case

In this section we shall prove[1] Proposition 31.5.

PROPOSITION 39.1. *Let f be a continuous concave function defined on Ω, and for some j with $1 \leqq j \leqq n$, assume that $f^j = \partial f/\partial x^j$ exists at each $x \in \Omega$ such that $x^j > 0$. Then f^j is continuous at each point x such that $x^j > 0$.*

[1] Very few of the tools developed in Sections 35, 37, and 38 will be used in the process, and sometimes only special cases—which could have been established more easily than the general cases—will be used. All in all, what is needed from those sections for the finite type case could have been developed separately in a few pages. We did not do this because we wished to avoid an unnecessary duplication, and because we consider the finite type case to be chiefly a stepping stone to the general one.

Proof. Without loss of generality let $j = 1$. Suppose f^1 is not continuous, say at y, where $y^1 > 0$. Let $x_k \to y$, $f^1(x_k) \to \alpha$, $\alpha \neq f^1(y)$ (possibly $\alpha = \pm\infty$). Without loss of generality let $x_k^1 > y^1/2$ for all k. Then for all $\gamma > -y^1/2$, we have

$$f(x_k + \gamma e_1) - f(x_k) \leqq f^1(x_k)\gamma. \tag{39.2}$$

If α is finite, let $k \to \infty$ and obtain from the continuity of f that

$$f(y + \gamma e_1) - f(y) \leqq \alpha\gamma$$

for all $\gamma > -y^1/2$. Hence, because $f^1(y)$ exists it must be equal to α, a contradiction. If $\alpha = \pm\infty$, let $\gamma = \mp y^1/4$; then (39.2) yields $f(x_k + \gamma e_1) - f(x_k) \to -\infty$, whereas continuity implies that it tends to $f(y + \gamma e_1) - f(y)$. This contradiction proves Proposition 39.1.

PROPOSITION 39.3. *Let $f \in \mathfrak{F}_0$, and let $u \in \mathfrak{U}_0$ be defined by $u_s = f$ for all $s \in I$. Then for all $a \in \Omega$ we have*

$$u_I(a) = f^*(a),$$

where f^* is the concavification[2] *of f.*

Proof. Assume first that $a > 0$ and that f is concave. Let

$$G = \{(\nu, x) \in E^1 \times \Omega: \nu \leqq f(x)\}.$$

Then G is convex, and $(f(a), a)$ is on the boundary of G. So there is a hyperplane containing $(f(a), a)$ that supports G, i.e. there is a $q \in E^n$ and a $\rho \in E^1$ such that $(\rho, q) \neq 0$ and

$$\rho\cdot\nu - q\cdot x \leqq \rho\cdot f(a) - q\cdot a \tag{39.4}$$

for all $(\nu, x) \in G$. If for any j, we would have $q^j < 0$, then by setting $\nu = 0$ and letting x^j be large and $x^i = 0$ for $i \neq j$, we would get a contradiction to (39.4). Hence, $q \in \Omega$. If $\rho < 0$, then, by fixing x and letting ν be a negative number with large absolute value, we again get a contradiction to (39.4). If $\rho = 0$ then by $(\rho, q) \neq 0$ and $q \in \Omega$ we get $q \geq 0$; hence since $a > 0$, we get $q\cdot a > 0$. But if we set $\nu = 0$ and $x = 0$ in (39.4), we get $0 \leqq -q\cdot a$, which is again a contradiction. We conclude that $\rho > 0$, which permits us to divide (39.4) by ρ and obtain a $p \in \Omega$ such that $\nu - p\cdot x \leqq f(a) - p\cdot a$ for all $(\nu, x) \in G$. If in

[2] See Section 36, in particular Proposition 36.2.

particular we set $v = f(x)$ and recall that $u(x, s) = f(x)$ for all s, we deduce that

$$u(x, s) - u(a, s) \leqq q \cdot (x - a).$$

Hence by Proposition 36.4, $u_I(a)$ is attained at $\mathbf{x} \equiv a$. Since f is concave, $f^* = f$, and so the proposition is proved in this case.

When f is concave but a is not necessarily >0, then we apply the case just proved to the subspace of E^n obtained by considering only those coordinates j for which a^j is positive, and obtain the desired result.

When f is not necessarily concave, then we apply Proposition 36.3, and deduce from the concave case just proved that

$$u_I(a) = u^*{}_I(a) = f^*(a).$$

This completes the proof of Proposition 39.3.

Lemma 39.5. Let $f \in \mathfrak{F}_1$. Then the concavification f^* is also in $\mathfrak{F}_1$.

Proof. Define $u \in \mathfrak{U}_1$ by $u(x, s) = f(x)$ for all $s \in S$. By Proposition 36.3, $u_s^* \in \mathfrak{U}_0$ for all $s \in S$, and hence $f^* \in \mathfrak{F}_0$. To prove that f^* obeys the differentiability condition (35.2), note that by Proposition 39.3, $u_I(x) = f^*(x)$ for all $x \in \Omega$. Hence by Proposition 38.1, $f^{*j}(x)$ exists whenever $x^j > 0$. Since f^* is continuous and concave, it follows from Proposition 39.1 that f^{*j} is continuous at each $x \in \Omega$ for which $x^j > 0$, and so (35.2) is verified. Finally, the fact that f^* is increasing follows from Lemma 37.12. This completes the proof of Lemma 39.5.

COROLLARY 39.6. *If $u \in \mathfrak{U}_1$, then $u^* \in \mathfrak{U}_1$.*

Proof. This is an immediate consequence of Lemma 39.5 and Proposition 36.3.

Let $f_1, \ldots, f_k$ be concave members of $\mathfrak{F}_1$. Denote the non-negative orthant of E^k by Ξ. For $y \in \Xi$ and $z \in \Omega$, define

$$(39.7) \quad g(y, z) = \max \left\{ \sum_{i=1}^{k} y^i f_i(x_i) : x_1, \ldots, x_k \in \Omega \text{ and } \sum_{i=1}^{k} y^i x_i \leqq z \right\}.$$

If we set $w = (y, z)$, then $w \in \Xi \times \Omega \subset E^{n+k}$. Thus $g = g(w)$ is a function of $k + n$ non-negative real variables.

Note that the inequality sign in the constraint $\Sigma_{i=1}^k y^i x_i \leqq z$ may be replaced by an equality unless $y = 0$ and $z \geq 0$.

It is easily seen that the max in the definition of g is attained. Indeed, if $y > 0$, then the constraint set is compact; and if one or more of the coordinates of y vanish, then we can ignore those coordinates and the corresponding x_i entirely, and the constraint set for the remaining x_i will still be compact. If all the y^i vanish—i.e. if $y = 0$—then, of course, $g(y, z) = 0$, and the max is achieved for any k-tuple of x_i in Ω.

LEMMA 39.8. *Let* $\{S_1, \ldots, S_k\}$ *be a partition of* I, *and define* u *in* $\mathfrak{U}_1$ *of finite type by*

$$u(x, s) = f_i(x) \quad \textit{when } s \in S_i.$$

For $S \subset I$, *define* $y_S \in \Xi$ *by* $y_S^i = \mu(S \cap S_i)$. *Then*

$$u_S(z) = g(y_S, z)$$

for all $z \in \Omega$.

Proof. Let $g(y_S, z)$ be attained at $(x_1, \ldots, x_k)$. Define $\mathbf{x}$ by

$$\mathbf{x}(s) = x_i \quad \text{for } s \in S \cap S_i.$$

Then

$$\int_S \mathbf{x} = \sum_{i=1}^{k} y_S^i x_i \leqq z,$$

and hence

$$u_S(z) \geqq \int_S u(\mathbf{x}) = \sum_{i=1}^{k} y_S^i f_i(x_i) = g(y_S, z).$$

To obtain the opposite inequality, let $u_S(z)$ be attained at $\mathbf{x}$. Define $(x_1, \ldots, x_k)$ by

$$x_i = \begin{cases} \dfrac{1}{\mu(S \cap S_i)} \displaystyle\int_{S \cap S_i} \mathbf{x}, & \text{if } \mu(S \cap S_i) \neq 0, \\ \text{arbitrary}, & \text{if } \mu(S \cap S_i) = 0. \end{cases}$$

Then by the concavity of f_i,

$$f_i(x_i) \geqq \frac{1}{\mu(S \cap S_i)} \int_{S \cap S_i} f_i(\mathbf{x}) = \frac{1}{y_S^i} \int_{S \cap S_i} u(\mathbf{x})$$

if $\mu(S \cap S_i) \neq 0$. Furthermore

$$\sum_{i=1}^{k} y_S^i x_i = \sum_{i=1}^{k} \mu(S \cap S_i) x_i = \int_S \mathbf{x} = z.$$

Hence

$$g(y_S, z) \geqq \sum_{i=1}^{k} y_S^i f_i(x_i) = \sum_{i=1}^{k} \int_{S \cap S_i} u(\mathbf{x}) = \int_S u(\mathbf{x}) = u_S(z).$$

This complete the proof of the lemma.

Lemma 39.9. *g is concave, non-decreasing, and continuous on* $\Xi \times \Omega$.

Proof. We first prove the concavity. Indeed, let (y_1, z_1) and (y_2, z_2) be in $\Xi \times \Omega$, and let $g(y_1, z_1)$ and $g(y_2, z_2)$ be taken on at $\{x_{11}, \ldots, x_{1k}\}$ and $\{x_{21}, \ldots, x_{2k}\}$ respectively. For $0 \leqq \alpha \leqq 1$, let

$$(y, z) = \alpha(y_1, z_1) + (1 - \alpha)(y_2, z_2),$$

and for each i define

$$x_i = \begin{cases} (\alpha y_1^i x_{1i} + (1 - \alpha) y_2^i x_{2i})/y^i, & \text{if } y^i > 0, \\ 0 & \text{otherwise.} \end{cases}$$

Then

$$\begin{aligned} \sum_{i=1}^{k} y^i x_i &= \sum_{i=1}^{k} [\alpha y_1^i x_{1i} + (1 - \alpha) y_2^i x_{2i}] \\ &= \alpha \sum_{i=1}^{k} y_1^i x_{1i} + (1 - \alpha) \sum_{i=1}^{k} y_2^i x_{2i} \\ &\leqq \alpha z_1 + (1 - \alpha) z_2 = z. \end{aligned}$$

So if we let $L = \{i\colon 1 \leqq i \leqq k \text{ and } y^i > 0\}$, we obtain from the definition of g and the concavity of the f_i that

$$\begin{aligned} g(y, z) &\geqq \sum_{i=1}^{k} y^i f_i(x_i) \\ &\geqq \sum_{i \in L} y^i \left[\frac{\alpha y_1^i}{y^i} f_i(x_{1i}) + \frac{(1 - \alpha) y_2^i}{y^i} f_i(x_{2i}) \right] \\ &= \alpha \sum_{i=1}^{k} y_1^i f_i(x_{1i}) + (1 - \alpha) \sum_{i=1}^{k} y_2^i f_i(x_{2i}) \\ &= \alpha g(y_1, z_1) + (1 - \alpha) g(y_2, z_2). \end{aligned}$$

This shows that g is indeed concave.

Next, we show that g is non-decreasing in $w = (y, z)$. Indeed, suppose $w_1 \geq w_0$ and $g(w_1) < g(w_0)$. If we draw a straight ray (half-line) starting at w_0 and passing through w_1, then this ray must always stay in $\Xi \times \Omega$. On the other hand, from the concavity of g it follows that, at a point on the ray sufficiently beyond w_1, g will be negative. But it is clear from its definition that g can never be negative. This demonstrates that g is non-decreasing.

Finally, we prove the continuity of g at each point $w_0 = (y_0, z_0)$ of $\Xi \times \Omega$. Note first that g is homogeneous of degree 1. Hence we may wholly restrict ourselves to the case in which $\Sigma_{i=1}^k y_0^i \leqq \frac{1}{2}$. In that case we may find a partition $\{S_1, \ldots, S_k\}$ of I and an $S \subset I$ such that

$$y_0^i = \mu(S \cap S_i).$$

If we define u in $\mathfrak{U}_1$ as in Lemma 39.8, then from that lemma it follows that

$$u_S(z) = g(y_0, z)$$

for all z in Ω. Hence from Proposition 37.13 it follows that $g(y, z)$ is continuous in z at (y_0, z_0).

To complete the proof of continuity it is sufficient to demonstrate

(39.10) For every ϵ there is a δ such that if $\|(y, z) - (y_0, z_0)\| \leqq \delta$, then $|g(y, z) - g(y_0, z)| < \epsilon$.

So let $\epsilon > 0$ be given. For each i there is an η such that $f_i(x) \leqq \epsilon\|x\|$ whenever $\|x\| > \eta$; w.l.o.g. we may choose the same η for all i. It then follows[3] that for all x and i, we have

(39.11) $$f_i(x) < \epsilon(\eta + \Sigma x).$$

Now for given (y, z), define $\hat{y} \in \Xi$ by $\hat{y}^i = \min(y^i, y_0^i)$ for all i. Let $g(y, z)$ be attained at $(x_1, \ldots, x_k)$. Since $\hat{y} \leq y$, it follows that $(x_1, \ldots, x_k)$ satisfies the constraints in the definition of $g(\hat{y}, z)$. Hence

$$g(\hat{y}, z) \geqq \sum_{i=1}^{k} \hat{y}_i f_i(x_i).$$

[3] Compare the proof of Lemma 37.9.

Therefore, using the monotonicity of g and (39.11),

$$\begin{aligned}(39.12)\quad g(y,z)-g(\hat{y},z) &\leqq \sum_{i=1}^{k} y^i f_i(x_i) - \sum_{i=1}^{k} \hat{y}^i f_i(x_i)\\ &\leqq \epsilon\eta \sum_{i=1}^{k} (y^i - \hat{y}^i) + \epsilon \sum_{j=1}^{n}\sum_{i=1}^{k} (y^i - \hat{y}^i)x_i^j\\ &\leqq \epsilon\eta \sum_{i=1}^{k} (y^i - \hat{y}^i) + \epsilon \sum_{j=1}^{n}\sum_{i=1}^{k} y^i x_i^j\\ &= \epsilon\eta \sum_{i=1}^{k} (y^i - \hat{y}^i) + \epsilon\Sigma z.\end{aligned}$$

Now from the definition of $\hat{y}$ it follows that

$$\|y-\hat{y}\| \leqq \|y-y_0\| \leqq \|(y,z)-(y_0,z_0)\|;$$

hence if in (39.10), δ is chosen sufficiently small, then the first term on the right side of (39.12) may be made less than ϵ, say. As for the second term, if δ is chosen less than $1/n$, say, then we will have

$$\epsilon\Sigma z \leqq \epsilon(1+\Sigma z_0).$$

Thus altogether we obtain

$$g(y,z)-g(\hat{y},z) \leqq \epsilon(2+\Sigma z_0).$$

Since g is monotonic and $y \geqq \hat{y}$, it follows that $g(y,z)-g(\hat{y},z) \geqq 0$. Hence

$$|g(y,z)-g(\hat{y},z)| \leqq \epsilon(2+\Sigma z_0).$$

Similarly we obtain

$$|g(y_0,z)-g(\hat{y},z)| \leqq \epsilon(2+\Sigma z_0).$$

Hence

$$|g(y,z)-g(y_0,z)| \leqq \epsilon(4+2\Sigma z_0).$$

This gives us (39.10) with a factor of $\epsilon(4+2\Sigma z_0)$, and (39.10) follows without difficulty. Thus the proof of continuity is complete, and with it the proof of Lemma 39.9.

PROPOSITION 39.13. *For each p with $1 \leqq p \leqq k+n$, $g^p = \partial g/\partial w^p$ exists and is continuous at each $w \in \Xi \times \Omega$ for which $w^p > 0$.*

Proof. First, we show that g is differentiable in y^1 whenever $y^1 > 0$. Indeed, let $y_0 \in \Xi$, $z_0 \in \Omega$, and let $g(y_0, z_0)$ be taken on at $\{x_{01}, \ldots,$

$x_{0k}\}$. Define a function h of the positive real variable y^1 by

$$h(y^1) = y^1 f_1(x_{01}y_0^1/y^1) + \sum_{i=2}^{k} y_0^i f_i(x_{0i}).$$

Since we wish to fix attention on the variable y^1, it is convenient to set $g_1(y^1) = g(y^1, y_0^2, \ldots, y_0^k, z_0)$; in particular, therefore, $g_1(y_0^1) = g(y_0, z_0)$. Now since

$$y^1(x_{01}y_0^1/y_1) + \sum_{i=2}^{k} y_0^i x_{0i} = \sum_{i=1}^{k} y_0^i x_{0i} = z,$$

it follows from the definitions of g and g_1 that

$$h(y^1) \leqq g_1(y^1) \tag{39.14}$$

for all $y^1 > 0$, and of course

$$h(y_0^1) = g_1(y_0^1). \tag{39.15}$$

Next, note that h is differentiable whenever $y^1 > 0$; this follows from the differentiability of $f_1(x_{01}y_0^1/y^1)$ as a function of y^1, which in turn follows from the fact that $f_1^j(x)$ exists whenever $x^j > 0$. Of course it may happen that some of the coordinates of x_{01} vanish, but then any change in y^1 does not affect the corresponding coordinates of $x_{01}y_0^1/y^1$, so that the differentiability of h is not affected. In fact, if we let $M = \{j \in N: x_{01}^j > 0\}$, then

$$h'(y^1) = \frac{d}{dy^1} h(y^1) = f_1(x_{01}y_0^1/y^1) + y^1 \sum_{j \in M} f_1(x_{01}y_0^1/y_1) \left(- \frac{x_{01}^j y_0^1}{(y^1)^2} \right).$$

In particular, we obtain

$$h'(y_0^1) = f_1(x_{01}) - \sum_{j \in M} x_{01}^j f_1^j(x_{01}).$$

On the other hand, since g is concave in w, it follows that g_1 is concave in y^1; hence there is a supporting line to its graph at y_0^1, i.e. there is a linear function $l(y^1)$ such that

$$l(y^1) \geqq g_1(y^1)$$

for all $y^1 > 0$, and

$$l(y_0^1) = g_1(y_0^1).$$

Recalling (39.14) and (39.15), we find that g_1 is "trapped" between the two differentiable functions l and h at y_0^1, and so must be differentiable. The differentiability of g_1 at y_0^1 is of course the same thing as the existence of the derivative $\partial g/\partial y^1$ at (y_0, z_0). A similar argument shows that all the derivatives $\partial g/\partial y^i$ exist whenever $y^i > 0$, for each i in $\{1, \ldots, k\}$.

The existence of the partial derivatives $\partial g/\partial z^j$ for $z^j > 0$ is an easy consequence of Proposition 38.1 and Lemma 39.8.

Combining the existence of the partial derivatives with the continuity and concavity of g (Lemma 39.9), and applying Proposition 39.1, we deduce the required continuity of the derivatives, and the proof is complete.[4]

Denote by H the set of all superadditive set functions in pNA that are homogeneous of degree 1 (see Section 26).

LEMMA 39.16. *If $u \in \mathfrak{U}_1$ and u is of finite type, then v (as defined in (30.1)) is in H.*

Proof. First we show that $v \in pNA$. By Corollary 39.6, we have $u^* \in \mathfrak{U}_1$, and certainly u^* is of finite type as well. Thus there is a finite set $\{f_1, \ldots, f_k\}$ of concave functions in $\mathfrak{F}_1$ such that each u_s^* is one of the f_i. If we now define g by (39.7), then from Lemma 39.8 we obtain $u_S^*(z) = g(y_S, z)$ for all $S \subset I$ and $z \in \Omega$, where y_S is defined by

$$y_S^i = \mu(S \cap S_i)$$

and S_i is defined by

$$S_i = \{s \in I: u_s^* = f_i\}.$$

Now by Proposition 36.3, $u_S^*(z) = u_S(z)$; hence $u_S(z) = g(y_S, z)$. If we write $\eta(S)$ instead of y_S, then we see that η is a k-dimensional vector of non-atomic measures on I, and we have

$$u_S(z) = g(\eta(S), z). \tag{39.17}$$

Now define an n-dimensional vector ζ of NA-measures by $\zeta(S) = \int_S \mathbf{a}$. Substituting $\zeta(S)$ for z in (39.17) and using the definition of v (30.1),

[4] The basic idea of this proof, to prove the differentiability of a function by "trapping" it between two differentiable functions, was adapted from Shapley (1964b). (See the lemma on p. 7, and its proof on pp. 8–9.)

we obtain

$$v(S) = u_S(\zeta(S)) = g(\eta(S), \zeta(S)).$$

Letting $\nu = (\eta, \zeta)$, we see that ν is a vector of non-negative measures in NA, and that

$$v = g \circ \nu. \tag{39.18}$$

Since g is continuous and non-decreasing (Lemma 39.9), and since for all p, $\partial g/\partial w^p$ exists and is continuous whenever $w^p > 0$, it follows from Proposition 10.17 that $g \circ \nu$, and hence v, is in pNA.

To show that v is homogeneous of degree 1, use the Weierstrass approximation theorem in $k + n$ dimensions to find a sequence $\{h_j\}$ of polynomials (in $k + n$ variables) such that $|h_j(w) - g(w)| \leqq 1/j$ for all w in the range of ν. For these polynomials it follows from the defining properties of the extension operator (in particular (21.1), (21.2), and (21.3)) that

$$(h_j \circ \nu)^*(\alpha\chi_S) = h_j(\alpha\nu(S)),$$

where * denotes the extension operator (Chapter IV). Letting $j \to \infty$ and using the continuity of the extension operator in the supremum norm (22.15), we deduce that

$$v^*(\alpha\chi_S) = (g \circ \nu)^*(\alpha\chi_S) = g(\alpha\nu(S)).$$

But since it is easily verified that g is homogeneous of degree 1, it follows that

$$g(\alpha\nu(S)) = \alpha g(\nu(S)) = \alpha v(S).$$

Hence $v^*(\alpha\chi_S) = \alpha v(S)$, and so v is homogeneous of degree 1.

We have demonstrated that $v \in pNA$ and that it is homogeneous of degree 1. Since its superadditivity is obvious, the proof of Lemma 39.16 is complete.

We are now ready for the

Proof of Proposition 31.5. This follows immediately from Lemma 39.16 and Theorem I.

40. Proof of Theorem J

The proof will proceed by reducing the general case to the finite type case. In the process, we shall also prove Proposition 33.2.

LEMMA 40.1. *Let* $u \in \mathcal{U}_1$ *and* $\mathbf{a}$ *be given, where* $\mathbf{a}$ *is* μ*-integrable. Let* $S \supset S_0$*, assume* $\int_{S_0} \mathbf{a} > 0$*, and let* $v(S) = u_S(\int_S \mathbf{a})$ *be attained at* $\mathbf{y}$*. Then there is a point* c *on the straight line connecting* $\int_{S_0} \mathbf{y}$ *with* $\int_{S_0} \mathbf{a}$*, such that*

$$(40.2) \qquad v(S) - v(S_0) = \int_{S \backslash S_0} [u(\mathbf{y}) + u'_{S_0}(c) \cdot (\mathbf{a} - \mathbf{y})]$$

Furthermore, if $\int_{S_0} \mathbf{y} > 0$*, then*

$$(40.3) \qquad u'_S\left(\int_S \mathbf{a}\right) = u'_{S_0}\left(\int_{S_0} \mathbf{y}\right).$$

Proof. Set $u_S = w$, $u_{S_0} = w_0$, $\int_S \mathbf{a} = b$, $\int_{S_0} \mathbf{a} = b_0$, and let $v(S) = w(b)$ be attained at $\mathbf{y}$. Set $V = S \backslash S_0$. For $0 \leqq \theta \leqq 1$, let

$$g(\theta) = w_0\left(\theta \int_{S_0} \mathbf{y} + (1 - \theta) b_0\right).$$

The function g is continuous on the closed unit interval $[0, 1]$, and is differentiable in the interior $(0, 1)$. Indeed the continuity follows from Proposition 37.13. To prove the differentiability, let

$$c_\theta = \theta \int_{S_0} \mathbf{y} + (1 - \theta) b_0.$$

Then since $b_0 = \int_{S_0} \mathbf{a} > 0$, it follows that $c_\theta > 0$ for all $\theta \in (0, 1)$. Applying Corollary 38.13, we deduce that $w_0^j(x)$ exists and is continuous at $x = c_\theta$ for each j and each θ in $(0, 1)$. We conclude that $g'(\theta)$ indeed exists for all $\theta \in (0, 1)$; moreover

$$g'(\theta) = w'(c_\theta) \cdot \Delta,$$

where $\Delta \in E^n$ is defined by

$$\Delta = \int_{S_0} \mathbf{y} - b_0 = \left[\int_S \mathbf{y} - \int_V \mathbf{y}\right] - \left[\int_S \mathbf{a} - \int_V \mathbf{a}\right] = \int_V (\mathbf{a} - \mathbf{y}).$$

Since g is continuous on $[0, 1]$ and differentiable in $(0, 1)$, the mean value theorem applies, and we deduce that $g(1) - g(0) = g'(\theta)$ for an appropriate θ. Setting $c = c_\theta$ for that θ, we obtain

$$(40.4) \qquad w_0\left(\int_{S_0} \mathbf{y}\right) - w_0(b_0) = w_0'(c) \cdot \Delta.$$

Now $w(b)$ is attained at $\mathbf{y}$; hence by Proposition 38.5,

$$(40.5) \qquad u(x, s) - u(\mathbf{y}(s), s) \leqq w'(b) \cdot (x - \mathbf{y}(s))$$

for almost all $s \in S$ and almost all $x \in \Omega$; in particular this is true for almost all $s \in S_0$. Since trivially we have $\int_{S_0} \mathbf{y} = \int_{S_0} \mathbf{y}$, it follows from Proposition 36.4 that $w_0(\int_{S_0} \mathbf{y})$ is attained at $\mathbf{y}|S_0$; that is, we have

$$(40.6) \qquad w_0\left(\int_{S_0} \mathbf{y}\right) = \int_{S_0} u(\mathbf{y}).$$

By applying Proposition 38.5 to S_0 we obtain

$$u(x, s) - u(\mathbf{y}(s), s) \leqq w_0'\left(\int_{S_0} \mathbf{y}\right)\cdot(x - \mathbf{y}(s))$$

whenever $s \in S_0$ and $\int_{S_0} \mathbf{y} > 0$. From this and (40.5) it follows easily that if $\int_{S_0} \mathbf{y} > 0$, then (40.3) holds (since for each j there must be an s in S_0 with $\mathbf{y}^j(s) > 0$). Moreover, (40.6) and (40.4) yield

$$w_0(b_0) = -w_0'(c)\cdot\Delta + \int_{S_0} u(\mathbf{y}),$$

which is equivalent to (40.2). This completes the proof of Lemma 40.1.

LEMMA 40.7. *Let $u \in \mathfrak{U}_1$ and $\mathbf{a}$ be given, where $\mathbf{a}$ is μ-integrable. Then for every $\epsilon > 0$ there is a $\delta > 0$ such that if m is a positive integer, $\hat{u} \in \mathfrak{U}_1$ is a δ-approximation to u, and*

$$S_1 \subset \cdots \subset S_m \subset S_{m+1} = I$$

is a sequence such that $\int_{S_1} \mathbf{a} \geqq \epsilon e$ and $u(S_{k+1}\backslash S_k) \leqq \delta$ for all k, then

$$(40.8) \qquad \sum_{k=1}^{m} |v(S_{k+1}) - \hat{v}(S_{k+1}) - (v(S_k) - \hat{v}(S_k))| < \epsilon,$$

where v and $\hat{v}$ are defined by

$$(40.9) \qquad v(S) = u_S\left(\int_S \mathbf{a}\right) \text{ and } \hat{v}(S) = \hat{u}_S\left(\int_S \mathbf{a}\right) \text{ for all } S.$$

Proof. Set $A_k = v(S_{k+1}) - \hat{v}(S_{k+1}) - (v(S_k) - \hat{v}(S_k))$. We start out by fixing attention on a single k. To simplify the notation by eliminating the need for a large number of subscripts, set $S_{k+1} = S$, $u_{S_{k+1}} = w$, $\int_{S_{k+1}} \mathbf{a} = b$, and let $v(S) = w(b)$ be attained at $\mathbf{y}$. Similarly, set $S_k = S_0$, $u_{S_k} = w_0$, $\int_{S_k} \mathbf{a} = b_0$, and let $v(S_0) = w_0(b_0)$ be attained at $\mathbf{y}_0$. Adopt a similar notation $\hat{v}$, $\hat{u}$, etc. Also set $V = S\backslash S_0$, $\Delta = \int_V (\mathbf{a} - \mathbf{y})$, and $\hat{\Delta} = \int_V (\mathbf{a} - \hat{\mathbf{y}})$.

Lemma 40.1, when applied both to u and to $\hat{u}$, yields points c and $\hat{c}$ in Ω such that

$$w(b) - w_0(b_0) = \int_V u(\mathbf{y}) + w_0'(c)\cdot\Delta$$

and

$$\hat{w}(b) - \hat{w}_0(b_0) = \int_V \hat{u}(\hat{\mathbf{y}}) + \hat{w}_0'(\hat{c})\cdot\hat{\Delta}.$$

We note also that

$$\hat{u}(\mathbf{y}(s), s) - \hat{w}'(b)\cdot\mathbf{y}(s) \leqq \hat{u}(\hat{\mathbf{y}}(s), s) - \hat{w}'(b)\cdot\hat{\mathbf{y}}(s) \tag{40.10}$$

for all $s \in S$; this follows from Proposition 38.5. Similarly

$$u(\hat{\mathbf{y}}(s), s) - w'(b)\cdot\hat{\mathbf{y}}(s) \leqq u(\mathbf{y}(s), s) - w'(b)\cdot\mathbf{y}(s).$$

Now let

$$\begin{aligned}
I_1 &= \int_V [u(\mathbf{y}) - \hat{u}(\mathbf{y})] \\
\hat{I}_1 &= \int_V [\hat{u}(\hat{\mathbf{y}}) - u(\hat{\mathbf{y}})] \\
I_2 &= [w'(b) - \hat{w}'(b)]\cdot\Delta \\
\hat{I}_2 &= [\hat{w}'(b) - w'(b)]\cdot\hat{\Delta} \\
I_3 &= [w_0'(c) - w'(b)]\cdot\Delta \\
\hat{I}_3 &= [\hat{w}_0'(\hat{c}) - \hat{w}'(b)]\cdot\hat{\Delta} \\
J_1 &= [\hat{w}_0'(\hat{c}) - w_0'(\hat{c})]\cdot\hat{\Delta} \\
J_2 &= [w_0'(\hat{c}) - w'(b)]\cdot\hat{\Delta}.
\end{aligned}$$

Note that

$$\hat{I}_3 = J_1 + J_2 - \hat{I}_2.$$

Then we have

$$\begin{aligned}
A_k &= w(b) - w_0(b_0) - (\hat{w}(b) - \hat{w}_0(b_0)) \\
&= \int_V u(\mathbf{y}) + w'(b)\cdot\Delta + (w_0'(c) - w'(b))\cdot\Delta \\
&\qquad - \int_V [\hat{u}(\hat{\mathbf{y}}) - \hat{w}'(b)\cdot\hat{\mathbf{y}} + \hat{w}'(b)\cdot\mathbf{a}] - (\hat{w}_0'(\hat{c}) - \hat{w}'(b))\cdot\hat{\Delta} \\
&\leqq \int_V u(\mathbf{y}) + w'(b)\cdot\Delta + I_3 \\
&\qquad\qquad - \int_V [\hat{u}(\mathbf{y}) - \hat{w}'(b)\cdot\mathbf{y} + \hat{w}'(b)\cdot\mathbf{a}] - \hat{I}_3 \\
&= I_1 + I_2 + I_3 - \hat{I}_3,
\end{aligned} \tag{40.11}$$

where the inequality follows from (40.10). Similarly

$$-A_k \leqq \hat{I}_1 + \hat{I}_2 + \hat{I}_3 - I_3 = \hat{I}_1 - I_3 + J_1 + J_2. \tag{40.12}$$

The proof of Lemma 40.7 will be completed by estimating the quantities I_i, $\hat{I}_i$, and J_i. These quantities can be made sufficiently small (in absolute value) to prove (40.8) if the δ appearing in the statement of the lemma is appropriately chosen. We now show how this is done.

It will be useful to make the following definition: a quantity is *uniformly small when δ is chosen sufficiently small* if for every $\epsilon_1 > 0$, it is less than ϵ_1 in absolute value, for appropriate choice of δ, uniformly in $\hat{u}$, S_0, and S (i.e. uniformly in $\hat{u}$, in the choice of the chain $\varnothing \subset S_1 \subset S_2 \subset \ldots$, and in the choice of a particular link in this chain).

Let $\alpha = \Sigma \int \mathbf{a}$. By Lemma 37.8, there is an integrable η, depending on u and ϵ only, such that if δ is sufficiently small then $\mathbf{y}(s) \leqq \eta(s)e$ for almost all $s \in S$. Choose η so that also $\mathbf{a}(s) \leqq \eta(s)$. From this it follows that

$$\|\Delta\| \leqq \int_V \eta, \quad \|\hat{\Delta}\| \leqq \int_V \eta. \tag{40.13}$$

Next, let D be the exceptional set in the definition of δ-approximation, i.e. the set in which we do not necessarily have $\|u_s - \hat{u}_s\| \leqq \delta$. Let ζ be an integrable real function such that $u(x, s) \leqq \zeta(s) + \Sigma x$ for all s in I and all x in Ω; such a ζ exists by Lemma 37.9. Then

$$\begin{aligned} |I_1| &\leqq \int_V |u(\mathbf{y}) - \hat{u}(\mathbf{y})| = \int_{V \setminus D} + \int_{V \cap D} \\ &\leqq \delta \int_{V \setminus D} (1 + \Sigma \mathbf{y}) + \int_{V \cap D} (\zeta + \Sigma \mathbf{y} + \sqrt{\Sigma \mathbf{y}}) \\ &\leqq \delta \int_{V \setminus D} (1 + n\eta) + \int_{V \cap D} (\zeta + 2n\eta). \end{aligned} \tag{40.14}$$

Similarly,

$$|\hat{I}_1| \leqq \delta \int_{V \setminus D} (1 + n\eta) + \int_{V \cap D} (\zeta + 2n\eta). \tag{40.15}$$

Next, we must estimate the terms that multiply Δ and $\hat{\Delta}$ in the expressions for I_2, $\hat{I}_2$, I_3, $\hat{I}_3$, J_1 and J_2. For this purpose note first that by (40.3), if $\int_{S_0} \mathbf{y} > 0$, then

$$I_3 = \left(w_0'(c) - w_0'\left(\int_{S_0} \mathbf{y}\right)\right) \cdot \Delta, \tag{40.16}$$

and

$$J_2 = \left(w_0'(\hat{c}) - w_0'\left(\int_{S_0} \mathbf{y}\right)\right) \cdot \hat{\Delta}; \tag{40.17}$$

and similarly if $\int_{S_0} \hat{\mathbf{y}} > 0$, then

$$\hat{I}_3 = \left(\hat{w}_0'(\hat{c}) - \hat{w}_0'\left(\int_{S_0} \hat{\mathbf{y}}\right)\right)\cdot\hat{\Delta}. \tag{40.18}$$

Now

$$c - \int_{S_0} \mathbf{y} = \theta \int_{S_0} \mathbf{y} + (1 - \theta)b_0 - \int_{S_0} \mathbf{y}$$
$$= (1 - \theta)\left(b_0 - \int_{S_0} \mathbf{y}\right) = -(1 - \theta)\Delta.$$

Hence

$$\left\| c - \int_{S_0} \mathbf{y} \right\| \leqq \|\Delta\|,$$

and similarly

$$\left\| \hat{c} - \int_{S_0} \hat{\mathbf{y}} \right\| \leqq \|\hat{\Delta}\|.$$

Combining this with (40.13) and with the fact that $\mu(V) \leqq \delta$, we obtain that, if δ is chosen sufficiently small, then

$$\left\| c - \int_{S_0} \mathbf{y} \right\| \text{ and } \left\| \hat{c} - \int_{S_0} \hat{\mathbf{y}} \right\| \text{ are uniformly small.} \tag{40.19}$$

Furthermore, we have

$$\hat{c} - \int_{S_0} \mathbf{y} = \hat{c} - c + c - \int_{S_0} \mathbf{y} = \theta(\hat{\Delta} - \Delta) + c - \int_{S_0} \mathbf{y}.$$

Hence

$$\left\| \hat{c} - \int_{S_0} \mathbf{y} \right\| \leqq \|\hat{\Delta}\| + \|\Delta\| + \left\| c - \int_{S_0} \mathbf{y} \right\|,$$

and combining this with (40.13) and (40.19) we obtain as above that if δ is chosen sufficiently small, then

$$\left\| \hat{c} - \int_{S_0} \mathbf{y} \right\| \text{ is uniformly small.} \tag{40.20}$$

We now make use of the assumption that $\int_{S_1} \mathbf{a} \geqq \epsilon e$. If we recall that $S_0 = S_k \supset S_1$ and that $b_0 = \int_{S_0} \mathbf{a}$, then from (40.13) and $\mu(V) \leqq \delta$ it follows that when δ is sufficiently small, $\int_{S_0} \mathbf{y} \geqq \frac{1}{2}\epsilon e$ and $\int_{S_0} \hat{\mathbf{y}} \geqq \frac{1}{2}\epsilon e$. Hence the vectors b, c, $\hat{c}$, $\int_{S_0} \mathbf{y}$ and $\int_{S_0} \hat{\mathbf{y}}$ are all in $A(\frac{1}{2}\epsilon, \alpha)$. So we may apply Propositions 38.7 and 38.14, and formulas (40.16) through (40.20), and deduce that if δ is chosen sufficiently small, then the terms multiplying Δ and $\hat{\Delta}$ in the definitions of I_2, $\hat{I}_2$, I_3, $\hat{I}_3$, J_1 and J_2 will be uniformly small. Taking into account formulas (40.11) through (40.15), we deduce that for any given $\epsilon_1 > 0$, if δ is chosen sufficiently small, then

$$(40.21) \qquad |A_k| \leqq \delta \int_V (1 + n\eta) + \int_{V \cap D} (\zeta + 2n\eta) + \epsilon_1 \int_V \eta;$$

here δ, η and ζ depend on u, $\mathbf{a}$, ϵ, and ϵ_1 only, and not on the choice of $\hat{u}$ or the S_k (providing, of course, that $\hat{u}$ and the S_k satisfy the conditions of the lemma). Writing V_k for V and recalling that $V_k = V = S \backslash S_0 = S_{k+1} \backslash S_k$, we deduce that the V_k are mutually disjoint, and their union is included in I; similarly, the $V_k \cap D$ are mutually disjoint, and their union is included in D. Hence from (40.21) we get

$$\sum_{k=1}^{m} |A_k| \leqq \delta \int (1 + n\eta) + \int_D (\zeta + 2n\eta) + \epsilon_1 \int \eta.$$

Using the integrability of η and ζ and the fact that $\mu(D) \leqq \delta$, we deduce that if δ and ϵ_1 are chosen small enough, then

$$\sum_{k=1}^{m} |A_k| \leqq \epsilon.$$

This completes the proof of Lemma 40.7.

PROPOSITION 40.22. *Let $u \in \mathfrak{U}_1$, and let $\mathbf{a}$ be μ-integrable. Then for each $\epsilon > 0$ there is a $\delta > 0$ such that if $\hat{u} \in \mathfrak{U}_1$ is a δ-approximation to u, and $\int_S \mathbf{a} \leqq \delta e$, then $\hat{v}(S) < \epsilon$, where $\hat{v}$ is defined by* (40.9).

Proof. Clearly $u_I(0) = 0$. By Proposition 37.13, u_I is continuous on Ω. Hence for δ sufficiently small, $\int_S \mathbf{a} \leqq \delta e$ yields

$$(40.23) \qquad u_I\left(\int_S \mathbf{a}\right) < \tfrac{1}{2}\epsilon.$$

On the other hand, if in Proposition 37.11 we substitute $\frac{1}{4}\epsilon$ for ϵ, then it follows that for δ sufficiently small, we have

$$\left| u_I\left(\int_S \mathbf{a}\right) - \hat{u}_I\left(\int_S \mathbf{a}\right) \right| < \tfrac{1}{4}\epsilon\left(1 + \Sigma \int_S \mathbf{a}\right) \leqq \tfrac{1}{4}\epsilon(1 + 1) = \tfrac{1}{2}\epsilon.$$

Combining this with (40.23), we get

$$\hat{v}(S) = \hat{u}_S\left(\int_S \mathbf{a}\right) \leqq \hat{u}_I\left(\int_S \mathbf{a}\right) < \tfrac{1}{2}\epsilon + \tfrac{1}{2}\epsilon = \epsilon.$$

This completes the proof of Proposition 40.22.

PROPOSITION 40.24. *Let* $u \in \mathcal{U}_1$, *let* $\mathbf{a}$ *be* μ-*integrable, and assume*

(40.25) *for all* $s \in I$, *either* $\mathbf{a}(s) > 0$ *or* $\mathbf{a}(s) = 0$.

Then for each $\epsilon > 0$ *there is a* $\delta > 0$ *such that if* $\hat{u} \in \mathcal{U}_1$ *is a* δ-*approximation to* u, *then*

$$\|v - \hat{v}\| < \epsilon,$$

where v *and* $\hat{v}$ *are defined by* (40.9).

Remark. Condition (40.25) says that for each s, either all coordinates of $\mathbf{a}(s)$ are positive or all vanish. This condition is implied both by (31.4) and by $n = 1$; we will use it to state (and prove) below a common generalization of Theorem J and Proposition 33.2.

Proof. Let δ_1 correspond to $\frac{1}{4}\epsilon$ in accordance with Proposition 40.22. Choose γ so that[1] $0 < \gamma < \int \mathbf{a}^1$ and so that $\int_S \mathbf{a}^1 \leqq \gamma$ implies $\int_S \mathbf{a} < \delta_1 e$; this is possible because of (40.25). Choose $\epsilon_1 > 0$ so that $\int_S \mathbf{a}^1 \geqq \gamma$ implies $\int_S \mathbf{a} > \epsilon_1 e$; this, again, is possible because of (40.25). Let

$$\epsilon_2 = \min(\epsilon_1, \tfrac{1}{4}\epsilon),$$

and choose δ_2 to correspond to ϵ_2 in accordance with Lemma 40.1. Let $\delta = \min(\delta_1, \delta_2)$.

Let $w = v - \hat{v}$, and let

$$\emptyset = S_0 \subset S_1 \subset \cdots \subset S_m \subset S_{m+1} = I$$

be a chain. It is always possible to insert finitely many additional sets $S_{01}, S_{02}, \ldots, S_{11}, S_{12}, \ldots\ldots, \ldots, S_{m1}, S_{m2}, \ldots$ into the chain so that $S_0 \subset S_{01} \subset S_{11} \subset \cdots \subset S_1 \subset S_{11} \subset S_{12} \subset \cdots \subset \cdots \subset S_m \subset S_{m1} \subset S_{m2} \subset \cdots \subset S_{m+1}$ and the measure of the difference between two neighboring sets is $< \delta$; that is, if we relabel the new sequence $U_0, \ldots, U_{p+1} = I$, then $\mu(U_{k+1} \backslash U_k) < \delta$ for all k. Furthermore, by Lyapunov's theorem in one dimension we may suppose w.l.o.g. that for one of the U_k, say for U_q, we have $\int_{U_q} \mathbf{a}^1 = \gamma$. Then

$$\sum_{k=0}^{m} |w(S_{k+1}) - w(S_k)| \leqq \sum_{k=0}^{p} |w(U_{k+1}) - w(U_k)| = \sum_{k=0}^{q-1} + \sum_{k=q}^{p}.$$

[1] If $\int \mathbf{a}^1 = 0$, then from (40.25) it follows that $\int \mathbf{a} = 0$, and the whole problem becomes trivial.

Lemma 40.7 and the fact that $\hat{u}$ is a δ-approximation—hence a fortiori a δ_2-approximation—to u yield

$$\sum_{k=q}^{p} \leqq \epsilon_2 \leqq \tfrac{1}{4}\epsilon.$$

Furthermore, since $\hat{u}$ is a δ-approximation to u, it is a fortiori a δ_1-approximation. Hence by the monotonicity of v and $\hat{v}$ and Proposition 40.22, we have

$$\begin{aligned}\sum_{k=0}^{q-1} |w(U_{k+1}) - w(U_k)| &= \sum_{k=0}^{q-1} |v(U_{k+1}) - v(U_k) - (\hat{v}(U_{k+1}) - \hat{v}(U_k))| \\ &\leqq \sum_{k=0}^{q-1} |v(U_{k+1}) - v(U_k)| + \sum_{k=0}^{q-1} |\hat{v}(U_{k+1}) - \hat{v}(U_k)| \\ &= \sum_{k=0}^{q-1} (v(U_{k+1}) - v(U_k)) + \sum_{k=0}^{q-1} (\hat{v}(U_{k+1}) - \hat{v}(U_k)) \\ &= v(U_q) + \hat{v}(U_q) < \tfrac{1}{4}\epsilon + \tfrac{1}{4}\epsilon = \tfrac{1}{2}\epsilon,\end{aligned}$$

because $\int_{U_q} \mathbf{a} < \delta_1 e$ (note that u is a 0-approximation, hence trivially a δ_1-approximation, to itself).

We conclude that

$$\sum_{k=0}^{m} |w(S_{k+1}) - w(S_k)| < \tfrac{1}{2}\epsilon + \tfrac{1}{4}\epsilon = \tfrac{3}{4}\epsilon,$$

and it follows that $\|v - \hat{v}\| = \|w\| < \epsilon$. This completes the proof of Proposition 40.24.

PROPOSITION 40.26. *Theorem J holds if* (31.4) *is replaced by* (40.25).

Proof. Recall that H is the set of all superadditive set-functions in pNA that are homogeneous of degree 1. By Proposition 35.6, for every δ there is a δ-approximation $\hat{u}$ to u that is of finite type. If $\hat{v}$ corresponds to the given $\mathbf{a}$ and to this $\hat{u}$ (in accordance with (40.9)), then by Proposition 40.24, for given ϵ we will have $\|\hat{v} - v\| < \epsilon$ when δ is sufficiently small. But by Lemma 39.16, $\hat{v} \in H$; thus v can be approximated in variation by members of H, i.e. it is in the closure of H. But H is closed (Proposition 27.12), and so we have proved that $v \in H$. Proposition 40.26 now follows from Theorem I.

Theorem J and Proposition 33.2 both follow immediately from Proposition 40.26.

41. Some Open Problems

A. POSITIVITY OF INITIAL RESOURCES

Propositions 31.5, 31.7, 33.2, and 40.26 give conditions under which the positivity condition (condition (31.4)) can be dispensed with. It is, however, possible that it can be dispensed with altogether, i.e. that Theorem J remains true if this condition is simply dropped, without substituting anything for it. We have not been able either to prove or disprove this possibility.

B. STRICT MONOTONICTY OF u

Our proofs make extensive use, especially in Section 37, of the assumption that each of the $u(x, s)$ be strictly increasing in x. It is, however, possible that a careful treatment might be able to dispense with this assumption, particularly under certain conditions (such as (31.4)). Compare Proposition 2.2 of Aumann and Perles (1965), which would also be considerably easier to prove if one would assume that the u-functions are strictly increasing in x.

C. ATTAINMENT OF THE MAX IN THE DEFINITION OF v

This question was treated via a number of examples in Subsection D of Section 33, where we showed that it is hopeless to try to extend our results to the case in which the max is not attained, at least in $v(I)$. The question arises, though, whether it is necessary to assume the asymptotic condition (31.2), or whether it would not be enough simply to assume that

(41.1) $v(I)$ is attained

or

(41.2) all the $v(S)$ are attained

or something of that nature.

Aesthetically, an "explicit" condition such as (31.2) is preferable to a condition like (41.1), since given a specific family of u-functions, it may be difficult to tell whether (41.1) holds. The mathematical question of whether (31.2) can be replaced by (41.1) remains, however.

Our method of proof is based on approximations by u's of finite type, this is based on the norm on $\mathfrak{F}_1$ defined in Section 35, and this in turn depends essentially on (31.2). Thus (31.2) is used not only to establish that the $v(S)$ are attained, but also directly in the proof. It appears that if one wishes to substitute (41.1) or (41.2), one would need an entirely new line of proof. We do not, of course, have a counterexample.

The most natural candidate for a replacement for (31.2) would seem to be neither (41.1) nor (41.2), but rather the stronger

(41.3) $\quad u_S(a)$ is attained for all $S \in \mathcal{C}$ and all $a \in \Omega$.

Condition (41.3) is equivalent to the condition that, for any $S \in \mathcal{C}$, the integral of the subgraphs[1] of $u(\cdot, s)$ over $s \in S$ be closed. Our problem is unsolved when any one of (41.1), (41.2), or (41.3) is substituted for (31.2).

In the extremely special case in which u is of finite type and all the $u(\cdot, s)$ are concave, the methods of Section 39 can probably be pushed through; that is to say, though we have not checked the details, we believe that in this case Theorem J can be proved without assumptions (31.2) and (31.4). In any event, (41.3) holds in this case, because the subgraph of the function u_S is a finite sum of closed subgraphs, and it may be verified that this must be closed.

[1] The *subgraph* of a function f is the set $\{(x, y): y \leqq f(x)\}$.

Chapter VII. The Diagonal Property

42. Introduction and Statement of Results

Let $v = f \circ \mu$, where μ is a vector of NA measures and f is continuously differentiable with $f(0) = 0$. Then according to Theorem B, v is in pNA, and

$$(\varphi v)(S) = \int_0^1 f_{\mu(S)}(t\mu(I))\, dt.$$

In Section 3 we have already pointed out a remarkable implication of this formula, namely that φv is completely determined by the behavior of f *near the diagonal* of μ only. In particular, if f vanishes in some neighborhood of the diagonal, then φv must vanish identically. Moreover, we know (refer to (3.3)) that φv is determined by its behavior "near the diagonal" even if $v = f \circ \mu$ near the diagonal only. In particular, if v is in $bv'NA$ and if $v(S) = 0$ for all S in some neighborhood of the diagonal of μ, then φv vanishes identically.

Proceeding formally, define $DIAG$ to be the set of all $v \in BV$ satisfying

(42.1) there is a positive integer k, a k-dimensional vector ζ of measures in NA^1, and a neighborhood U in E_k of the diagonal $[0, \zeta(I)]$ such that if $\zeta(S) \in U$ then $v(S) = 0$.

Note that $DIAG$ is a symmetric subspace of BV. Let Q be a symmetric subspace of BV, φ a value on Q. We shall say that the pair (Q, φ) enjoys the *diagonal property*, and that φ is a *diagonal value*, if we have $\varphi v = 0$ for all $v \in Q \cap DIAG$. We might paraphrase this by saying that if v vanishes in a neighborhood of some "diagonal," then its value also vanishes. Strictly speaking, the diagonal property is a property of the pair (Q, φ), but, when no confusion can result, we shall speak of it as applying to Q or φ alone.

This property appears to be basic in the study of values; it is reflected also in (7.7) and in Theorem H (Chapter IV). In this chapter we shall prove that the diagonal property is enjoyed by all the values we have defined thus far—i.e. the values on $bv'NA$ (and pNA), MIX, and $ASYMP$. It is an open question whether or not the diagonal property follows from the definition of value, i.e. whether all values are diagonal. In any case, it certainly appears that the diagonal property is enjoyed by any value that is in any way connected with random order considerations.

Define $pNAD$ to be the closure of $pNA + DIAG$; it may be thought of as consisting of set functions that "behave almost like" those of pNA "close to the diagonal." If one thinks of pNA as expressing certain differentiability properties (cf.[1] Theorems B, C, and H), then $pNAD$ expresses the same differentiability properties, but on the diagonal only (cf. Proposition 44.22 below). One might say that the definition of $pNAD$ is obtained from that of pNA by ignoring sets S that are not relevant to value considerations, or at least to "random order" value considerations.

In line with this we shall prove an extension of Theorem F, according to which $pNAD \subset ASYMP$ (Corollary 43.14). One cannot expect to prove $pNAD \subset MIX$, since $pNAD$ (and in fact $DIAG$) contains functions that are not in ORD, whereas MIX is by definition a subset of ORD; but we *shall* prove (Proposition 43.2) that $pNAD \cap ORD \subset MIX$. We shall also prove (Proposition 43.13) that there is a unique diagonal value on $pNAD$ that is continuous in the variation norm; this is a "diagonal analogue" of the uniqueness of the value on pNA (Theorem B). Of course, the continuous diagonal value must coincide with the asymptotic value, and on $pNAD \cap ORD$ it must coincide with the mixing value (Proposition 43.13).

We shall also prove analogues (Propositions 44.22 and 44.28) of Theorems H and I. In these analogues (or rather generalizations), the hypothesis $v \in pNA$ is replaced by[2]

$$v \in pNAD \cap pNA'. \tag{42.2}$$

Since $v \in pNA'$, the extension v^* is defined;[3] this is needed in both

[1] It is true that Theorem H refers to the diagonal only, but as far as the differentiability properties are concerned, similar results could easily be proved off the diagonal as well.

[2] Recall that pNA' is the closure of pNA in the supremum norm. See Section 22.

[3] Proposition 22.16.

theorems. In the conclusions, we are no longer justified in referring to "the" value φv; however, the conclusions remain true if the value is replaced by the continuous diagonal value.

Finally, we shall prove Proposition 31.7. This will be done by first showing (Propositions 45.1 and 46.1) that under the conditions of that Proposition, $v \in pNAD \cap ORD$, and the continuous diagonal value coincides with the unique point in the core (Proposition 45.1). This then yields the "asymptotic" part of Proposition 31.7; finally, in Section 46 we shall demonstrate $v \in ORD$ and deduce the "mixing" part as well.

43. The Diagonal Property for $bv'NA$, MIX, and $ASYMP$

PROPOSITION 43.1. *The unique value on $bv'NA$ enjoys the diagonal property.*

Proof. Apply Proposition 8.32 to deduce that if $v \in bv'NA$ satisfies (42.1), then $v \in pNA$. Then apply Proposition 7.6 with $f \doteq 0$ to deduce that $\varphi v = 0$. This completes the proof of Proposition 43.1.

PROPOSITION 43.2. *$pNAD \cap ORD \subset MIX$, and the mixing value, regarded as a value on MIX, is a diagonal value.*

Proof. Let $v \in pNAD \cap ORD$. Let $\{\epsilon_1, \epsilon_2, \ldots,\}$ be a sequence tending to 0; then for each j we may find a decomposition of v (depending on j),

$$v = v_1 + v_2 + v_3,$$

where $v_1 \in pNA$, $v_2 \in DIAG$, and $\|v_3\| \leqq \epsilon_j$. Since $v_1 \in pNA \subset MIX$ (Theorem E), there is in NA^1 a measure μ_{v_1} that corresponds to v_1 in accordance with the definition of mixing value. For each j, let k, ζ, and U (which, of course, depend on j) correspond to v_2 in accordance with (42.1), and let

$$\mu^j = \mu_{v_1} + \frac{1}{k} \sum_{i=1}^{k} \zeta_i.$$

Let

$$\mu_v = \sum_{j=1}^{\infty} \frac{1}{2^j} \mu^j.$$

Let $\mu \in NA^1$ be such that $\mu_v \ll \mu$, let $\{\Theta_1, \Theta_2, \ldots\}$ be a μ-mixing sequence, and let $\mathcal{R}$ be a measurable order. Now fix j for the time being; note that $\mu_{v_1} \ll \mu$ and that $\zeta_i \ll \mu$ for all i. Pick a neighborhood V of the diagonal $[0, \zeta(I)]$ with $V \subset U$, and a number $\epsilon > 0$ such that an "ϵ-thickening" of V is still in U, i.e.

$$x \in V \text{ and } \|y - x\| < \epsilon \Rightarrow y \in U. \tag{43.3}$$

Next, pick a number $\delta > 0$ such that for all $S \subset I$,

$$\mu(S) < \delta \Rightarrow \|\zeta(S)\| < \epsilon; \tag{43.4}$$

this is possible because $\zeta_i \ll \mu$ for each i. Finally, pick a finite sequence $s_1, \ldots, s_l$ in I such that

$$s_l \,\mathcal{R}\, s_{l-1} \,\mathcal{R}\, \cdots \,\mathcal{R}\, s_1,$$

and for every s in I there is an s_k such that

$$\|\mu(I(s; \mathcal{R})) - \mu(I(s_k; \mathcal{R}))\| < \delta; \tag{43.5}$$

this is possible because μ is non-atomic (see also Lemmas 12.15 and 12.14).

For each k, apply Lemma 15.12 with $\xi = \zeta_i$ and $T = I(s_k; \mathcal{R})$, obtaining

$$\zeta(\Theta_m I(s_k; \mathcal{R})) \to \mu(I(s_k; \mathcal{R}))\zeta(I)$$

as $m \to \infty$. It follows that $\zeta(\Theta_m I(s_k; \mathcal{R})) \in V$ for all sufficiently large m, say for $m > m_0$; and since there are only finitely many k, we may choose m_0 independent of k (though not of $\mathcal{R}$ or of μ). But then for $m > m_0$, it follows from (43.3), (43.4), (43.5), and the fact that Θ_m preserves μ-measure that $\zeta(\Theta_m I(s; \mathcal{R})) \in U$ for all s. In particular, we may substitute $\Theta_m^{-1}s$ for s, and deduce that

$$\zeta(I(s; \Theta_m\mathcal{R})) = \zeta(\Theta_m I(\Theta_m^{-1}s; \mathcal{R})) \in U$$

whenever $m > m_0$. It follows that for all $s \in I$ and $m > m_0$,

$$v_2(I(s; \Theta_m\mathcal{R})) = 0. \tag{43.6}$$

For fixed $m > m_0$, let $\mathcal{Q} = \Theta_m\mathcal{R}$, and let W be in the field (*not* σ-field) $H(\mathcal{Q})$ generated by the initial segments $I(s; \mathcal{Q})$ (cf. the proof of Proposition 12.8). Then W can be written in the form

$$W = \bigcup_{q=1}^{p} [t_q, s_q)_{\mathfrak{Q}},$$

i.e. as a finite union of disjoint $\mathfrak{Q}$-intervals, where

$$s_p \,\mathfrak{Q}\, t_p \,\mathfrak{Q} \cdots \mathfrak{Q}\, s_1 \,\mathfrak{Q}\, t_1. \tag{43.7}$$

Since $v_2 + v_3 = v - v_1 \in ORD$, it follows that $\varphi(v_2 + v_3; \mathfrak{Q})$ exists, and so by (12.2) and (43.6),

$$\begin{aligned} |\varphi(v_2 + v_3; \mathfrak{Q})(W)| &= \Big| \varphi(v_2 + v_3; \mathfrak{Q}) \Big(\bigcup_{q=1}^{p} [t_q, s_q)_{\mathfrak{Q}}\Big) \Big| \\ &= \Big| \sum_{q=1}^{p} [\varphi(v_2 + v_3; \mathfrak{Q})(I(s_q; \mathfrak{Q})) - \varphi(v_2 + v_3; \mathfrak{Q})(I(t_q; \mathfrak{Q}))] \Big| \\ &= \Big| \sum_{q=1}^{p} [(v_2 + v_3)(I(s_q; \mathfrak{Q})) - (v_2 + v_3)(I(t_q; \mathfrak{Q}))] \Big| \\ &= \Big| \sum_{q=1}^{p} [v_3(I(s_q; \mathfrak{Q})) - v_3(I(t_q; \mathfrak{Q}))] \Big|. \end{aligned}$$

By (43.7), the last expression is the variation of v_3 over a subchain of a certain chain, and so it is $\leqq \|v_3\|$. Thus we have

$$|\varphi(v_2 + v_3; \Theta_m \mathfrak{R})(W)| \leqq \|v_3\|$$

whenever $W \in H(\Theta_m \mathfrak{R})$; but since the field $H(\Theta_m \mathfrak{R})$ generates the σ-field $\mathfrak{C}$ (by (12.3)), it follows from a standard approximation argument[1] that

$$|\varphi(v_2 + v_3; \Theta_m \mathfrak{R})(S)| \leqq \|v_3\| \tag{43.8}$$

for all $S \in \mathfrak{C}$; of course this holds for all $m > m_0$. From this and the fact that $v_1 \in MIX$ (Theorem E), it follows that for all $S \in \mathfrak{C}$,

$$\begin{aligned} &\limsup_{m \to \infty} \varphi(v; \Theta_m \mathfrak{R})(S) \\ &\qquad \leqq \lim \varphi(v_1; \Theta_m \mathfrak{R})(S) + \limsup \varphi(v_2 + v_3; \Theta_m \mathfrak{R})(S) \\ &\qquad \leqq (\varphi v_1)(S) + \|v_3\| \leqq (\varphi v_1)(S) + \epsilon_j, \end{aligned}$$

where φv_1 is the mixing value (or, for that matter, the unique value on

[1] See Halmos (1950), p. 56, Theorem D. Compare the end of the proof of Proposition 12.8.

pNA). Similarly

$$\liminf_{m\to\infty} \varphi(v; \Theta_m\mathcal{R})(S) \geqq (\varphi v_1)(S) - \epsilon_j.$$

Hence

$$\limsup \varphi(v; \Theta_m\mathcal{R})(S) - \liminf \varphi(v; \Theta_m\mathcal{R})(S) \leqq 2\epsilon_j.$$

Now the left side of this inequality is independent of j; so we may let $j \to \infty$ and conclude that $\lim_{m\to\infty} \varphi(v; \Theta_m\mathcal{R})(S)$ exists. This shows that $pNAD \cap ORD \subset MIX$.

To prove the second assertion of the proposition, let $v \in MIX \cap DIAG$. Since $MIX \subset ORD$ and $DIAG \subset pNAD$, it follows that $v \in pNAD \cap ORD$, and so (43.8) applies. Since $v \in DIAG$, we may set $v_2 = v$, $v_1 = v_3 = 0$. Letting $m \to \infty$ in (43.8), we deduce

$$|(\varphi v)(S)| = |(\varphi v_2)(S)| \leqq \|v_3\| = 0,$$

where φ is the mixing value; thus $\varphi = 0$. This completes the proof of Proposition 43.2.

Corollary 43.9. *Let $w + r \in pNAD \cap ORD$, where $w \in DIAG$. Then*

$$\|\varphi(w + r)\| \leqq 2\|r\|,$$

where φ is the mixing value.

Proof. For given $\epsilon > 0$, let S be such that

$$\|\varphi(w + r)\| \leqq |\varphi(w + r)(S)| + |\varphi(w + r)(I\backslash S)| + \epsilon. \tag{43.10}$$

Since $\varphi(w + r) \in FA$ (Lemma 15.2), such an S exists. In the previous proof, set $v_1 = 0$, $v_2 = w$, $v_3 = r$. Letting $m \to \infty$ in (43.8), we deduce

$$|\varphi(w + r)(S)| \leqq \|r\|$$
$$|\varphi(w + r)(I\backslash S)| \leqq \|r\|,$$

and so the corollary follows from (43.10).

Remark. By going over the proof of (43.8), one could easily sharpen the conclusion in Corollary 43.9 to $\|\varphi(w + r)\| \leqq \|r\|$; this sharper form is, however, not needed in the sequel.

Proposition 43.11. *$DIAG \subset ASYMP$, and the asymptotic value, regarded as a value on $ASYMP$, enjoys the diagonal property.*

Proof. Follows immediately from Corollary 18.10.

COROLLARY 43.12. $pNAD \subset ASYMP$.

Proof. $pNA \subset ASYMP$ (Theorem F), $DIAG \subset ASYMP$ (Proposition 43.11), and $ASYMP$ is closed (Theorem F).

A value is called *continuous* if it is continuous in the variation norm.

PROPOSITION 43.13. *There is a unique continuous diagonal value on* $pNAD$; *on* $pNAD$ *it coincides with the asymptotic value, and on* $pNAD \cap ORD$, *with the mixing value.*

Proof. The asymptotic value is a value on $pNAD$ (Corollary 43.12); it is continuous (Proposition 18.1) and enjoys the diagonal property (Proposition 43.11). To prove the uniqueness, let φ be a continuous diagonal value. Then φ is determined on pNA by the uniqueness of the value on pNA (Theorem B), and on $DIAG$ it must vanish identically by the diagonal property; hence it is determined on $pNA + DIAG$, and so by continuity on its closure, namely $pNAD$.

To prove the last clause of the proposition, let $v \in pNAD \cap ORD$. For given $\epsilon > 0$, set

$$v = v_1 + v_2 + v_3,$$

where $v_1 \in pNA$, $v_2 \in DIAG$, and $\|v_3\| \leqq \epsilon$. Denoting the asymptotic value by φ^A and the mixing value by φ^M, we then have (using Theorems E and F, Propositions 18.1 and 43.11, and Corollary 43.9)

$$\begin{aligned}\|(\varphi^A - \varphi^M)(v)\| &= \|(\varphi^A - \varphi^M)(v_2 + v_3)\| \\ &\leqq \|\varphi^A(v_2)\| + \|\varphi^A(v_3)\| + \|\varphi^M(v_2 + v_3)\| \\ &\leqq \|v_3\| + 2\|v_3\| \leqq 3\epsilon.\end{aligned}$$

Letting $\epsilon \to 0$ we deduce $\varphi^A v = \varphi^M v$, as was to be proved. This completes the proof of Proposition 43.13.

44. Analogues of Theorems H and I

Throughout this section, φ will denote the unique continuous diagonal value on $pNAD$. In the next few lemmas we shall make free use of the notations and terminology of Chapter IV, in particular of Sections 22 and 23. We begin with a generalization of Lemma 22.1.

LEMMA 44.1. *Let ξ be a finite-dimensional vector of measures in NA. Let $g_1, \ldots, g_m$ be ideal set functions with*

$$g_1 \leqq \cdots \leqq g_m.$$

Then there are sets $T_1, \ldots, T_m$ in $\mathcal{C}$ with

$$T_1 \subset \cdots \subset T_m$$

such that for all i,

$$\xi(T_i) = \int g_i \, d\xi.$$

Proof. The proof is exactly analogous to that of Lemma 22.1.

LEMMA 44.2. *Let $v \in BV \cap pNA'$. Then $v^* \in IBV$, and $\|v^*\| = \|v\|$.*

Remark. When we write $\| \quad \|$, we are, of course, referring to the variation norm. The supremum norm is denoted by $\| \quad \|'$, and the equation $\|v^*\|' = \|v\|'$ has already been established (see (22.14)).

Proof. Let Ω be a chain

$$0 = g_0 \leqq g_1 \leqq \cdots \leqq g_m \leqq g_{m+1} = \chi_I$$

of ideal set functions. For a given ϵ, let $\xi_1, \ldots, \xi_m$ be vectors of measures in NA, and $\delta_1, \ldots, \delta_m$ positive numbers, such that for all i,

$$\text{(44.3)} \qquad \left\|\int (f - g_i) \, d\xi_i\right\| \leqq \delta_i \Rightarrow |v^*(f) - v^*(g_i)| < \epsilon;$$

the existence of such ξ_i and δ_i follows from the continuity of v^* in the NA-topology (see (22.11)). Let ξ be the vector $(\xi_1, \ldots, \xi_m)$, and define $T_1, \ldots, T_m$ in accordance with Lemma 44.1. Then

$$\int (\chi_{T_i} - g_i) \, d\xi = 0$$

for all i, and hence $\int (\chi_{T_i} - g_i) \, d\xi_i = 0 \leqq \delta_i$ for all i. Hence

$$\text{(44.4)} \qquad |v(T_i) - v^*(g_i)| = |v^*(\chi_{T_i}) - v^*(g_i)| < \epsilon.$$

Setting $T_0 = \emptyset$ and $T_{m+1} = I$, we deduce

$$\begin{aligned} \|v^*\|_\Omega &= \sum_{i=0}^{m} |v^*(g_{i+1}) - v^*(g_i)| \\ &\leqq \sum_{i=0}^{m} |v(T_{i+1}) - v(T_i)| + 2(m+1)\epsilon \\ &\leqq \|v\| + 2(m+1)\epsilon. \end{aligned}$$

Letting $\epsilon \to 0$, we deduce $\|v^*\|_\Omega \leqq \|v\|$, and hence

$$\|v^*\| = \sup_\Omega \|v^*\|_\Omega \leqq \|v\|. \tag{44.5}$$

Since the inequality $\|v\| \leqq \|v^*\|$ is obvious, the proof of Lemma 44.2 is complete.

LEMMA 44.6. *Let* $T_1, \ldots, T_m$ *be disjoint measurable subsets of* $(0, 1)$. *With each* t *in each* T_i, *let there be associated a family* $\mathcal{H}^t$ *of closed intervals*[1] *in* $(0, 1)$, *one of whose endpoints is* t; *assume moreover, that each* $\mathcal{H}^t$ *contains arbitrarily short intervals. Let*

$$\mathcal{H}_i = \bigcup_{t \in T_i} \mathcal{H}^t, \text{ and } \mathcal{H} = \bigcup_{i=1}^{m} \mathcal{H}_i.$$

Then for each $\epsilon > 0$, *there is a finite family* $\mathcal{S}$ *of mutually disjoint intervals in* $\mathcal{H}$, *such that if* S_i *is the union of the intervals in* $\mathcal{S} \cap \mathcal{H}_i$, *then*

$$\lambda(S_i + T_i) < \epsilon,$$

where λ *is Lebesgue measure and "+" denotes the symmetric difference.*

Proof. Let us call T_i *pure left* if, for each t in T_i, $\mathcal{H}^t$ contains arbitrarily short intervals whose left endpoint is t. Define *pure right* analogously, and call T_i *pure* if it is either pure left or pure right (or both).

First we prove the lemma in the case in which each T_i is pure, proceeding by induction on m. The case in which $m = 1$ and T_i is pure left is proved in Titchmarsh (1939), §11.41, Lemmas 1 and 2, pp. 356–357; the pure right case, of course, follows from the pure left case by symmetry arguments.

Now assume that the lemma has been proved for $m - 1$; and let $T_1, \ldots, T_m$, and the sets $\mathcal{H}^t$, $\mathcal{H}_i$ and $\mathcal{H}$ satisfy the hypotheses of the lemma. Let $\mathcal{H}_* = \mathcal{H}_1 \cup \cdots \cup \mathcal{H}_{m-1}$. Applying the induction hypothesis (for ϵ/m instead of ϵ), we obtain a family $\mathcal{S}_*$ of mutually disjoint intervals in $\mathcal{H}_*$, such that for $i = 1, \ldots, m - 1$, if S_i is the union of the intervals in $\mathcal{S}_* \cap \mathcal{H}_i$, then

$$\lambda(S_i + T_i) < \epsilon/m. \tag{44.7}$$

[1] All intervals in this lemma and its proof are understood to have positive length.

Let $T_* = T_1 \cup \cdots \cup T_{m-1}$, and $S_* = S_1 \cup \cdots \cup S_{m-1}$. Note that

$$S_* + T_* \subset (T_1 + S_1) \cup \cdots \cup (T_{m-1} + S_{m-1}),$$

and hence

$$\lambda(S_* + T_*) \leqq \frac{m-1}{m}\epsilon < \epsilon.$$

Next, since T_m is pure, we may assume w.l.o.g. that it is pure left. Let $T_{**} = T_m \backslash S_*$, and for $t \in T_{**}$, let $\mathfrak{H}^t_{**}$ be the set of all intervals in $\mathfrak{H}^t$ whose left-hand endpoint is t and which do not intersect S_*. $\mathfrak{H}^t_{**}$ is non-empty and contains arbitrarily small intervals, because S_* is closed and $\mathfrak{H}^t$ contains arbitrarily small intervals whose left-hand endpoint is t. Let $\mathfrak{H}_{**} = \bigcup_{t \in T_m} \mathfrak{H}^t_{**}$. Apply the case $m = 1$ (with ϵ/m instead of ϵ) to T_{**} and $\mathfrak{H}_{**}$, obtaining a finite family $\mathfrak{S}_m$ of disjoint intervals in $\mathfrak{H}_{**}$ such that

$$\lambda(S_m + T_{**}) < \epsilon/m,$$

where S_m is the union of the intervals in $\mathfrak{S}_m$. Let $\mathfrak{S} = \mathfrak{S}_* \cup \mathfrak{S}_m$. Since $S_* \cap S_m = \varnothing$, the members of $\mathfrak{S}$ are disjoint. Furthermore, it may be verified that

$$S_m + T_m \subset (S_m + T_{**}) \cup (S_* \backslash T_*);$$

hence

$$\lambda(S_m + T_m) < \lambda(S_m + T_{**}) + \lambda(S_* + T_*) < \frac{\epsilon}{m} + \frac{m-1}{m}\epsilon = \epsilon.$$

This, together with (44.7), completes the proof of the lemma in the case in which all the T_i are pure.

In the general case, let T_i^L, for each i, be the set of all t in T_i for which $\mathfrak{H}_t$ contains arbitrarily small intervals whose left-hand endpoint is t; let $T_i^R = T_i \backslash T_i^L$. Let $\mathfrak{H}_i^L = \bigcup_{t \in T_i^L} \mathfrak{H}^t$, and $\mathfrak{H}_i^R = \bigcup_{t \in T_i^R} \mathfrak{H}^t$. Then T_i^L is pure left and T_i^R is pure right, so we may apply the case just proved to the system consisting of $T_1^L, \ldots, T_m^L, T_1^R, \ldots, T_m^R$, and the $\mathfrak{H}^t$. If we use $\frac{1}{2}\epsilon$ instead of ϵ, this yields a finite family $\mathfrak{S}$ of mutually disjoint intervals in $\mathfrak{H}$, such that if S_i^L and S_i^R is the union of the intervals in $\mathfrak{S} \cap \mathfrak{H}_L^i$ and $\mathfrak{S} \cap \mathfrak{H}_R^i$ respectively, then

$$\lambda(S_i^L + T_i^L) < \tfrac{1}{2}\epsilon \quad \text{and} \quad \lambda(S_i^R + T_i^R) < \tfrac{1}{2}\epsilon.$$

Since

$$S_i + T_i \subset (S_i^L + T_i^L) \cup (S_i^R + T_i^R),$$

it follows that

$$\lambda(S_i + T_i) < \tfrac{1}{2}\epsilon + \tfrac{1}{2}\epsilon = \epsilon,$$

and the proof of the lemma is complete.

LEMMA 44.8. *Let f be a non-negative extended real-valued*[2] *function on* $(0, 1)$. *For each positive integer k and each i with $0 \leqq i \leqq k^2$, define*

$$T_{ik} = \begin{cases} \{t: i/k \leqq f(t) < (i+1)/k\} & \text{for } i < k^2 \\ \{t: k \leqq f(t)\} & \text{for } i = k^2. \end{cases}$$

Then

$$\sum_{i=0}^{k^2} \frac{i}{k} \lambda(T_{ik}) \to \int_0^1 f(t)dt$$

as $k \to \infty$.

Proof. This follows easily from any of the standard definitions of the Lebesgue integral.

Let $v^* \in IBV$. For each $\delta > 0$, define the δ-*norm* $\|v^*\|_\delta$ to be

$$\sup \sum_{i=0}^{m} |v^*(g_{i+1}) - v^*(g_i)|,$$

where the sup is over all chains

$$0 = g_0 \leqq g_1 \leqq \cdots \leqq g_m \leqq g_{m+1} = \chi_I$$

such that for all g_i and all $s, s' \in I$, we have

$$|g_i(s) - g_i(s')| < \delta. \tag{44.9}$$

The restriction (44.9) means that the g_i are "close" to the diagonal; indeed, if $g_i(s) = g_i(s')$ for all s, s' in I, then g_i is of the form $t\chi_I$, and so is on the diagonal. Thus $\|v^*\|_\delta$ is the sup of the variation of v^* over chains that always remain in a δ-neighborhood of the diagonal. Note that $\|v^*\|_\delta \leqq \|v^*\|$; and so if $v \in BV \cap pNA'$, then by Lemma 44.2, it follows that

$$\|v^*\|_\delta \leqq \|v\|. \tag{44.10}$$

[2] I.e. f may take the value $+\infty$ as well as finite non-negative values.

The next lemma generalizes the hypothesis as well as sharpens the conclusion of Lemma 23.1.

LEMMA 44.11. *Let $v \in BV \cap pNA'$, and let $S \in \mathcal{C}$. Then $|\partial v^*(t)|^+$ is integrable over* $[0, 1]$, *and for all $\delta > 0$,*

$$\int_0^1 |\partial v^*(t)|^+ \, dt \leqq \|v^*\|_\delta.$$

Proof. Fix $\delta > 0$. Let k be a positive integer. Define a partition $\{T_0, T_1, \ldots, T_{k^2}\}$ of $(0, 1)$ by

$$(44.12) \qquad T_i = \begin{cases} \{t: i/k \leqq |\partial v^*(t)|^+ < (i+1)/k\} & \text{for } i < k^2 \\ \{t: k \leqq |\partial v^*(t)|^+\} & \text{for } i = k^2. \end{cases}$$

For each $t \in (0, 1)$ there are numbers τ, arbitrarily small in absolute value, such that

$$(44.13) \qquad \begin{gathered} \tau \neq 0,\ |\tau| < \delta,\ |t + \tau| \in (0, 1), \text{ and} \\ |\partial v^*(t)|^+ - \left| \frac{v^*(t\chi_I + \tau\chi_S) - v^*(t\chi_I)}{\tau} \right| < \frac{1}{k}. \end{gathered}$$

With each t and τ satisfying (44.13), associate the interval whose endpoints are t and $t + \tau$; let $\mathcal{H}^t$ be the family of intervals so defined. Now apply Lemma 44.6 to the system defined by $\{T_0, T_1, \ldots, T_{k^2}\}$ and the families $\mathcal{H}^t$. This yields a family $\mathcal{G}$ of intervals satisfying the conclusions of that lemma (for given ϵ). Denote the intervals of $\mathcal{G}$ by $U_1, \ldots, U_p$, where for all h, the right endpoint of U_h is left of the left endpoint of U_{h+1}; this is possible because the U_h are disjoint. Each U_h has endpoints t and $t + \tau$ satisfying (44.13); denote them by t_h and $t_h + \tau_h$, respectively. Now construct a chain Ω of ideal sets

$$0 = g_0 \leqq g_1 \leqq \cdots \leqq g_{2p+1} = \chi_I$$

by letting

$$g_{2h-1} = t_h\chi_I \quad \text{and} \quad g_{2h} = t_h\chi_I + \tau_h\chi_S$$

if $\tau > 0$, and

$$g_{2h-1} = t_h\chi_I + \tau_h\chi_S \quad \text{and} \quad g_{2h} = t_h\chi_I$$

if $\tau < 0$. Note that since $|\tau_h| < \delta$ for all h, the chain Ω satisfies the condition (44.9), and hence $\|v^*\|_\delta \geqq \|v^*\|_\Omega$. To evaluate $\|v^*\|_\Omega$, for each i relabel the intervals constituting $\mathcal{G} \cap \mathcal{H}_i$ by $U_{i1}, \ldots, U_{iq}$ (where q

depends on i); these are some of the U_h, and when i varies, we get all of the U_h. If $U_{ij} = U_h$, let $t_{ij} = t_h$ and $\tau_{ij} = \tau_h$. Then $t_{ij} \in T_i$, and so by Lemma 44.6 and (44.12), we have

$$\begin{aligned}
\|v^*\|_\delta \geqq \|v^*\|_\Omega &\geqq \sum_{h=1}^{p} |v^*(t_h\chi_I + \tau_h\chi_S) - v^*(t_h\chi_I)| \\
&= \sum_{h=1}^{p} \left| \frac{v^*(t_h\chi_I + \tau_h\chi_S) - v^*(t_h\chi_I)}{\tau_h} \right| \lambda(U_h) \\
&= \sum_i \sum_j \left| \frac{v^*(t_{ij}\chi_I + \tau_{ij}\chi_S) - v^*(t_{ij}\chi_I)}{\tau_{ij}} \right| \lambda(U_{ij}) \\
&= \sum_i \sum_j \frac{i-1}{k} \lambda(U_{ij}) = \sum_i \frac{i-1}{k} \lambda(S_i) \\
&= \sum_i \frac{i}{k} \lambda(S_i) - \frac{1}{k} \sum_i \lambda(S_i) \geqq \sum_i \frac{i}{k} \lambda(S_i) - \frac{1}{k},
\end{aligned}$$

where $S_i = \cup_j U_{ij}$ is as in Lemma 44.6. From Lemma 44.6 we obtain

$$\lambda(S_i + T_i) < \epsilon;$$

combining this with the previous inequality, we obtain

$$\|v^*\|_\delta \geqq \sum_i \frac{i}{k} \lambda(T_i) - \epsilon \sum_i \frac{i}{k} - \frac{1}{k} = \sum_i \frac{i}{k} \lambda(T_i) - \epsilon \frac{k^2(k^2+1)}{2k} - \frac{1}{k}.$$

Letting $\epsilon \to 0$, we deduce

$$\|v^*\|_\delta \geqq \sum_i \frac{i}{k} \lambda(T_i) - \frac{1}{k}.$$

But as $k \to \infty$, we have by Lemma 44.8 that[3]

$$\sum_i \frac{i}{k} \lambda(T_i) \to \int_0^1 |\partial v^*(t)|^+ \, dt.$$

Since $\frac{1}{k} \to 0$ as $k \to \infty$, we deduce

$$\|v^*\|_\delta \geqq \int_0^1 |\partial v^*(t)|^+ \, dt.$$

This completes the proof of Lemma 44.11.

[3] The T_i depends on k as well as on i.

LEMMA 44.14. *Let $w + r \in BV \cap pNA'$, where $w \in DIAG$. Then for $\delta > 0$ sufficiently small,*

$$\|(w + r)^*\|_\delta \leqq \|r\|.$$

Proof. Since w is in $DIAG$, it satisfies (42.1). Let k, ζ, and U correspond to w in accordance with (42.1). Let δ be such that

$$(\exists t)(\|\zeta(S) - te\| < \delta) \Rightarrow \zeta(S) \in U; \tag{44.15}$$

this is true for all sufficiently small δ. Let $v = w + r$. For given $\epsilon > 0$, let Ω be a chain

$$0 = g_0 \leqq g_1 \leqq \cdots \leqq g_m \leqq g_{m+1} = \chi_I$$

of ideal set functions satisfying (44.9), such that

$$\|v^*\|_\Omega > \|v^*\|_\delta - \epsilon. \tag{44.16}$$

Since $v \in pNA'$, we may find a polynomial in measures $f \circ \nu$, where ν is a vector of measures in NA^+, such that

$$\|v - f \circ \nu\|' < \epsilon/4m; \tag{44.17}$$

from this and (22.14) it follows that $\|(v - f \circ \nu)^*\|' < \epsilon/4m$, and hence

$$\|(v - f \circ \nu)^*\|_\Omega < 2(m + 1)\epsilon/4m \leqq \epsilon. \tag{44.18}$$

Now apply Lemma 44.1 to the vector measures $\xi = (\zeta, \nu)$, obtaining a chain Γ of sets

$$\emptyset = T_0 \subset T_1 \subset \cdots \subset T_m \subset T_{m+1} = I$$

in $\mathcal{C}$ such that for all i

$$\xi(T_i) = \int g_i \, d\xi.$$

It follows that for all i, $\nu(T_i) = \int g_i \, d\nu$, and so

$$(f \circ \nu)^*(g_i) = f(\int g_i \, d\nu) = (f \circ \nu)(T_i);$$

hence

$$\|(f \circ \nu)^*\|_\Omega = \|f \circ \nu\|_\Gamma. \tag{44.19}$$

Next, for each i, since g_i satisfies (44.9), there is a number t_i such that

$$|(g_i - t_i\chi_I)(s)| < \delta$$

for all s in I. Hence for each component ζ_p of ζ,

$$|\zeta_p(T_i) - t_i| = |\int g_i \, d\zeta_p - t_i| = |\int (g_i - t_i\chi_I) \, d\zeta_p| \leqq \int |g_i - t_i\chi_I| \, d\zeta_p < \delta.$$

Therefore $\|\zeta(T_i) - t_i e\| < \delta$, and so by (44.15), $\zeta(T_i) \in U$; hence $w(T_i) = 0$, and so

$$\|w\|_\Gamma = 0. \tag{44.20}$$

Note also that by (44.17),

$$\|v - f \circ \nu\|_\Gamma < 2(m+1)\epsilon/4m \leqq \epsilon. \tag{44.21}$$

Combining (44.16), (44.18), (44.19), (44.20), and (44.21), we get

$$\begin{aligned}\|v^*\|_\delta &\leqq \epsilon + \|v^*\|_\Omega \leqq \epsilon + \|(v - f \circ \nu)^*\|_\Omega + \|(f \circ \nu)^*\|_\Omega \\ &\leqq 2\epsilon + \|f \circ \nu\|_\Gamma \leqq 2\epsilon + \|f \circ \nu - v\|_\Gamma + \|v - w\|_\Gamma \\ &+ \|w\|_\Gamma \leqq 3\epsilon + \|r\|_\Gamma + 0 \leqq 3\epsilon + \|r\|.\end{aligned}$$

Letting $\epsilon \to 0$, we obtain the desired result. This completes the proof of Lemma 44.14.

The following proposition is an analogue of Theorem H. (Recall that φ is the unique continuous diagonal value on $pNAD$.)

PROPOSITION 44.22. *Let $v \in pNAD \cap pNA'$. Then for each $S \in \mathcal{C}$, the derivative $\partial v^*(t, S)$ exists for almost all t in* [0, 1], *and is integrable over* [0, 1] *as a function of t; and*

$$(\varphi v)(S) = \int_0^1 \partial v^*(t, S) \, dt.$$

Proof. The proof follows the ideas of the proof of Theorem H (Section 23). Define

$$\Delta_v(t) = \limsup_{\tau \to 0} \frac{v^*(t\chi_I + \tau\chi_S) - v^*(t\chi_I)}{\tau} - \liminf_{\tau \to 0} \frac{v^*(t\chi_I + \tau\chi_S) - v^*(t\chi_I)}{\tau}$$

(cf. (23.5)), and

$$\Delta_v = \int_0^1 \Delta_v(t) \, dt.$$

From Lemma 44.11 we then obtain

$$0 \leqq \Delta_v \leqq 2 \int_0^1 |\partial v^*(t)|^+ \, dt \leqq 2\|v^*\|_\delta$$

for all $\delta > 0$ (cf. (23.6)); furthermore

$$\Delta_{v+w} \leqq \Delta_v + \Delta_w$$

whenever $v, w \in BV \cap pNA'$. Now let $v \in pNAD \cap pNA'$; for given $\epsilon > 0$, let $v = q + w + r$, where $q \in pNA$, $w \in DIAG$, and $\|r\| < \epsilon$. From Theorem H it follows that $\Delta_q = 0$; hence since $w + r = v - q \in BV \cap pNA'$, we get, using Lemma 44.14, that for δ sufficiently small,

$$0 \leqq \Delta_v \leqq \Delta_q + \Delta_{w+r} \leqq 0 + 2\|(w + r)^*\|_\delta \leqq \|r\| < \epsilon.$$

Letting $\epsilon \to 0$, we deduce $\Delta_v = 0$. Hence $\Delta_v(t) = 0$ for almost all t, i.e. $\partial v^*(t)$ exists[4] a.e. for all v. Whenever it exists we have $|\partial v^*(t)| = |\partial v^*(t)|^+$, and hence by Lemmas 44.11 and 44.2, we have, for all $\delta > 0$,

$$\int_0^1 |\partial v^*(t)| \, dt \leqq \|v^*\|_\delta \leqq \|v^*\| = \|v\|; \tag{44.23}$$

in particular, this implies the integrability of $\partial v^*(t)$.

Now let

$$\theta v = \int_0^1 \partial v^*(t) \, dt = \int_0^1 \partial v^*(t,S) \, ds;$$

then θv is linear in v, and by (44.23),

$$|\theta v| \leqq \|v^*\|_\delta \leqq \|v\| \tag{44.24}$$

for all $\delta > 0$. By Theorem H, $(\varphi q)(S) = \theta q$; therefore by (44.24) and Lemma 44.14,

$$\begin{aligned} |(\varphi v)(S) - \theta v| &\leqq |(\varphi q)(S) - \theta q| + |\varphi(w + r)(S) - \theta(w + r)| \\ &\leqq 0 + |\varphi(w + r)(S)| + |\theta(w + r)| \\ &\leqq |\varphi(w + r)(S)| + \|(w + r)^*\|_\delta \\ &\leqq |\varphi(w + r)(S)| + \|r\| < |\varphi(w + r)(S)| + \epsilon. \end{aligned} \tag{44.25}$$

Now φ coincides with the asymptotic value on $pNAD$ (Proposition 43.13); furthermore, since w and $w + r$ are both in $pNAD$, so is r

[4] As in Section 23, we write $\partial v^*(t)$ for $\partial v^*(t,S)$.

(though it may not be in pNA'). Since $\|\varphi\| = 1$ for the asymptotic value (Proposition 18.1), it follows that

$$(44.26) \qquad \|\varphi r\| \leqq \|r\| < \epsilon.$$

On the other hand, $w \in DIAG$ and φ has the diagonal property (Proposition 43.11); hence $\varphi w = 0$. Combining this with (44.25) and (44.26), we obtain

$$|\varphi v(S) - \theta v| \leqq \|\varphi(w + r)\| + \epsilon \leqq \|\varphi w\| + \|\varphi r\| + \epsilon \leqq 2\epsilon;$$

hence letting $\epsilon \to 0$, we deduce $\varphi v(S) = \theta v$, as was to be proved. This completes the proof of Proposition 44.22.

The following is an analogue of Proposition 27.8.

PROPOSITION 44.27. *If*[5] $w \in pNA'$, *then every member of the core of* w *is in* NA.

Proof. The proof follows the ideas of that of Proposition 27.8. What is needed is the existence of an NA^+ measure ν such that if $T_1, T_2, \ldots$ is a sequence of sets in $\mathcal{C}$ with $\nu(T_i) \to 0$, then (27.10) holds, i.e.

$$w(T_i) \to 0 \quad \text{and} \quad w(I \backslash T_i) \to w(I);$$

this may be called[6] the *continuity of* w *w.r.t.* ν *at* $\emptyset$ *and* I. To establish this, let $w_1, w_2, \ldots$ be a sequence in pNA such that

$$\|w - w_j\|' \to 0.$$

Let ν_j in NA^1 be such that $w_j \ll \nu_j$ for all j. Set $\nu = \Sigma_{i=1}^{\infty} \nu_j/2^j$. If $\nu(T_i) \to 0$, then $\nu_j(T_i) \to 0$ for all j, and hence as $i \to \infty$,

$$w_j(T_i) \to 0 \quad \text{and} \quad w_j(I \backslash T_i) \to w_j(I).$$

For given $\epsilon > 0$, pick j such that $\|w - w_j\|' < \epsilon/2$, and let N be such that whenever $i > N$,

$$|w_j(T_i)| < \epsilon/2 \quad \text{and} \quad |w_j(I \backslash T_i) - w_j(I)| < \epsilon/2.$$

[5] The same conclusion holds when $w \in pNAD$.

[6] It is clear how this definition may be generalized to cover continuity of w w.r.t. ν at an arbitrary S. Compare the discussion of Example 33.11, where continuity of a set function at S is defined without referring to ν. It may be seen that continuity at S is implied by continuity at S w.r.t. some ν in NA^+; and this, in turn, is implied by absolute continuity.

Then whenever $i > N$,

$$|w(T_i)| < \epsilon \quad \text{and} \quad |w(I \backslash T_i) - w(I)| < \epsilon,$$

i.e. (27.10) holds. Thus the desired continuity property is established. The remainder of the proof is exactly as in Proposition 27.8.

The following proposition is an analogue of Theorem I.

PROPOSITION 44.28. *Let v be a superadditive set function in $pNAD \cap pNA'$ that is homogeneous of degree* 1. *Then there is a unique point in the core of v, namely φv.*

Proof. The proof follows the lines of the proof of Theorem I rather closely. Results analogous to Lemmas 27.2, 27.4, 27.5, and Corollary 27.3 are readily established, the only difference in the proofs being that references to Theorem H must be replaced by references to Proposition 44.22. Thus it is established that φv is in the core of v, and that it is the only member of the core of v that is in NA. But by Proposition 44.27, there are no other members of the core of v, and so the proof of Proposition 44.28 is complete.

45. The Asymptotic Value of a Transferable Utility Economy

As in the last section, φ will continue to denote the unique continuous diagonal value (i.e. the asymptotic value) on $pNAD$.

We now return to the world of Chapter VI; all the notations will be as they were there. In this section we shall prove:

PROPOSITION 45.1. *Let* **a** *be μ-integrable, let $u \in \mathfrak{U}_1$, and let v be given by* (30.1). *Then v is well-defined and is in $pNAD$, and the core of v consists of the single point φv.*

This immediately[1] yields the asymptotic part of Proposition 31.7.

Just as we used Theorem I to prove Theorem J, we shall use Proposition 44.28 to prove Proposition 45.1. Thus we must show that the v of Proposition 45.1 is well-defined, that it is in $pNAD \cap pNA'$, and that it is superadditive and homogeneous of degree 1.

Let $\mathfrak{D}$ be any subset of $\mathcal{C}$ containing $\emptyset$ and I. A real-valued function

[1] Proposition 43.13.

w on $\mathfrak{D}$ with $w(\emptyset) = 0$ is called a $\mathfrak{D}$-*function.* Monotonicity, bounded variation, and the variation norm are defined for $\mathfrak{D}$-functions just as they are for set-functions. Thus a $\mathfrak{D}$-function w is *monotonic* if $S_1 \supset S_2$ and $S_1, S_2 \in \mathfrak{D}$ imply $w(S_1) \geqq w(S_2)$; w is *of bounded variation* if it is the difference of monotonic functions; and in that case, its (*variation*) *norm* $\|w\|$ is defined by

$$\|w\| = \inf(w_1(I) + w_2(I)), \tag{45.2}$$

where the inf is taken over all monotonic $\mathfrak{D}$-functions w_1 and w_2 such that $w = w_1 - w_2$.

LEMMA 45.3. *If w is a $\mathfrak{D}$-function of bounded variation, then*

$$\|w\| = \sup \sum_{i=1}^{k} |w(S_i) - w(S_{i-1})|,$$

where the sup is taken over all chains

$$\emptyset = S_0 \subset \cdots \subset S_k = I$$

of S_i in $\mathfrak{D}$. Furthermore, the inf in the definition (45.2) *of $\|w\|$ is attained.*

Proof. The proof is similar to that of Proposition 4.1.

LEMMA 45.4. *Let w be a $\mathfrak{D}$-function of bounded variation. Then there exists a set function v in BV such that $v|\mathfrak{D} = w$ and*

$$\|v\| = \|w\|.$$

Proof. First let w be monotonic. Define v by

$$v(S) = \sup\{w(T)\colon T \in \mathfrak{D} \quad \text{and} \quad T \subset S\}.$$

Clearly v is monotonic, and $v(I) = w(I)$. This completes the proof in case w is monotonic.

In the general case, using Lemma 45.3, let w_1 and w_2 be monotonic $\mathfrak{D}$-functions such that

$$w = w_1 - w_2$$

and $w_1(I) + w_2(I) = \|w\|$. Let v_1 and v_2 be monotonic set functions such that $v_1|\mathfrak{D} = w_1$, $v_2|\mathfrak{D} = w_2$, and $v_1(I) = w_1(I)$, $v_2(I) = w_2(I)$. Define $v = v_1 - v_2$. Then

$$\|v\| \leqq \|v_1\| + \|v_2\| = w_1(I) + w_2(I) = \|w\|.$$

But again from Lemma 45.3 it is clear that $\|v\| \geqq \|w\|$, since v is an extension of w. This completes the proof of Lemma 45.4.

Throughout the remainder of this section, $\mathbf{a}$ will be a fixed μ-integrable function from I to Ω, u will be a fixed member of $\mathfrak{U}_1$, and v will be the set-function corresponding to $\mathbf{a}$ and u (i.e. $v(S) = u_S(\int_S \mathbf{a})$); that it is well-defined follows from Proposition 36.1. We shall use the notation $\hat{u}$ for δ-approximations to u (though $\hat{u}$ is not fixed throughout the discussion), and for given $\hat{u}$, we will denote by $\hat{v}$ the set function corresponding to $\mathbf{a}$ and $\hat{u}$ (i.e. $\hat{v}(S) = \hat{u}_S(\int_S \mathbf{a})$).

LEMMA 45.5. *For each $\delta > 0$, define*

$$\mathfrak{D}_\delta = \left\{S \in \mathcal{C}: \left\| \int_S \mathbf{a} - \mu(S) \int \mathbf{a} \right\| < \delta\right\}.$$

Then for every $\epsilon > 0$ there is a $\delta > 0$ such that if $\hat{u} \in \mathfrak{U}_1$ is a δ-approximation to u, then

$$\|v|\mathfrak{D}_\delta - \hat{v}|\mathfrak{D}_\delta\| < \epsilon.$$

Proof. The proof is similar to that of Proposition 40.24, the restriction to $\mathfrak{D}_\delta$ taking the place of condition (40.25). W.l.o.g.[2] let $\int \mathbf{a} = e$. Let δ_1 correspond to $\frac{1}{4}\epsilon$ in accordance with Proposition 40.22; w.l.o.g. let $\delta_1 < 1$. Let

$$\epsilon_2 = \min(\tfrac{1}{2}\delta_1, \tfrac{1}{2}\epsilon),$$

and choose δ_2 to correspond to ϵ_2 in accordance with Lemma 40.7. Let $\delta = \min(\frac{1}{8}\delta_1, \delta_2)$.

Let $w = v - \hat{v}$, and let

$$\emptyset = S_0 \subset S_1 \subset \cdots \subset S_m \subset S_{m+1} = I$$

be a chain in $\mathfrak{D}_\delta$. By Lyapunov's theorem (Proposition 1.3), it is always possible to insert finitely many additional sets

$$S_{01}, S_{02}, \ldots, S_{11}, S_{12}, \ldots, \ldots, S_{m1}, S_{m2}, \ldots$$

in $\mathfrak{D}_\delta$ into the chain so that

$$S_0 \subset S_{01} \subset S_{11} \subset \cdots \subset S_1 \subset S_{11} \subset S_{12} \subset \cdots \subset \cdots \subset S_m \subset S_{m1} \subset S_{m2} \subset \cdots \subset S_{m+1}$$

[2] Any commodity j for which $\int \mathbf{a}^j = 0$ may simply be excluded from consideration.

and the measure of the difference between two neighboring sets is $< \delta$; that is, if we relabel the new sequence $U_0, \ldots, U_{p+1} = I$, then $\mu(U_{k+1} \backslash U_k) < \delta$ for all k. Furthermore, by Lyapunov's theorem (in n dimensions[3]) we may suppose w.l.o.g. that for one of the U_k, say for U_q, we have $\int_{U_q} \mathbf{a}^1 = \frac{3}{4}\delta_1$. From $U_q \in \mathfrak{D}_\delta$ it follows that for all j,

$$\left| \int_{U_q} \mathbf{a}^j - \int_{U_q} \mathbf{a}^1 \right| < 2\delta \leqq \tfrac{1}{4}\delta_1,$$

and hence

$$\int_{U_q} \mathbf{a}^j < \tfrac{3}{4}\delta_1 + \tfrac{1}{4}\delta_1 = \delta_1 \tag{45.6}$$

and

$$\int_{U_q} \mathbf{a}^j > \tfrac{3}{4}\delta_1 - \tfrac{1}{4}\delta_1 = \tfrac{1}{2}\delta_1 \geqq \epsilon_2. \tag{45.7}$$

Now

$$\sum_{k=0}^{m} |w(S_{k+1}) - w(S_k)| \leqq \sum_{k=0}^{p} |w(U_{k+1}) - w(U_k)| = \sum_{k=0}^{q-1} + \sum_{k=q}^{p}.$$

From (45.7), Lemma 40.7, and the fact that $\hat{u}$ is a δ-approximation—hence a fortiori a δ_2-approximation—to u, we then obtain

$$\sum_{k=q}^{p} \leqq \epsilon_2 \leqq \tfrac{1}{2}\epsilon.$$

Furthermore, since u is a δ-approximation to u, it is a fortiori a δ_1-approximation. Hence by the monotonicity of v and $\hat{v}$ and by (45.6) and Lemma 40.22, we have, exactly as in the proof of Proposition 40.24, that

$$\sum_{k=0}^{q-1} \leqq v(U_q) + \hat{v}(U_q) < \tfrac{1}{4}\epsilon + \tfrac{1}{4}\epsilon = \tfrac{1}{2}\epsilon.$$

We conclude that

$$\sum_{k=0}^{m} |w(S_{k+1}) - w(S_k)| < \tfrac{1}{2}\epsilon + \tfrac{1}{2}\epsilon = \epsilon,$$

and it follows that

$$\|v|\mathfrak{D}_\delta - \hat{v}|\mathfrak{D}_\delta\| = \|w|\mathfrak{D}_\delta\| < \epsilon.$$

This completes the proof of Lemma 45.5.

[3] The one-dimensional Lyapunov theorem is not sufficient because we must make sure that $U_q \in \mathfrak{D}_\delta$.

COROLLARY 45.8. $v \in pNAD$.

Proof. Let ϵ be given, and let δ and $\hat{u}$ correspond to ϵ in accordance with Lemma 45.5. By Lemma 45.4, there then exists a set-function r in BV such that $r|\mathfrak{D}_\delta = (v - \hat{v})|\mathfrak{D}_\delta$ and

$$\|r\| = \|(v - \hat{v})|\mathfrak{D}_\delta\| < \epsilon.$$

Then

$$v - r = (v - \hat{v} - r) + \hat{v}.$$

Now $(v - \hat{v} - r)|\mathfrak{D}_\delta = 0$, and hence $v - \hat{v} - r \in DIAG$; and by Lemma 39.16, $\hat{v} \in pNA$. Thus

$$v - r \in pNA + DIAG.$$

Since $\|r\| < \epsilon$ and ϵ was chosen arbitrarily small, it follows that v is in the closure of $pNA + DIAG$. This completes the proof of Corollary 45.8.

Let H' be the set of all superadditive set functions in pNA' that are homogeneous of degree 1.

PROPOSITION 45.9. *H' is closed in the supremum norm.*

Proof. The proof is exactly analogous to that of Proposition 27.12.

PROPOSITION 45.10. $v \in H'$.

Proof. Let $\epsilon > 0$ be given, let δ correspond to $\epsilon/(1 + \Sigma\int \mathbf{a})$ in accordance with Proposition 37.11, and let $\hat{u} \in \mathfrak{U}_1$ be a δ-approximation to u. Then for all $S \in \mathcal{C}$ we have

$$|v(S) - \hat{v}(S)| = \left| u_S\left(\int_S \mathbf{a}\right) - \hat{u}_S\left(\int_S \mathbf{a}\right) \right| < \epsilon \frac{1 + \Sigma \int_S \mathbf{a}}{1 + \Sigma \int \mathbf{a}} \leqq \epsilon.$$

Hence

$$\|v - \hat{v}\|' = \sup_S |v(S) - \hat{v}(S)| < \epsilon.$$

Since an appropriate $\hat{u}$ can be found (Proposition 35.6) and since $\hat{v} \in H$ (Lemma 39.16), it follows that v is in the closure of H in the supremum norm. But this is certainly included in the sup-closure of H', and so by Proposition 45.9 in H'. The proof of Proposition 45.10 is complete.

Proposition 45.1 follows immediately from Proposition 44.28, Corollary 45.8, and Proposition 45.10.

46. The Mixing Value of a Transferable Utility Economy

A u in $\mathfrak{U}_1$ and a μ-integrable $\mathbf{a}$ will continue to be given throughout this section, and v will be defined by (30.1); all other conventions of Chapter VI will be maintained as well. Also, φ will continue to denote the unique continuous diagonal value on $pNAD$ (Proposition 43.13). The main result of this section may then be stated as follows:

PROPOSITION 46.1. $v \in ORD$.

Together with Proposition 45.1, this shows that $v \in pNAD \cap ORD$; then from Propositions 43.13 and 45.1 it follows that v has a mixing value, and it equals the unique point in the core. This proves the "mixing" part of Proposition 31.7, and so completes the proof of that proposition (the "asymptotic" part was proved in the previous section). Thus it remains only to prove Proposition 46.1.

LEMMA 46.2. *For every ϵ there is a β such that for all S with $\mu(S) \geqq \epsilon$ and all b with $\int \mathbf{a} \geqq b \geqq \epsilon e$, we have*[1]

$$\|u_S'(b)\| \leqq \beta.$$

Proof. Let $\boldsymbol{\eta}$ correspond to ϵ and $\alpha = \Sigma \int \mathbf{a}$ in accordance with Lemma 37.8. From the integrability of $\boldsymbol{\eta}$ it follows that

$$\sup \left\{ \int_T \boldsymbol{\eta} \colon \mu(T) \leqq \delta \right\} \to 0$$

as $\delta \to 0$; hence there is a δ such that

$$\mu(T) \leqq \delta \Rightarrow \int_T \boldsymbol{\eta} < \epsilon/2. \tag{46.3}$$

For each s in I, u_s^j is continuous at each point of the compact set

$$\{y \in \Omega \colon y \leqq \boldsymbol{\eta}(s) \quad \text{and} \quad y^j \geqq \epsilon/2\};$$

hence u_s^j is bounded there. If we let $\boldsymbol{\beta}^j(s)$ be the maximum of u_s^j there, then $\boldsymbol{\beta}^j(s)$ is measurable. From this it follows that there is a number β^j

[1] Recall that $u_S'(b)$ denotes the vector of partial derivatives $(u_S^1(b), \ldots, u_S^n(b))$.

such that in every set U of measure $\geqq \delta$ there is a non-null set of s with $\boldsymbol{\beta}^j(s) \leqq \beta^j$. Set

$$\beta = \max_j \beta^j.$$

Suppose now that S and b are as in the lemma, and let $u_S(b)$ be attained at $\mathbf{y}$. For each j, let

$$U^j = \{s \in S: \mathbf{y}^j(s) \geqq \epsilon/2\}.$$

Then we assert that for all j,

$$\mu(U^j) > \delta. \tag{46.4}$$

Indeed, if not, then for some j, for all $s \in S$ except possibly a set T of s's of measure $\leqq \delta$, we will have $\mathbf{y}^j(s) < \epsilon/2$; and for that set we will have $\mathbf{y}^j(s) \leqq \boldsymbol{\eta}(s)$. Hence by (46.3),

$$\epsilon \leqq b^j \leqq \int_S \mathbf{y}^j \leqq \int_{S\setminus T} \epsilon/2 + \int_T \boldsymbol{\eta} < \epsilon/2 + \epsilon/2,$$

an absurdity; this establishes (46.4).

From (46.4) it follows that for a nonnull set of s's in U^j, $\boldsymbol{\beta}^j(s) \leqq \beta^j \leqq \beta$; hence since $\mathbf{y}(s) \leqq \boldsymbol{\eta}(s)$ and $\mathbf{y}^j(s) \geqq \epsilon/2$, it follows that

$$u^j_\prime(\mathbf{y}(s), s) \leqq \beta.$$

Since $U^j \subset S$, it follows from this and Proposition 38.5 that $u^j_S(b) \leqq \beta$, as was to be proved. This completes the proof of Lemma 46.2.

LEMMA 46.5. *For every ϵ there is an integrable function ζ such that for all S and T with $S \supset T$ and $\int_T \mathbf{a} \geqq \epsilon e$, we have*

$$v(S) - v(T) \leqq \int_{S\setminus T} \zeta. \tag{46.6}$$

Proof. Let $v(S)$ be attained at $\mathbf{y}$. From (40.2) with $S_0 = T$ we obtain

$$v(S) - v(T) = \int_{S\setminus T} [u(\mathbf{y}) + u'_T(c)\cdot(\mathbf{a} - \mathbf{y})],$$

where

$$c = \theta \int_T \mathbf{y} + (1 - \theta) \int_T \mathbf{a}$$

for some $\theta \in [0, 1]$. From Lemma 37.8 (with a different ϵ) we deduce that $\mathbf{y}$ is bounded uniformly[2] by an integrable function, say $\boldsymbol{\eta}e$; hence,

[2] In S and T, but not in ϵ.

from Corollary 37.10, it follows that there is also such a function for $u(\mathbf{y})$. Thus it remains only to show that $u'_T(c)$ is uniformly bounded. But by Lemma 46.2, this would follow if we could show that all coordinates of c are uniformly bounded away from 0; and indeed it would suffice to show this for $\int_T \mathbf{y}$, since we already know that $\int_T \mathbf{a} \geqq \epsilon e$.

Suppose first that S and T are such that

$$\int_{S\backslash T} \eta \leqq \epsilon/2. \tag{46.7}$$

Then

$$\int_T \mathbf{y} = \int_S \mathbf{y} - \int_{S\backslash T} \mathbf{y} \geqq \int_S \mathbf{a} - \int_{S\backslash T} \eta e \geqq \epsilon e - \epsilon e/2 = \epsilon e/2;$$

this gives us the desired uniform lower bound, it being understood that the "uniformness" is subject to (46.7). Thus the theorem is proved in this case; i.e., subject to (46.7), there is a ζ satisfying (46.6). But now even when S and T do not satisfy (46.7), we may construct a chain

$$T = S_0 \subset S_1 \subset \cdots \subset S_k = S$$

such that

$$\int_{S_{i+1}\backslash S_i} \eta \leqq \epsilon/2$$

for any i. If one then applies what we have just proved to the pair (S_{i+1}, S_i), one deduces

$$v(S_{i+1}) - v(S_i) \leqq \int_{S_{i+1}\backslash S_i} \zeta;$$

and summing over i, one obtains (46.6). This completes the proof of Lemma 46.5.

Corollary 46.8. *For every ϵ there is an integrable function ζ such that if S and T are such that $S \supset T$ and for all j,*

$$\int_S \mathbf{a}^j > 0 \Rightarrow \int_T \mathbf{a}^j \geqq \epsilon,$$

then

$$v(S) - v(T) \leqq \int_{S\backslash T} \zeta.$$

Proof. Let $N = \{1, \ldots, n\}$. For each $L \subset N$, define

$$\Omega^L = \{a \in \Omega : a^j = 0 \text{ for } j \notin L\}.$$

By restricting the utility functions $u(x, s)$ to $x \in \Omega^L$ and considering only s for which $\mathbf{a}(s) \in \Omega^L$, we obtain an economy whose commodities are those indexed by $j \in L$ only rather than all $j \in N$. Lemma 46.5 applies to this economy, and yields a function ζ_L. The proof of Corollary 46.8 is concluded by setting $\zeta = \max_{L \subset N} \zeta_L$.

A set function w is said to be *continuous*[3] if for all $S \in \mathcal{C}$, all non-decreasing sequences $\{S_i\}$ such that $\cup S_i = S$, and all non-increasing sequences $\{S_i\}$ such that $\cap S_i = S$, we have

$$\lim w(S_i) = w(S).$$

LEMMA 46.9. *v is continuous.*

Proof. Every measure is continuous, because of the countable additivity. Continuity is preserved under multiplication, addition, and passage to the limit w.r.t. the supremum norm; hence each member of pNA' is continuous. But by Proposition 45.10, $v \in pNA'$, so the proof of Lemma 46.9 is complete.

Proof of Proposition 46.1. The proof is similar to that of Proposition 12.8, where we showed that $AC \subset ORD$. Let $\mathcal{R}$ be a measurable order; it will be fixed throughout the proof. Recall that $S \subset I$ is an *initial set* if

$$s \in S \quad \text{and} \quad s \mathcal{R} t \quad \text{imply} \quad t \in S.$$

All initial sets are measurable, i.e. are in $\mathcal{C}$ (Lemma 12.14). Therefore the σ-field generated by the initial sets is contained in $\mathcal{C}$; but since it contains all the initial segments $I(s; \mathcal{R})$, it must equal $\mathcal{C}$ (by (12.3)). Now let $K(\mathcal{R})$, or $\mathcal{K}$ for short, denote the field (not σ-field) generated by the initial sets; if we define an *$\mathcal{R}$-convex* set to be the set difference between two initial sets, then the members of $\mathcal{K}$ are precisely the finite unions of disjoint $\mathcal{R}$-convex sets. Whether or not $v \in ORD$, there is always a unique finitely additive ν on $\mathcal{K}$ such that

$$\nu(S) = v(S) \tag{46.10}$$

for all initial sets S. We shall show that ν can be extended to a countably additive measure $\varphi^{\mathcal{R}} v$ on the σ-field generated by $\mathcal{K}$; since the initial segments $I(s; \mathcal{R})$ are initial sets, (46.10) yields (12.2), and so we obtain $v \in ORD$.

In order to obtain the desired extension, we must[4] prove that ν is bounded and countably additive on $\mathcal{K}$. Since ν is indeed bounded, namely

[3] Cf. the definition of *continuity at S* in connection with Example 33.11; also the discussion of continuity in the proof of Proposition 44.27.

[4] See, for example, Dunford and Schwartz (1958), III 5.9, p. 136.

by $v(I)$, it remains only to establish the countable additivity; that is, if $U_1, U_2, \ldots$ is an infinite sequence of pairwise disjoint sets in $\mathcal{K}$ whose union U is also in $\mathcal{K}$, we must show that

$$\nu(U) = \sum_{i=1}^{\infty} \nu(U_i).$$

If we set $W_p = U \backslash \cup_{i=1}^{p} U_i$, then W_p is a non-increasing sequence of sets in $\mathcal{K}$ with empty intersection, and what we must prove is that

$$\nu(W_p) \to 0 \tag{46.11}$$

as $p \to \infty$.

The idea of the proof is quite simple. One can think of the players entering a room in the order $\mathcal{R}$, each one bringing along his initial bundle $\mathbf{a}(s)\mu(ds)$ and dumping it into the middle of the room. The pile gradually grows. At certain points of time, commodities that were not previously there will be added to the pile; but there can be only finitely many such "critical" points—no more than the number of commodities. Away from the "neighborhood"[5] of a critical point, ν is absolutely continuous w.r.t. μ, because of Corollary 46.8; therefore, on that part of W_p that is not near the critical points, ν tends to 0. But we can choose the neighborhood H_k of each critical point as small as we wish, and because of the continuity of v (Lemma 46.9), $\nu(H_k)$ will then be as small as we wish. Therefore $\nu(H^*)$ will also be small, where H^* is the union of all the neighborhoods of the critical points; and so ν will also be small on that part of W_p that is in H^*, as desired.

To formalize this argument, define, for each $b \in \Omega$,

$$L(b) = \{j: b^j > 0\}.$$

Let α be the initial resource measure, i.e. the vector measure on $\mathcal{C}$ defined by

$$\alpha(S) = \int_S \mathbf{a};$$

w.l.o.g. assume that all the α^j are in NA^1. Now the initial sets S are totally ordered under inclusion; hence the vectors $\alpha(S)$ are totally

[5] The "neighborhood" here is one-sided; it is a set of points coming "after" the critical point.

ordered under the relation $\geqq$ on Ω; hence the sets $L(\alpha(S))$ are totally ordered under inclusion. Thus we can find a sequence

$$\emptyset = L^0 \subset L^1 \subset \cdots \subset L^n = \{1, \ldots, n\},$$

where $|L^j| = j$, such that for all initial sets S, $L(\alpha(S))$ is one of the L^j. (Of course it is understood that not all the L^j need actually occur among the $L(\alpha(S))$; under (31.4), for example, only L^0 and L^n occur.) Renumber the goods so that for all j,

$$L^j = \{1, \ldots, j\}.$$

For each j, let J^j be the union of all initial sets S such that $\alpha^{j+1}(S) = 0$; clearly J^j is itself an initial set. If $s \in J^j$, then s is in some initial set S with $\alpha^{j+1}(S) = 0$. Then $I(s; \mathcal{R}) \subset S$, and hence $\alpha^{j+1}(S) = 0$; so by Lemma 12.14 (with $v = \alpha^{j+1}$),

$$\alpha^{j+1}(J^j) = \sup_{s \in J^j} \alpha^{j+1} I(s; \mathcal{R}) = 0.$$

For $m = 1, 2, \ldots$, let

$$J^j_m = J(1/m; \alpha^j, \mathcal{R})$$

i.e. J^j_m is the intersection of all initial segments of α^j-measure $> 1/m$. For fixed j, the J^j_m form a non-increasing sequence of initial sets all of which contain J^{j-1}; hence

$$\cap_m J^j_m \supset J^{j-1}.$$

But by Lemma 12.15,

$$\alpha^j(\cap_m J^j_m) = \lim_m \alpha^j(J^j_m) = 0 = \alpha^j(J^{j-1});$$

hence by the definition of J^{j-1},

$$\cap_m J^j_m = J^{j-1}.$$

Hence by Lemma 46.9,

(46.12) $$v(J^j_m) \to v(J^{j-1})$$

as $m \to \infty$.

Now let $\delta > 0$ be given; to prove (46.11), we must show that for p sufficiently large,

(46.13) $$\nu(W_p) < \delta.$$

Choose an m such that for each j,

$$(46.14)\qquad v(J^j_m) - v(J^{j-1}) < \delta/2n;$$

since there are only finitely many j, the existence of such an m follows from (46.12). For this m, let

$$\begin{aligned} H^j &= J^j_m \backslash J^{j-1} \\ H^* &= \textstyle\bigcup_j H^j \\ I^* &= I \backslash H^* \\ I^j &= (J^j - J^{j-1}) \cap I^*, \end{aligned}$$

where $J_{-1} = \emptyset$. All the sets so defined are in $\mathcal{K}$. By (46.14), we have

$$(46.15)\qquad \nu(H^*) \leqq \sum_j \nu(H^j) < n(\delta/2n) = \delta/2.$$

Next, let ζ correspond to $\epsilon = 1/m$ in accordance with Corollary 46.8. If V is an $\mathcal{R}$-convex subset of I^j, then $V = S \backslash T$, where S and T are initial sets that are included in J^j and include J^{j-1} and all the J^k_m with $k \leqq j$. Hence both $\alpha^k(S)$ and $\alpha^k(T)$ are $\geqq 1/m$ for $k \leqq j$, and both vanish otherwise; so from Corollary 46.8 we deduce

$$\nu(V) \leqq \int_V \zeta.$$

Since both sides of this inequality are (finitely) additive in V, we deduce the same inequality for any subset of I^j that is in $\mathcal{K}$; and hence also for any subset of I^* in $\mathcal{K}$, since any such set can be partitioned into subsets of the individual I^j. From this it follows that

$$(46.16)\qquad \nu(W_p \cap I^*) \leqq \int_{W_p \cap I^*} \zeta.$$

But since $\{W_p\}$ is non-increasing and $\bigcap W_p = \emptyset$, the right side of (46.16) tends to 0; hence for p sufficiently large,

$$\nu(W_p \cap I^*) < \delta/2.$$

Combining this with (46.15) and $I^* = I \backslash H^*$, we obtain

$$\nu(W_p) = \nu(W_p \cap I^*) + \nu(W_p \cap H^*) < \delta.$$

The proof of Proposition 46.1, and so also of Proposition 31.7, is thus complete.

Chapter VIII. Removal of the Standardness Assumption

47. Prelude

In this chapter we shall thoroughly investigate how much of what we have done remains valid when the Standardness Assumption (2.1) is removed or modified. Though the situation is rather complicated, the upshot is that, for most of the results, one can either substitute a weaker form of (2.1), or dispense with it altogether.

The Standardness Assumption is used in five key places in this book: in the proofs of Proposition 6.1 and Lemma 12.5; implicitly in the definitions of the mixing and asymptotic values (Sections 14 and 18); and again implicitly in the quotations from Aumann and Perles (1965) in Section 36. It is also used in Appendix C. We will consider each of these uses separately.

Proposition 6.1 asserts that if $\mu \in NA^1$ and $f \in bv$, and if φ is a value on a space containing $f \circ \mu$, then

$$(47.1) \qquad \varphi(f \circ \mu) = \mu.$$

This is the key to the proofs of uniqueness of the value on pNA and on $bv'NA$, and also to the proof of Theorem D (the "Random Order Impossibility Principle," Section 12). We shall show below[1] that, without the Standardness Assumption, the underlying space can be chosen so that Proposition 6.1 and Theorem D are false, and there are many values both on pNA and on $bv'NA$. As a result, the diagonal formula (3.1) loses its force, since it was asserted to hold for *the* value on pNA; similarly for Theorems H, I, and J.

On the other hand, Proposition 6.4, which is a sufficient condition for a value, is not affected by the removal of the Standardness Assumption; the existence proofs for the value are based on Proposition 6.4, and they are not affected either. Thus even without (2.1), there exists a value on $bv'NA$ (and a fortiori on pNA); in fact, there exists a value

[1] Propositions 48.4, 48.12, 48.22, and 48.24 and Corollaries 48.5 and 48.26.

that obeys (47.1), and for this value one will get the diagonal formula (3.1), and Theorems H, I, and J.

Theorems C and G are not affected at all by removal of the Standardness Assumption.

Proposition 6.1—and (47.1) in particular—may be considered an expression of the symmetry axiom (2.2) (see the proof of Proposition 6.1). The reason that Proposition 6.1 cannot be proved without the Standardness Assumption is that it is then possible to construct underlying spaces that have very few automorphisms, so that the symmetry axiom no longer says very much. In a sense, when there are few automorphisms it becomes easier to fulfill the conditions for a value. Thus where there was only one value before (on pNA and $bv'NA$), there are now many; and where it was impossible to define a value of a certain type (Theorem D), it now becomes possible.

This suggests that to restore the requisite "tightness" to the system, one could try replacing the symmetry axiom (2.2) by (47.1), or adding (47.1) as an additional axiom. If this is done (in the absence of (2.1)), then the uniqueness of the value on pNA and $bv'NA$ is restored; thus Theorems A, B, H, I, and J again become true.

Lemma 12.5 asserts that if $\mathcal{R}$ is a measurable order on I, then I has a denumerable $\mathcal{R}$-dense subset. It is used in the treatment of the Random Order Impossibility Principle (Sections 12 and 13) and again in the treatment of the mixing value (Sections 15, 43, and 46). To prove Lemma 12.5, it is however not necessary to use the full force of (2.1); rather, it is sufficient to assume that

(47.2) $\mathcal{C}$ is isomorphic to a subset[2] of $([0, 1], \mathcal{B})$.

To prove Theorem D, it is sufficient to assume (47.1) and (47.2). To get a mixing value, though, this is not sufficient; for this purpose one needs μ-mixing sequences, whose existence is guaranteed by Proposition 14.3. Even under (47.2) though, $(I, \mathcal{C})$ can be chosen (Proposition 48.13) so that there are no μ-mixing sequences for any μ. Thus, the definition of mixing value becomes vacuous,[3] and the entire theory of mixing

[2] This is equivalent to saying that $\mathcal{C}$ is countably generated and for any two points of I, there is a member of $\mathcal{C}$ containing one but not the other.

[3] Any set function will vacuously satisfy the definition of mixing value; in particular, it is no longer possible to speak of *the* mixing value of v.

values loses its meaning. Proposition 14.3 is used also in the proof of Proposition 13.4; the latter, too, is false without (2.1).

The Standardness Assumption enters the analysis of *ASYMP* via the notion of a "separating" sequence of partitions; in a general measurable space, such a sequence may not exist. In this case, though, (47.2) is sufficient; with it, the asymptotic value can be defined and Theorem F proved.

In Section 36, several theorems are quoted from Aumann and Perles (1965). These theorems depend on the theory of integrals of set-valued functions (Aumann, 1965), this in turn depends on a selection theorem of von Neumann (1949, p. 448, Lemma 5), and this in turn depends on (2.1). In this instance, though, (2.1) can be dispensed with entirely, since von Neumann's theorem can be generalized (Aumann, 1969) so as not to depend on standardness.

Appendix C uses the full force of the Standardness Assumption.

Proofs of the assertions made in this section regarding the non-uniqueness of the value, etc., will be given in Section 48. As we said above, these proofs will revolve around the construction of spaces on which there are few automorphisms. In particular, we will construct a subspace of ([0, 1], $\mathcal{B}$) with the cardinality of the continuum all of whose automorphisms are trivial, in the sense that they differ from the identity on a denumerable set only.

48. Fugue

In this section we present three examples of underlying spaces such that the value on pNA (and a fortiori on $bv'NA$) is not unique. In each case the idea is to have sufficiently few automorphisms so that the symmetry axiom cannot take full effect. The first two examples will be relatively simple from the measure-theoretic and set-theoretic viewpoint; but because the spaces still have non-trivial automorphisms, the subsequent construction of more than one value on pNA is not immediate. In the last example, the space has no non-trivial automorphisms at all; the construction of such a space is relatively involved from the measure-theoretic and set-theoretic viewpoint, but once it is constructed, the existence of many values on pNA is easy to show.

Throughout this section, c will denote the cardinality of the continuum.

We motivate the first of our examples by reexamining the argument used in Section 3 as an intuitive support for $\varphi(f \circ \mu) = \mu$: "In the game $f \circ \mu$, the payoff to a coalition depends only on its μ-measure. There would therefore seem to be no reason to 'discriminate' between coalitions having equal μ-measure, given assumption (2.1), and it seems natural to conjecture that they should get the same value. Because of the non-atomicity and the normalization condition (2.3), this implies that the value equals the μ-measure." The first portion of this argument is less than convincing in the context of an arbitrary measure space because it ignores the fact that sets having the same μ-measure may yet be distinguishable because of the underlying structure of the space I. The simplest example of this phenomenon is that in which some individual points have positive μ-measure (see Milnor and Shapley, 1961, and Hart, 1973). This violates the non-atomicity, but the first part of the argument does not really depend on non-atomicity; this is used only in the last part. Another example of this phenomenon, in which the measure *is* non-atomic, is as follows:

Example 48.1. Let $I = I_1 \cup I_2$, where $I_1 \cap I_2 = \emptyset$, I_1 is the cartesian product of denumerably many copies of the discrete space $\{0, 1\}$, and I_2 is the cartesian product of c copies of $\{0, 1\}$. The σ-field $\mathcal{C}$ on I is that generated by the product σ-fields on I_1 and I_2, respectively.

For the above reasons, it seems reasonable to believe that for the underlying space of this example, there are values on pNA and on $bv'NA$ that do not obey (47.1). This is indeed the case.[1] The crux of the proof is the following lemma:

LEMMA 48.2. *In Example* 48.1, *each automorphism* Θ *of* $(I, \mathcal{C})$ *leaves both* I_1 *and* I_2 *fixed.*

Proof. We first establish

(48.3) Every non-empty measurable subset of I_2 has 2^c elements.

To prove (48.3) let $I_2 = \times_\beta \Omega_\beta$, where each Ω_β is a copy of $\{0, 1\}$, and β ranges over an index set C of cardinality c. We claim that each

[1] And is equivalent to the non-uniqueness of the value; see Proposition 48.4 and its proof.

measurable subset S of I_2 is of the form $S' \times \times_{\beta \in C \setminus D} \Omega_\beta$, where D is a denumerable subset of C, and $S' \subset \times_{\beta \in D} \Omega_\beta$. This is certainly true of the sets that generate the σ-field of I_2, for which D contains only one element and $S' = \{0\}$ or $S' = \{1\}$. Furthermore, if our claim holds for a given S, then it also holds for its complement $I_2 \setminus S$; and if it holds for a sequence $S_1, S_2, \ldots$, then it also holds for its union $\cup_i S_i$. A standard argument then yields the truth of our claim, and (48.3) follows.

Now if Θ is an automorphism, then $I_2 \cap \Theta(I_1)$ is measurable; it must be of cardinality $\leq c$ because it is included in $\Theta(I_1)$, and be either empty or of cardinality $\geq 2^c$ because it is a measurable subset of I_2; so it is empty, i.e. $\Theta(I_1) \subset I_1$. Similarly $\Theta^{-1}(I_1) \subset I_1$, i.e. $I_1 \subset \Theta(I_1)$; so

$$I_1 = \Theta(I_1)$$

and hence also $I_2 = \Theta(I_2)$. This completes the proof of Lemma 48.2.

PROPOSITION 48.4. *In Example* 48.1, *there is more than one value on* pNA.

Proof. For $v \in pNA$, consider the set functions v^1 and v^2 on the underlying spaces I_1 and I_2, respectively, defined by

$$\begin{aligned} v^1(S) &= v(S), & S &\subset I_1 \\ v^2(S) &= v(S \cup I_1) - v(I_1), & S &\subset I_2. \end{aligned}$$

These set functions can be thought of as games derived from the game v by placing I_1 "before" I_2, i.e. letting the players in I_1 come to a settlement without taking I_2 into account, and only afterwards letting the players in I_2 bargain with each other for the remainder of the worth in the game. For $i = 1, 2$, let NA_i denote the space of non-atomic measures on I_i, and pNA_i the closure in the variation norm of polynomials[2] in NA_i measures. Denote by φ_i the value defined on pNA_i by Proposition[3] 7.6.

We claim that $v^i \in pNA_i$ for $i = 1, 2$. This follows from the fact that if v_n is a polynomial in NA measures, then the v_n^i are polynomials in NA_i measures, and if $v_n \to v$ in the variation norm, then $v_n^i \to v^i$ in the variation norm. Now define

$$\varphi v = \varphi_1 v^1 + \varphi_2 v^2.$$

[2] By "polynomial," we mean "polynomial with vanishing constant term."

[3] I.e., the existence part of Theorem B. As noted in Section 47, this does not use (2.1).

In this case it is immediate that φ is well defined, since the definition is directly in terms of v, and not in terms of a representation of v. All the conditions in the definition of value, including the positivity, follow from the corresponding conditions for the φ_i; for the symmetry, one must also use Lemma 48.2. So φ is a value on pNA.

If we take I_2 first and I_1 second, i.e. define

$$\begin{aligned} v^1(S) &= v(S) & S &\subset I_2 \\ v^2(S) &= v(S \cup I_2) - v(I_2), & S &\subset I_1 \end{aligned}$$

and proceed similarly, we get another value φ. The two values are different; indeed if μ and ν in NA^1 are concentrated on I_1 and I_2, respectively, then the value of $\mu\nu$ is ν in the first case and μ in the second. This completes the proof of Proposition 48.4.

COROLLARY 48.5. *In Example* 48.1, *there is more than one value on* $bv'NA$.

Proof. Any value φ on pNA obeys $\|\varphi v\| \leqq \|v\|$; this follows from the internality of pNA (Propositions 4.7 and 7.19; these propositions do not use (2.1)). Now define a value on $s'NA$ in accordance with the existence part of Theorem A; this value also obeys $\|\varphi v\| \leqq \|v\|$. Using Remark 8.23 and Proposition 4.6, we may combine the values on pNA and on $s'NA$ to yield a value on $bv'NA$; and so the non-uniqueness on $bv'NA$ follows from the non-uniqueness on pNA. This completes the proof of Corollary 48.5.

We may remark that the two values constructed in the proof of Proposition 48.4 are not the only values on pNA in Example 48.1. There is, of course, the value constructed in Proposition 7.6, i.e. the one obeying the diagonal formula. This is different from both values constructed in the proof of Proposition 48.4; for the set function $\mu\nu$ considered in that proof, the "diagonal value" yields $(\mu + \nu)/2$. There are, moreover, many other values, for example, all convex combinations of the above values, but we will not describe them further here.

Let us recall that a measurable space is called *separable* if its σ-field has a denumerable generating subset. Does Proposition 48.4 depend essentially on the non-separability of the underlying space, by all odds a rather pathological feature? Proposition 48.4 depends on the extreme inhomogeneity of the player space, the fact that there are no automorphisms that can "mix" I_1 and I_2. Can a similar inhomogeneity occur even when the player space is separable?

The answer is yes,[4] though the inhomogeneity will not be as clear cut as in Example 48.1. Before presenting the example, we introduce some notation and terminology, and prove a lemma.

From now on, *negligible*[5] means "of cardinality less than c," and *non-negligible* means "of cardinality c." These words will also be used adverbially; thus "A and B intersect non-negligibly" means "$|A \cap B| = c$." The first non-negligible ordinal number will be denoted Γ.

Remark 48.6. A Borel subset of $[0, 1]$ is negligible if and only if it is denumerable.

Proof. Follows from Proposition 1.1.

LEMMA 48.7. *There is a subset I_2 of the real interval* $(1, 2]$ *that includes no non-denumerable Borel subset of* $(1, 2]$ *but intersects every such set.*

Proof. There are continuum many non-denumerable Borel subsets of $(1, 2]$; denote them by B_α, where α runs over all ordinal numbers less than Γ. Similarly, let the points of $(1, 2]$ be x_α, where α runs over the same set of ordinal numbers. Define two sets $\{y_\alpha\}$ and $\{z_\alpha\}$, where α again runs over all ordinals $<\Gamma$, as follows: y_1 is the first point in B_1 (i.e. the x_α with smallest α), and z_1 is the second point in B_1. If y_β and z_β have been defined for all ordinals less than a given denumerable ordinal α, define y_α to be the first point in B_α that is larger than all the points y_β and z_β already defined, and z_α to be the second such point; they exist because B_α is non-negligible (Remark 48.6) and only negligibly many points have already been defined. The proof is completed by setting $I_2 = \{y_\alpha\}_{\alpha<\Gamma}$.

Example 48.8. Let I_1 be the unit interval $[0, 1]$, I_2 the subset of $(1, 2]$ provided by Lemma 48.7, and $I = I_1 \cup I_2$. Define $\mathcal{C}$ to be the set of subsets of I of the form $I \cap U$, where U is a Borel subset of $[0, 2]$.

In discussing Example 48.8, we will use the notion of "subspace." Let $(J, \mathfrak{D})$ be a measurable space, and let S be a (not necessarily measurable) subset of J. If we define $\mathcal{E}$ to be the set of all sets of the form

[4] See Lemma 48.9.

[5] Under the continuum hypothesis, "negligible" and "countable" are the same.

$S \cap T$, where $T \in \mathfrak{D}$, then $\mathcal{E}$ is called the subspace σ-field, and $(S, \mathcal{E})$ is a *subspace* of $(J, \mathfrak{D})$; when no confusion can result, we sometimes refer to this subspace simply as S, and it is to be understood that the σ-field to be attached to S is the subspace σ-field. Note that if $J \supset S \supset T$, then T considered as a subspace of S is the same as T considered as a subspace of J.

We are now ready to establish the basic inhomogeneity property of Example 48.8. This property is similar to the corresponding property of Example 48.1 (see Lemma 48.2), though it is not quite as clear-cut as there.

LEMMA 48.9. *If* Θ *is an automorphism of the player space* $(I, \mathcal{C})$ *in Example* 48.8, *then both* $\Theta I_1 \backslash I_1$ *and* $\Theta^{-1} I_1 \backslash I_1$ *are denumerable.*

Proof. The measurable spaces involved in this proof are $[0, 2]$ with its Borel subsets and the subspaces I, I_1, and ΘI_1 of $[0, 2]$. From the fact that ΘI_1 may also be considered a subspace of I, it follows that Θ is an isomorphism from I_1 onto ΘI_1. Let Φ be the inclusion mapping from ΘI_1 into $[0, 2]$. It then follows that $\Phi\Theta$ is a one-one measurable function from I_1 into $[0, 2]$, whose range is ΘI_1. Applying Proposition 1.2, we deduce that ΘI_1 is a Borel set. Hence $\Theta I_1 \backslash I_1$ is a Borel set, but since it is included in I_2, it cannot be non-denumerable; this proves that $\Theta I_1 \backslash I_1$ is denumerable. If we substitute Θ^{-1} for Θ, we get the second half of the lemma, and this completes its proof.

We next show that there are some NA^1 measures on $(I, \mathcal{C})$, i.e. that $pNA \neq \{0\}$ (and $bv'NA \neq \{0\}$). If J is a real interval, we abuse our notation slightly by letting $\mathcal{B}$ be the σ-field of Borel sets in J. A subspace $(K, \mathcal{E})$ of $(J, \mathcal{B})$ is called *absolutely thick* (in[6] J) if K intersects every non-denumerable Borel set in J. Thus from Lemma 48.7 it follows that the $(I, \mathcal{C})$ of Example 48.8 is absolutely thick (in $[0, 2]$).

LEMMA 48.10. *If* $(K, \mathcal{E})$ *is absolutely thick in the interval* J, *there is a one-one correspondence between the* NA *measures* μ *on* $(J, \mathcal{B})$ *and the* NA *measures* μ' *on* $(K, \mathcal{E})$, *such that*

$$\mu(S) = \mu'(S \cap K) \tag{48.11}$$

for all $S \in \mathcal{B}$.

[6] The definition is easily extended to an arbitrary measurable space.

Proof. We use the self-explanatory notations $NA(J)$ and $NA(K)$. For each μ in $NA(J)$, define an element $f(\mu) = \mu'$ of $NA(K)$ by (48.11). Then μ' is well defined: for if S and T in $\mathcal{B}$ are such that $S \cap K = T \cap K$, then[7] $\emptyset = (S \cap K) + (T \cap K) = (S + T) \cap K$; hence $S + T$ is denumerable, and hence $\mu(S) = \mu(T)$. Conversely, for each μ' in $NA(K)$, define an element $g(\mu') = \mu$ of $NA(J)$ by (48.11); then μ is trivially well defined. Moreover, $g \circ f$ and $f \circ g$ are the identities on $NA(J)$ and $NA(K)$, respectively. Hence both f and g are one-one and onto, and so Lemma 48.10 is proved.

PROPOSITION 48.12. *In Example* 48.8, *there is more than one value on* pNA *and on* $bv'NA$.

Proof. The proof is exactly like those of Proposition 48.4 and Corollary 48.5; instead of Lemma 48.2, one uses Lemma 48.9. To show that there are measures on I_2 (which is needed in the proof), one notes that I_2 is absolutely thick in [1, 2], and uses Lemma 48.10.

Our final example is the furthest one can go in the way of inhomogeneity, as long as one restricts oneself to subspaces[8] of $([0, 1], \mathcal{B})$.

PROPOSITION 48.13. *There is an absolutely thick subspace of* $([0, 1], \mathcal{B})$, *for all of whose automorphisms* Θ, *the set* $\{x: \Theta x \neq x\}$ *is denumerable.*

We precede the proof of Proposition 48.13 by two lemmas.

LEMMA 48.14. *Let* $(I, \mathcal{C})$ *be a subspace of* $([0, 1], \mathcal{B})$. *Then every measurable function* f *from* $(I, \mathcal{C})$ *into itself can be extended to a measurable function* g *from* $([0, 1], \mathcal{B})$ *into itself.*

Proof. We wish to find a measurable function g from $([0, 1], \mathcal{B})$ into itself such that $g|I = f$. First note that f is measurable also when viewed as a function from $(I, \mathcal{C})$ to $([0, 1], \mathcal{B})$; indeed, if $T \in \mathcal{B}$, then $I \cap T \in \mathcal{C}$, so

$$f^{-1}(T) = f^{-1}(I \cap T) \in \mathcal{C}.$$

For each non-negative integer n and each j with $1 \leqq j \leqq 2^n$, define

$$I_n^j = f^{-1}[(j - 1)/2^n, j/2^n);$$

[7] Recall that "+" denotes the symmetric difference.

[8] A measurable space $(I, \mathcal{C})$ is called *regular* if any two points are separated, i.e. if for $x, y \in I$, there is an S in $\mathcal{B}$ such that $x \in S$ and $y \notin S$. Any space that is both separable and regular is isomorphic to a subspace of $([0, 1], \mathcal{B})$ (see Mackey (1957), Theorem 2.1).

then $I_n^j \in \mathcal{C}$. We will define Borel subsets J_n^j of [0, 1] that for fixed n are pairwise disjoint, and such that $J_n^j \cap I = I_n^j$. The definition is inductive. For $n = 0$, $J_n^j = [0, 1]$. Next, assume the J_m^j have been defined for $m < n$. Let i be even; note that

$$(48.15) \qquad I_n^i \cup I_n^{i-1} = I_{n-1}^{i/2}.$$

Let K $(=K_n^i)$ be a Borel set for which $K \cap I = I_n^i$; there is such a K because $I_n^i \in \mathcal{C}$. Define

$$J_n^i = J_{n-1}^{i/2} \cap K$$
$$J_n^{i-1} = J_{n-1}^{i/2} \backslash K.$$

The disjointness of all the 2^n sets J_n^j follows from that of the J_{n-1}^j. Furthermore,

$$J_n^i \cap I = J_{n-1}^{i/2} \cap K \cap I = J_{n-1}^{i/2} \cap I_n^i = I_n^i,$$

and by (48.15),

$$J_n^{i-1} \cap I = (J_{n-1}^{i/2} \backslash K) \cap I = I_{n-1}^{i/2} \backslash I_n^i = I_n^{i-1}.$$

This completes the inductive construction of the J_n^j.

Define functions f_n on the half-open interval [0, 1) by

$$f_n(x) = j/2^n \quad \text{when } x \in J_n^j.$$

The f_n satisfy the following conditions:

$$(48.16) \qquad f_n \text{ is measurable.}$$
$$(48.17) \qquad |f_n(x) - f(x)| < 1/2^n \qquad \text{for } x \in I\backslash\{1\}.$$
$$(48.18) \qquad |f_n(x) - f_m(x)| < 1/2^n \qquad \text{whenever } n < m.$$

By (48.18), the sequence $\{f_n(x)\}_{n=0}^{\infty}$ is a Cauchy sequence, and so converges; by (48.17), the limit is $f(x)$ when $x \in I\backslash\{1\}$; and by (48.16), the limit is measurable. Setting

$$f^*(x) = \begin{cases} \lim_{n\to\infty} f_n(x) & \text{if } x \in [0, 1), \\ f(1), & \text{if } x = 1 \text{ and } 1 \in I, \\ 0, & \text{if } x = 1 \text{ and } 1 \notin I, \end{cases}$$

we complete the proof of Lemma 48.14.

LEMMA 48.19. *Let* $(I, \mathcal{C})$ *be a subspace of* $([0, 1], \mathcal{B})$. *Then every automorphism* Θ *of* $(I, \mathcal{C})$ *can be extended to an automorphism* Φ *of* $([0, 1], \mathcal{B})$.

Proof. Both Θ and Θ^{-1} are measurable functions from $(I, \mathcal{C})$ into itself, and so can be extended to measurable functions from $([0, 1], \mathcal{B})$ to itself; call the extensions g and $\hat{g}$, respectively. Let

$$J = \{x \in [0, 1]: x = \hat{g}(g(x))\}.$$

Then J is a Borel subset of $[0, 1]$, and $J \supset I$. Let $K = g(J)$. Then $\hat{g}(K) = \hat{g}(g(J)) = J$. Furthermore if $y \in K$, then there is an x in J such that $y = g(x)$; hence

$$g(\hat{g}(y)) = g(\hat{g}(g(x))) = g(x) = y.$$

Conversely, if $y \in [0, 1]$ is such that $g(\hat{g}(y)) = y$, then setting $x = \hat{g}(y)$ we find

$$\hat{g}(g(x)) = \hat{g}(g(\hat{g}(y))) = \hat{g}(y) = x;$$

hence $x \in J$, and so $y = g(x) \in K$. Thus

$$K = \{y \in [0, 1]: y = g(\hat{g}(y))\};$$

hence K too is a Borel subset of $[0, 1]$, and $K \supset I$. Letting $h = g|J$ and $h = \hat{g}|K$, we find that $\hat{h}h$ and $h\hat{h}$ are the identities on J and K, respectively, and so, by a familiar argument, both are one-one and onto; since both are measurable, it follows that h is an isomorphism from the subspace of $([0, 1], \mathcal{B})$ determined by J onto that determined by K. Thus we have extended Θ to an isomorphism between two subspaces of $([0, 1], \mathcal{B})$, both of which are defined by Borel sets.

It remains only to define the extension Φ outside of these subspaces. Denoting $[0, 1]\backslash X$ by X', we have the following rather tiresome list of cases:

(i) I' is at most denumerable.

(ii) I' is non-denumerable, but both J' and K' are at most denumerable.

(iii) Either J' or K' or both are non-denumerable.

In case (i), I is itself Borel, so if we let Φ be the identity on I', we obtain the desired extension. In case (ii), let L be a subset of $J\backslash I$ of cardinality $\aleph_0$. Then $h|J\backslash L$ is an isomorphism from the subspace of $([0, 1], \mathcal{B})$ determined by the Borel set $J\backslash L$ to that determined by the Borel set $K\backslash h(L)$, and this isomorphism extends Θ. Since h is one-one, $|h(L)| = \aleph_0$; therefore $|J' \cup L| = |K' \cup h(L)| = \aleph_0$, and any one-one correspondence between $J' \cup L$ and $K' \cup h(L)$ will complete the extension of Θ to $([0, 1], \mathfrak{B})$. In case (iii), suppose w.l.o.g. that K' is non-denumerable. Then $h|J \cap K$ is an isomorphism from the subspace of $([0, 1], \mathfrak{B})$ determined by the Borel set $J \cap K$ to that determined by the Borel set $h(J \cap K)$, and this isomorphism extends Θ. The complement of $(J \cap K)$ is $J' \cup K'$, a non-denumerable Borel set; moreover $(h(J \cap K))'$ is also a Borel set, and it is non-denumerable because it contains $(h(J))' = K'$. Hence there is an isomorphism between the subspaces of $([0, 1], \mathfrak{B})$ determined by $(J \cap K)'$ and $(h(J \cap K))'$, and this isomorphism completes the extension of Θ to $([0, 1], \mathfrak{B})$. This completes the proof of Lemma 48.19.

Proof of Proposition 48.13. Let $\mathfrak{F}$ be the set of all automorphisms of $([0, 1], \mathfrak{B})$; it is known that $|\mathfrak{F}| = c$. For each Φ in $\mathfrak{F}$, write

$$A_\Phi = \{x : \Phi x \neq x\}.$$

Let $\mathfrak{H}$ be the set of all Φ in $\mathfrak{F}$ such that A_Φ is non-negligible. Since it is easy to construct c different members[9] of $\mathfrak{H}$, it follows that $|\mathfrak{H}| = c$.

Recall that Γ is the first non-negligible ordinal number. Denote the members of $\mathfrak{H}$ by Φ_α, where α runs over all ordinals $< \Gamma$. There are c non-denumerable[10] Borel subsets of $[0, 1]$; denote them B_ω, where again α runs over all ordinals $< \Gamma$. Inductively define a family $\{D_\alpha\}_{\alpha < \Gamma}$ by choosing D_α to be the first B_β (i.e. the B_β with the smallest β) not previously chosen that intersects A_{Φ_α} non-negligibly; there always is such a B_β since A_{Φ_α} intersects c Borel sets non-negligibly, and fewer than c Borel sets B_β were previously chosen. We assert that

(48.20) D_α runs over all non-denumerable Borel subsets of $[0, 1]$.

Indeed, suppose that one of the B_β is not a D_α. Then the cardinality of the set of Φ in $\mathfrak{H}$ such that $|A_\Phi \cap B_\beta| = c$ is at most that of β, i.e. it is $< c$. But $\mathfrak{H}$ includes c automorphisms Φ that leave *no* point fixed,[11] i.e. such that $A_\Phi = [0, 1]$, and hence $|A_\Phi \cap B_\beta| = |B_\beta| = c$. This contradiction proves (48.20).

[9] E.g. the "rotations": first map $[0, 1]$ isomorphically onto the circle, and then rotate the circle.

[10] I.e. non-negligible; see Lemma 48.6.

[11] E.g. the rotations.

Now define a family $\{x_\alpha\}_{\alpha<\Gamma}$ inductively as follows: Choose x_1 in D_1. If x_β has been chosen for all $\beta < \alpha$, choose x_α in $D_\alpha \cap A_{\Phi_\alpha}$ to be different from the $\Phi_\alpha^{-1}(x_\beta)$ as well as the $\Phi_\beta(x_\beta)$ for all $\beta < \alpha$; this is always possible, because 2α is negligible whereas $D_\alpha \cap A_{\Phi_\alpha}$ is non-negligible.

Let $I = \{x_\alpha\}_{\alpha<\Gamma}$, and let $(I, \mathcal{C})$ be the subspace of $([0, 1], \mathcal{B})$ determined by I. Since $x_\alpha \in D_\alpha$, it follows from (48.20) that I is absolutely thick. Next, let Θ be an automorphism of $(I, \mathcal{C})$; then by Lemma 48.19, Θ can be extended to an automorphism Φ of $([0, 1], \mathcal{B})$. If Φ is in $\mathcal{K}$, then it is one of the Φ_α. Then $\Phi_\alpha(x_\alpha) \neq x_\alpha$, because $x_\alpha \in A_{\Phi_\alpha}$. Furthermore, from $x_\alpha \neq \Phi_\alpha^{-1}(x_\beta)$ for all $\beta < \alpha$, it follows that $\Phi_\alpha(x_\alpha) \neq x_\beta$ for all $\beta < \alpha$. Finally, when $\gamma > \alpha$, the choice of x_γ dictates that $\Phi_\alpha(x_\alpha) \neq x_\gamma$. Hence $\Phi_\alpha(x_\alpha) \notin I$, and hence $\Theta(x_\alpha) = \Phi(x_\alpha) \notin I$, in contradiction to our choice of Θ as an automorphism of $(I, \mathcal{C})$. Hence Φ cannot be in $\mathcal{K}$, i.e. $A_\Phi = \{x\colon \Phi x \neq x\}$ is negligible. But since Φ is Borel-measurable, A_Φ is a Borel set; so if it is negligible, it must be denumerable. But clearly the set $\{x\colon \Theta x \neq x\} \subset A_\Phi$, so it too is denumerable; and thus the proof of Proposition 48.13 is complete.

Example 48.21. *Let $(I, \mathcal{C})$ be the subspace of* $([0, 1], \mathcal{B})$ *whose existence is asserted by Proposition* 48.13.

Proposition 48.22. *In Example* 48.21, *there is more than one value on pNA, on $bv'NA$, and even on BV.*

Proof. Let μ be an NA^1 measure on $(I, \mathcal{C})$; the existence of more than one such measure is guaranteed by Lemma 48.10. Define

$$\varphi v = v(I)\mu.$$

Then from Proposition 48.13 it follows that φv is invariant under all automorphisms of $(I, \mathcal{C})$, and the symmetry axiom (2.2) follows easily. The other conditions for the value are easily verified. This completes the proof of Proposition 48.22.

An esthetic flaw in the value φ defined in the above proof is that the value of a measure is usually a different measure. This certainly does not contradict our definitions; however, considerations of "dummies" and "strategic equivalence" (cf. von Neumann and Morgenstern (1953), p. 245) might make it not unreasonable to add to the definition of value a "projection" axiom, namely that

$$\text{(48.23)} \qquad \text{if } v \in FA, \text{ then } \varphi v = v.$$

PROPOSITION 48.24. *In Example* 48.21, *there is more than one value on* pNA *satisfying* (48.23).

Proof. Let Θ be an automorphism of $(I, \mathcal{C})$ and let $v \in pNA$; we first show that

$$\text{(48.25)} \qquad \Theta_* v = v.$$

Indeed, if $v \in NA^1$, then (48.25) follows from Proposition 48.13; hence (48.25) is true also if v is a power of an NA^1 measure; and hence, by the linearity of Θ_*, also if v is a linear combination of such powers. Finally, we note that $\|\Theta_* v\| = \|v\|$, whence Θ_* is continuous; thus (48.25) follows for all of pNA.

Now let $\mathcal{R}$ be a measurable order on $(I, \mathcal{C})$; any measurable order on $([0, 1], \mathcal{B})$ induces such an order[12]. Then $\varphi^{\mathcal{R}} v \in NA$; this follows from Propositions 12.8 and $pNA \subset AC$ (Corollary 5.3), neither of which depend on the Standardness Assumption. Applying (48.25) both to v and $\varphi^{\mathcal{R}} v$, we obtain

$$\varphi^{\mathcal{R}}(\Theta_* v) = \varphi^{\mathcal{R}} v = \Theta_*(\varphi^{\mathcal{R}} v);$$

hence $\varphi^{\mathcal{R}}$ obeys the symmetry axiom (2.2). The other conditions for value are easily verified, and it follows that $\varphi^{\mathcal{R}}$ is a value. Since different $\mathcal{R}$ may be found for which $\varphi^{\mathcal{R}}$ are different, and since $\varphi^{\mathcal{R}}$ satisfies (48.23), the proof of Proposition 48.24 is complete.

COROLLARY 48.26. *In Example* 48.21, *there is more than one value on* $bv'NA$ *obeying* (48.23).

Proof. The proof is like that of Corollary 48.5.

[12] This follows from the fact that all initial sets are measurable; see Lemma 12.14, which does not depend on the Standardness Assumption.

Appendix A. Finite Games and Their Values

The value for non-atomic games is a generalization, or analogue, of a concept introduced by Shapley (1953a) for finite games. Though formally most of this book is independent of finite game considerations,[1] conceptually the connection is important. In this appendix we review some basic facts concerning finite games and their values.

A finite game in coalitional (or characteristic function) form, or simply *finite game* for short, is a pair (N, v), where N is a finite set and v is a real-valued function on the family 2^N of all subsets of N, with $v(\emptyset) = 0$. We shall also refer to the function v itself as a *finite game* or a *game on* N. The members of N are called *players*, the members of 2^N *coalitions*. For $S \subset N$, $v(S)$ is called the *worth* of S; it is interpreted as the total payoff that the coalition S can obtain for its members.

If S is any finite set, an *S-vector* is a real-valued function on S; the set of all S-vectors is denoted E^S. E^S can be thought of as a euclidean space of dimension $|S|$, whose points have coordinates indexed by the members of S rather than by numbers. By an obvious identification, E^S can also be regarded as a subspace of E^T, whenever $S \subset T \subset N$. If $x \in E^S$ and $T \subset S$, we write

$$x(T) = \sum_{i \in T} x(i).$$

Let N be a finite set and let $n = |N|$. The games on N form a euclidean space of dimension $2^n - 1$, which we denote G^N. A *permutation* of N is a 1-1 function from N onto itself; if Θ is such a permutation, $v \in G^N$, and $x \in E^N$, define $\Theta_* v$ in G^N by $(\Theta_* v)(S) = v(\Theta S)$ and $\Theta_* x$ in E^N by $(\Theta_* x)(i) = x(\Theta i)$. Given v in G^N, a *null player* of v is a member i of N such that $v(S \cup \{i\}) = v(S)$ for all $S \subset N$. A *value* on G^N is then a function φ from G^N to E^N such that

(A.1) φ is linear,

[1] They appear formally only in Chapter III and in subsequent work on the asymptotic value.

and for all games $v \in G^N$ and permutations Θ of N we have

(A.2) $$\varphi(\Theta_* v) = \Theta_*(\varphi v),$$

(A.3) $$(\varphi v)(N) = v(N), \quad \text{and}$$

(A.4) $$\text{if } i \text{ is a null player of } v, \text{ then } (\varphi v)(i) = 0.$$

Conditions (A.1), (A.2), (A.3), and (A.4) are called the linearity, symmetry, efficiency, and dummy axioms, respectively.

PROPOSITION A.5. *There is one and only one value on G^N; it is given by*

(A.6) $$(\varphi v)(i) = \sum_{S \subset N \backslash \{i\}} \gamma_S [v(S \cup \{i\}) - v(S)]$$

where $\gamma_S = |S|!(n - |S| - 1)!/n!$.

Remark A.7. Another form of (A.6) is usually more amenable to calculations and is more closely related to the general approach of this book, particularly that of Chapter II. By an *order* on N, we mean a total order, so that there are $n!$ distinct orders[2] on N. If $\mathcal{R}$ is such an order, and if $i \in N$, we denote by $P_i^{\mathcal{R}}$ the set of players preceding i in $\mathcal{R}$. Then (A.6) is equivalent to

(A.8) $$(\varphi v)(i) = \frac{1}{n!} \sum_{\mathcal{R}} [v(P_i^{\mathcal{R}} \cup \{i\}) - v(P_i^{\mathcal{R}})],$$

where the sum runs over all orders $\mathcal{R}$ on N. To see this, we merely fix S and i and count the number of orders $\mathcal{R}$ for which $P_i^{\mathcal{R}} = S$; since the members of S and of $(N\backslash S)\backslash\{i\}$ can be permuted at will, this number comes to $|S|!(n - |S| - 1)!$, which is just $n!\gamma_S$.

Formula (A.8) may be interpreted in terms of the following procedure for imputing a payoff to the players. Choose an order on N at random, in such a way that each of the $n!$ possibilities has equal probability. Allow the players to enter a room one by one in the chosen order. On entry, each player is paid the amount that his entry contributes to the worth of the coalition in the room. The expected payment to player i under this random procedure is then given directly by (A.8).

[2] Note that we keep separate the ideas of order and permutation; in the infinite case the distinction is quite significant. The set N is not intrinsically ordered, though it is often convenient to identify N with the set of integers from 1 to n.

Proof of Proposition A.5. It is easily verified that the function φ defined by (A.8) satisfies (A.1) through (A.4); this shows that there is a value on G^N. To prove that it is unique, define for each non-empty $T \subset N$ a game $u_T \in G^N$ by

$$u_T(S) = \begin{cases} 1 & \text{if } S \supset T, \\ 0 & \text{otherwise.} \end{cases}$$

From (A.2) it follows that $(\varphi u_T)(i) = (\varphi u_T)(j)$ whenever i and j are in T. Hence by (A.3) and (A.4),

$$(\varphi u_T)(i) = \begin{cases} 1/|T| & \text{for } i \in T \\ 0 & \text{for } i \notin T, \end{cases}$$

thus determining φ on all the games u_T. There are $2^n - 1$ different games of this kind; if we could show that they are linearly independent, it would follow that they form a basis for the $(2^n - 1)$-dimensional linear space G^N, and thereby determine φ on all of G^N, which is what we wish to prove.

Suppose therefore, contrariwise, that a nontrivial linear combination of the u_T vanishes. Let T_0 have minimum cardinality among all the T such that u_T has a nonvanishing coefficient in this linear combination. Then we can write

$$u_{T_0} = \sum \alpha_T u_T, \tag{A.9}$$

where the sum runs over a family of sets T for which $T_0 \not\supset T$, and so for which $u_T(T_0) = 0$. If we evaluate both sides of (A.9) at T_0, we find $1 = 0$, an absurdity. This completes the proof of Proposition A.5.

A game v on N is *monotonic* if $T \supset S$ implies $v(T) \geqq v(S)$ for all S, $T \subset N$. The analogue for infinite games of the following simple fact plays an important role in Chapter 1.

COROLLARY A.10. *If v is monotonic, then*

$$(\varphi v)(i) \geqq 0 \textit{ for all } i \textit{ in } N.$$

Proof. Immediate from the non-negativity of the coefficients in (A.6) or (A.8).

A finite game some of whose players are null can also be considered a game on a smaller set, with some or all of the null players removed.

We now show that the value is essentially unaffected by such a change in the underlying player set. Call M a *carrier* of $v \in G^N$ if $v(S) = v(S \cap M)$ for all $S \subset N$. It may be verified that M is a carrier of v if and only if $N \backslash M$ consists of null players. For $M \subset N$ denote by v^M the restriction of $v \in G^N$ to the subsets of M; thus, v^M is the member of G^M such that $v^M(S) = v(S)$ for $S \subset M$.

COROLLARY A.11. *Let φ^M be the unique value on G^M. If M is a carrier of v, then*

$$\varphi^M v^M = \varphi v | M.$$

Proof. Any $w \in G^M$ has an extension w_N in G^N defined by $w_N(S) = w(S \cap M)$; intuitively w_N is obtained from w by adding null players. Define φ' on G^M by $\varphi' w = \varphi w_N | M$. Then φ' obeys (A.1) through (A.4), with M substituted for N. Therefore $\varphi' = \varphi^M$ and for $v \in G^N$ we have

$$\varphi^M v^M = \varphi' v^M = \varphi (v^M)_N | M;$$

but since M is a carrier of v, it follows that $(v^M)_N = v$, so the proof of Corollary A.11 is complete.

NOTES

1. The set function v is known in the game theory literature by the name "characteristic function"; hence also the term "characteristic function form," since games are often presented in various other forms, not relevant to the considerations in this book. These terms are somewhat misleading and in conflict with mathematical usage, however, in that v "characterizes" only certain coalitional aspects of the overall competitive situation, suppressing e.g. the strategic aspects. "Coalitional form" is more descriptive, but may suggest something more general than a simple numerical measure of each coalition's worth. Perhaps the terms[3] "worth function" or "coalitional worth function" for v would help keep ideas straight in a general account; in the present volume we have found the neutral term "set function" sufficient.

2. The concept of value described in this Appendix is due to Shapley (1953a); there are however a number of minor differences between the earlier treatment (which we shall refer to as "[S]" in this note) and that of the present Appendix. *First*, [S] restricts attention to superadditive games (i.e. those satisfying $v(S \cup T) \geqq v(S) + v(T)$ when $S \cap T = \emptyset$). This restriction entails only minor adjustments to the statements of the results and their proofs. *Second*, instead of a finite set N, [S] works with an arbitrary set U of players,

[3] For obvious reasons we avoid the word "value," which was undoubtedly the original inspiration for the conventional letter "v."

taking G^U to consist of the (superadditive) games on U that have finite carriers. The finite dimensionality of G^N is exploited in the proof of Proposition A.5; in order to show that the games u_T form a basis, it is only necessary to count them and show their linear independence. The corresponding proof in [S] derives an explicit representation for $v \in G^U$ in terms of the u_T. Given Proposition A.5, the existence and uniqueness of a value on G^U follows easily from Corollary A.11 and the observation that if there were two different values on G^U then they would disagree on some game with finite carrier. *Third*, the present axioms (A.3) and (A.4) are combined in [S] into a single axiom which says that $(\varphi v)(M) = v(M)$ for any carrier M; this is natural in [S] in view of the possibly infinite underlying space of players, but it tends to obscure the fact that two rather different intuitive ideas are involved. *Fourth*, the linearity axiom (A.1) is replaced in [S] by an "additivity" axiom: $\varphi(v + w) = \varphi v + \varphi w$, which comes to the same thing in the presence of the other axioms.

Formula (A.8) and Corollary A.10 are given in [S], while Corollary A.11 is subsumed in the main theorem of [S], which in effect asserts (A.6) for all finite carriers of v.

3. Two players i and j are called *substitutes* if $v(S \cup \{i\}) = v(S \cup \{j\})$ for all $S \subset I \backslash \{i, j\}$. The symmetry axiom (A.2) is equivalent (in the presence of the other axioms) to the statement: if i and j are substitutes, then $(\varphi v)(i) = (\varphi v)(j)$.

The reader should note that (A.2) does not refer to games having symmetries, i.e. games satisfying $v = \Theta_* v$ for a non-trivial permutation Θ. But (A.2) is equivalent (in the presence of the other axioms) to the condition

$$\text{(A.12)} \qquad \text{if } \Theta_* v = v \text{ then } \varphi(\Theta_* v) = \varphi v.$$

Indeed, this need only be asserted for a few permutations Θ, as above, since the only use of (A.2) in the proof is in evaluating the games u_T.

4. A "dummy" in v is a player i such that $v(S \cup \{i\}) = v(S) + v(\{i\})$ whenever $i \notin S$, or equivalently, such that for some additive set function α, i is a null player in the game $v + \alpha$. (It is easily seen that the value for finite games is relatively invariant under the addition of additive games, thus $\varphi(v + \alpha) = \varphi v + \alpha$.) Thus every null player is a dummy, but not conversely. An alternative version of (A.4) states that if i is a dummy then $(\varphi v)(i) = v(\{i\})$; hence (A.4) has come to be known as the dummy axiom.

5. Formula (A.6) can also be given a direct probability interpretation; it expresses player i's expected marginal contribution to a coalition chosen "at random," according to the following scheme: First one of the n integers 0, 1, . . . , $n - 1$ is chosen with probability $1/n$; call the choice s. Then one of the $\binom{n-1}{s}$ subsets of $N \backslash \{i\}$ of size s is chosen with probability $1 \Big/ \binom{n-1}{s}$; call the choice S. Finally, i is paid the amount $v(S \cup \{i\}) - v(S)$. Since the probability of S is just γ_S, i's expectation is precisely $(\varphi v)(i)$.

A non-probabilistic model that yields the value has been given by Harsanyi (1959). In Harsanyi's model, each set S of players is imagined to form a "syndicate" that declares a "dividend" $d(S)$ (positive or negative), distributing evenly among its members the difference between its worth, $v(S)$, and the total dividend declarations of all its subsyndicates. Thus, if $s = |S|$, $t = |T|$, we have

$$d(S) = \frac{1}{s}\left[v(S) - \sum_{T \subset S, T \neq S} td(T)\right]. \tag{A.13}$$

It can then be shown that the sum of all dividends received by a player under this scheme is exactly his value.

To see this, define φ by $(\varphi v)(i) = \Sigma_{S \ni i}\, d(S)$. Then (A.1) and (A.2) are immediate, (A.3) follows from (A.13) with $S = N$, and (A.4) follows from the fact that if any player in S is null then $d(S) = 0$, which can be established by a simple calculation.

In establishing a solution concept like the value, two methods are complementary. A deductive approach, from "plausible" axioms, works with necessary conditions and tends to show that there is *at most one* intuitively acceptable definition. A constructive approach, like those outlined here and in Remark A.7, works with a "plausible" model and endeavors to show that there is *at least one* intuitively satisfactory definition. Each method helps to justify and clarify the other.

Appendix B. ϵ-Monotonicity

For motivation of this Appendix, see the discussion following Proposition 4.15, and also Proposition 7.19.

For $v \in BV$, define the *downward variation* of v to be

$$\sup \sum_i \max\{(v(S_i) - v(S_{i+1})), 0\},$$

where the sup is taken over all chains

$$\emptyset = S_0 \subset \cdots \subset S_k = I.$$

v is said to be *ϵ-monotonic* if its downward variation is $\leqq \epsilon$. Note that a set function is monotonic if and only if it is 0-monotonic.

The chief result we wish to prove here is

PROPOSITION B.1. *Let Q be a subspace of BV with $NA^+ \subset Q \subset AC$; let $\epsilon > 0$; and let Q^ϵ denote the set of ϵ-monotonic elements of Q. Then*

$$Q = Q^\epsilon - Q^\epsilon$$

(in fact, $Q = Q^\epsilon - Q^+$).

For the proof we need the following.

LEMMA B.2. *For any $v \in AC$ and $\epsilon > 0$, there exists $\mu \in NA^+$ and a real number $c > 0$ such that $v + c\mu$ is ϵ-monotonic.*[1] *Moreover, μ (but not c) may be chosen independently of ϵ.*

Proof. By definition of AC, there is a $\mu \in NA^+$ such that for all $\epsilon > 0$ there is a $\delta > 0$ such that for any subchain Λ of any chain Ω

$$\text{(B.3)} \qquad \|\mu\|_\Lambda \leqq \delta \Rightarrow \|v\|_\Lambda \leqq \epsilon.$$

For any $S \subset T$ with $v(S) \neq v(T)$, define the ratio

$$\rho(S, T) = \frac{|v(T) - v(S)|}{\mu(T \backslash S)};$$

[1] That ϵ cannot be taken to be 0 can be seen by taking $v = -\sqrt{\lambda}$, where λ is Lebesgue measure on $([0, 1], \mathcal{B})$.

$\rho(S, T)$ may be infinite, and is always positive. Let Λ' denote the set of links $(S, T) \in \Lambda$ such that $\rho(S, T) > \epsilon/\delta$. We claim that

(B.4) $$\|\mu\|_{\Lambda'} < \delta.$$

Suppose not. Let $\eta > 0$ satisfy $\eta < \delta$, and interpolate sets R_j:

$$S = R_0 \subset R_1 \subset \cdots \subset R_k = T$$

between the S, T of each link in Λ' in such a way that $\mu(R_j \backslash R_{j-1}) \leqq \eta$, $j = 1, \ldots, k$. Let Λ^* denote the set of all links $\{R_{j-1}, R_j\}$ obtained in this way; clearly Λ^* is a subchain of some chain Ω^* that "refines" Ω. Moreover

(B.5) $$\|v\|_{\Lambda^*} \geqq \|v\|_{\Lambda'} \quad \text{and} \quad \|\mu\|_{\Lambda^*} = \|\mu\|_{\Lambda'}.$$

Now arrange the links of Λ^* according to their ρ values. Choosing those of highest ρ value, we may, since (B.4) is false, select enough of them to make a subchain $\Lambda^{*\prime}$ with

$$\delta - \eta \leqq \|\mu\|_{\Lambda^{*\prime}} \leqq \delta.$$

Then by definition of δ,

$$\|v\|_{\Lambda^{*\prime}} \leqq \epsilon.$$

Hence

(B.6) $$\frac{\epsilon}{\delta - \eta} \geqq \frac{\|v\|_{\Lambda^{*\prime}}}{\|\mu\|_{\Lambda^{*\prime}}}.$$

Since we used the highest ρ-values in $\Lambda^{*\prime}$, we have

(B.7) $$\frac{\|v\|_{\Lambda^{*\prime}}}{\|\mu\|_{\Lambda^{*\prime}}} \geqq \frac{\|v\|_{\Lambda^*}}{\|\mu\|_{\Lambda^*}}.$$

By (B.5), we have

(B.8) $$\frac{\|v\|_{\Lambda^*}}{\|\mu\|_{\Lambda^*}} \geqq \frac{\|v\|_{\Lambda'}}{\|\mu\|_{\Lambda'}}.$$

If ρ_0 denotes the *smallest* ρ-value used in Λ', then we have

(B.9) $$\frac{\|v\|_{\Lambda'}}{\|\mu\|_{\Lambda'}} \geqq \rho_0 > \frac{\epsilon}{\delta}.$$

Now (B.6), (B.7), (B.8), (B.9) yield

$$\frac{\epsilon}{\delta - \eta} \geqq \rho_0 > \frac{\epsilon}{\delta},$$

and as $\eta \to 0$ we obtain a contradiction. This proves (B.4). From (B.3) and (B.4) we obtain

$$\|v\|_{\Lambda'} \leqq \epsilon. \tag{B.10}$$

Now define

$$w = v + \frac{\epsilon}{\delta}\mu.$$

For $(S, T) \in \Lambda \backslash \Lambda'$, we have

$$\begin{aligned} w(T) - w(S) &= v(T) - v(S) + \frac{\epsilon}{\delta}\mu(T\backslash S) \\ &\geqq -|v(T) - v(S)| + \frac{\epsilon}{\delta}\mu(T\backslash S) \\ &= \left(-\rho(S, T) + \frac{\epsilon}{\delta}\right)\mu(T\backslash S) \\ &\geqq 0. \end{aligned}$$

Hence

$$\begin{aligned} \sum_{(S,T)\in\Lambda} |\min(0, w(T) - w(S))| &= \sum_{\Lambda'} \left|\min(0, v(T) - v(S) + \frac{\epsilon}{\delta}\mu(T\backslash S))\right| \\ &\leqq \sum_{\Lambda'} |v(T) - v(S)| \\ &= \|v\|_{\Lambda'} \leqq \epsilon \qquad \text{by (B.10)}. \end{aligned} \tag{B.11}$$

On the other hand, the downward variation of w is the supremum of the left side of (B.11), over all subchains Λ; hence it is $\leqq \epsilon$. This completes the proof of the lemma.

Proof of Proposition B.1. Let $v \in Q$. Choose μ, c by the lemma so that $w = v + c\mu$ is ϵ-monotonic. Then $w \in Q^\epsilon$ and $c\mu \in Q^+ \subset Q^\epsilon$, and $v = w - c\mu$. This completes the proof of the proposition.

Appendix C. The Mixing Value of Absolutely Continuous Set Functions

In this appendix, we shall give an alternative characterization of the mixing value for $v \in AC$. For motivation, see the discussion of the measure μ_v in Section 14 (before the statement of Proposition 14.3).

PROPOSITION C.1. *Let $v \in AC$. A necessary and sufficient condition that the set function φv be the mixing value of v is that for all μ in NA^1 with $\mu \gg v$, condition* (14.2) *holds.*

The proof of Proposition C.1 is the object of this section. It will require a number of lemmas, including some of independent interest.

LEMMA C.2. *Let $\mathcal{R}$ be an order on I, and assume that I has a countable $\mathcal{R}$-dense subset. Then there is a real-valued function Θ on I such that for all s and t.*

$$s \mathcal{R} t \Leftrightarrow \Theta s > \Theta t. \tag{C.3}$$

Proof. This is essentially Proposition 5 of Debreu (1964, p. 291). There, I is an arbitrary set, without any measurable structure.

LEMMA C.4. *Let $\mathcal{R}$ be a measurable order on the underlying space $(I, \mathcal{C})$. Then there is a Borel subset D of the real line, such that if $\mathfrak{D}$ is the family of Borel subsets of D, then $(I, \mathcal{C})$ is isomorphic to $(D, \mathfrak{D})$ under an isomorphism Θ that preserves order, i.e. that satisfies* (C.3).

Proof. Because of Lemma 12.5, we can apply Lemma C.2 to this situation, obtaining a real-valued Θ obeying (C.3). Define

$$D = \Theta(I). \tag{C.5}$$

Clearly Θ is a one-one transformation from I onto D. Let $\mathcal{E}$ be the σ-field of subsets of D consisting of all intersections of D with Borel sets. Now for an arbitrary real α, the set

$$J = \Theta^{-1}(D \cap (-\infty, \alpha))$$

has the property that

$$s \in J,\ s \mathcal{R} s' \Rightarrow s' \in J.$$

Hence by Lemma 12.14, J is measurable. The sets $D \cap (-\infty, \alpha)$ generate $\mathcal{E}$, and each of their inverse images is measurable; hence Θ is a measurable transformation from $(I, \mathcal{C})$ onto $(D, \mathcal{E})$. Then it follows from Proposition 1.2 that Θ is an isomorphism from $(I, \mathcal{C})$ onto $(D, \mathcal{E})$, and D is a Borel subset of the real line. Hence if $\mathfrak{D}$ is the family of Borel subsets of D, then $\mathfrak{D} = \mathcal{E}$, and the lemma is proved.

LEMMA C.6. *On* $(I, \mathcal{C})$, *let* $\mathcal{R}$ *be a measurable order, and let* ξ *be a measure in* NA^1. *Let* λ *be Lebesgue measure on* $[0, 1]$. *Then there is an isomorphism* Ξ *from* $(I, \mathcal{C})$ *onto* $([0, 1], \mathcal{B})$ *such that*

$$\xi(S) = \lambda(\Xi(S))$$

for all $S \subset I$, *and such that there is a* $U \subset I$ *of* ξ*-measure* 0 *such that*

$$s \mathcal{R} t \Rightarrow \Xi s > \Xi t$$

whenever s *and* t *are in* $I \backslash U$.

Remark. The lemma says that the "ordered measure spaces" $(I, \mathcal{C}, \xi, \mathcal{R})$ and $([0, 1], \mathcal{B}, \lambda, >)$ are "almost isomorphic," in the sense that there is a one-one mapping from one onto the other that preserves measurability and measure, and that "almost" preserves the order. Of course, it follows that any two "ordered measure spaces" are "almost isomorphic" in this sense, provided that the underlying measurable spaces are isomorphic to $([0, 1], \mathcal{B})$, that the measures are in NA^1, and that the orders are measurable.

Proof. By the previous lemma, we may assume w.l.o.g. that I is a Borel subset of the real line, that $\mathcal{C}$ is the family of Borel subsets of I, and that the order is the natural order $>$ ("greater than"). All measurable spaces appearing in this proof will consist of a Borel subset X of the real line E^1, together with the family of Borel subsets of X; thus identifying such a space by X only can cause no confusion.

Since $I \subset E^1$, we may extend ξ from I to a measure ξ' on all of E^1 by defining

$$\xi'(S) = \xi(S \cap I).$$

Let us define a mapping Ψ of E^1 into $[0, 1]$ by

$$\Psi(s) = \xi'((-\infty, s)). \tag{C.7}$$

Since ξ is in NA^1, so is ξ'; hence Ψ, the cumulative distribution, is a monotonic non-decreasing continuous function. Hence if $s \in E^1$, then from the continuity and monotonicity of Ψ, and the fact that the continuous image of a connected set is connected, it follows that $\Psi((-\infty, s))$ is an interval with end-points 0 and $\Psi(s)$. Hence from (C.7) we obtain

$$\lambda(\Psi((-\infty, s))) = \Psi(s) = \xi'((-\infty, s)). \tag{C.8}$$

Next, let W be the set of points where Ψ fails to be univalent, i.e.

$$W = \{s \in E^1 : \exists t \neq s \text{ with } \Psi(t) = \Psi(s)\}.$$

Because of the monotonicity and continuity of Ψ, W is a disjoint union of nondegenerate closed intervals, each of which is carried into a single point under Ψ, and each of which is of ξ-measure 0. Since they are non-degenerate and disjoint, there are at most denumerably many of them; hence $\Psi(W)$ is denumerable, and

$$\xi'(W) = 0, \tag{C.9}$$

$$\lambda(\Psi(W)) = 0. \tag{C.10}$$

Let Ψ^* be the restriction of Ψ to $E^1 \backslash W$, and let ξ^* be the restriction of ξ' to the Borel subsets of $E^1 \backslash W$. Then Ψ^* is an isomorphism from the measurable space $E^1 \backslash W$ to the measurable space $\Psi(E^1) \backslash \Psi(W)$. Furthermore for all $s \in E^1$ we have

$$\Psi^*((-\infty, s) \backslash W) = \Psi((-\infty, s) \backslash W) = \Psi((-\infty, s)) \backslash \Psi(W).$$

Therefore from (C.10) it follows that

$$\lambda(\Psi^*((-\infty, s) \backslash W)) = \lambda(\Psi((-\infty, s))),$$

and hence from (C.9) and (C.8) we obtain

$$\begin{aligned} \xi^*((-\infty, s) \backslash W) &= \xi'((-\infty, s) \backslash W) \\ &= \xi'((-\infty, s)) = \lambda(\Psi((-\infty, s))) = \lambda(\Psi^*((-\infty, s) \backslash W)). \end{aligned} \tag{C.11}$$

Now since Ψ^* is an isomorphism, $\lambda\Psi^*$ is a measure on $E^1 \backslash W$. Formula (C.11) says that this measure coincides with ξ^* for all sets of the form

$(-\infty, s)\backslash W$, which generate all the measurable subsets of $E^1\backslash W$; hence

$$\xi^*(S) = \lambda(\Psi^*(S)) \tag{C.12}$$

for all $S \subset E^1\backslash W$.

From (C.9) it follows that $\xi^*(I\backslash W) = 1$; hence $I\backslash W$ is non-denumerable, and so contains a non-denumerable Borel set U' of ξ-measure 0. Let $V' = \Psi^*(U')$; by (C.12), V' is of Lebesgue measure 0. Now from (C.12) and (C.9) it follows that

$$\lambda(\Psi^*(I\backslash W)) = \xi^*(I\backslash W) = \xi'(I\backslash W) = \xi'(I) = \xi(I) = 1;$$

hence $[0, 1]\backslash\Psi^*(I\backslash W)$ is of Lebesgue measure 0. Let

$$U = U' \cup (I \cap W),$$
$$V = V' \cup ([0, 1]\backslash\Psi^*(I\backslash W));$$

then U is a non-denumerable Borel subset of I whose ξ-measure is 0 (by (C.9)), V is a non-denumerable Borel subset of [0, 1] whose λ-measure is 0, and Ψ^* is an isomorphism from $I\backslash U$ onto $[0, 1]\backslash V$. Let Θ be an arbitrary isomorphism from the measurable space U onto the measurable space V. Define a one-one mapping Ξ from I onto [0, 1] by

$$\Xi(s) = \begin{cases} \Theta(s) & \text{when } s \in U \\ \Psi^*(s) & \text{when } s \in I\backslash U. \end{cases}$$

Then the demands of the lemma are satisfied, and so our proof is complete.

Corollary C.13. *Let $\mathcal{R}$ be a measurable order on I, and let $\xi \in NA^+$. Let S and T be measurable subsets of I of equal ξ-measure. Then there is an automorphism Ξ of I such that*

$$\Xi(S) = T$$

and such that there is a $U \subset I$ of ξ-measure 0 such that

$$s \,\mathcal{R}\, t \Leftrightarrow \Xi s \,\mathcal{R}\, \Xi t$$

whenever s and t are in $S\backslash U$, and whenever s and t are in $(I\backslash S)\backslash U$.

Proof. Assume first that

$$0 < \xi(S) = \xi(T) < \xi(I).$$

Consider the following two "ordered measure spaces": First, the space formed from S, its measurable subsets, the order $\mathcal{R}$ restricted to S, and the NA^1 measure $\xi(\cdot)/\xi(S)$; and, for the second, the space formed in a similar way from T. Because of Lemma C.6 and the remark following its statement, these two "ordered measure spaces" are "almost isomorphic"; let Ξ_1 be the appropriate "almost isomorphism." Similarly, the "ordered measure spaces" formed from $I\backslash S$ and $I\backslash T$ in a similar way are "almost isomorphic"; let Ξ_2 be the appropriate "almost isomorphism." If we combine Ξ_1 and Ξ_2, we obtain an automorphism of I that satisfies the required conditions.

If $0 = \xi(S) = \xi(T)$, then we substitute an arbitrary (non-order-preserving) isomorphism of S onto T for Ξ_1; similarly if $\xi(S) = \xi(T) = \xi(I)$, we substitute an arbitrary (non-order-preserving) isomorphism of $I\backslash S$ onto $I\backslash T$ for Ξ_2. This completes the proof of Corollary C.13.

LEMMA C.14. *Let $u \ll \mu$, where $\mu \in NA^+$, let $T \subset I$ be of μ-measure* 0, *and let $\mathcal{R}$ and $\mathcal{R}'$ be measurable orders such that*

$$s \,\mathcal{R}\, t \Leftrightarrow s \,\mathcal{R}'\, t$$

whenever both s and t are in $I\backslash T$. Then

$$\varphi^{\mathcal{R}} u = \varphi^{\mathcal{R}'} u.$$

Proof. For $s \notin T$, we have

$$\begin{aligned} I(s; \mathcal{R}) &= \{t: s \,\mathcal{R}\, t\} = \{t \notin T: s \,\mathcal{R}\, t\} \cup \{t \in T: s \,\mathcal{R}\, t\} \\ &= \{t \notin T: s \,\mathcal{R}'\, t\} \cup \{t \in T: s \,\mathcal{R}\, t\} \\ &= (\{t: s \,\mathcal{R}'\, t\} \backslash \{t \in T: s \,\mathcal{R}'\, t\}) \cup \{t \in T: s \,\mathcal{R}\, t\} \\ &\cong_\mu I(s; \mathcal{R}'), \end{aligned}$$

where $\cong_\mu$ means "equal except possibly for a set of μ-measure 0." Since both $u \ll \mu$ and $\varphi^{\mathcal{R}'} u \ll \mu$ (Proposition 12.8), it follows that for $s \in I\backslash T$,

$$\begin{aligned} (\varphi^{\mathcal{R}} u)(I(s; \mathcal{R})) = u(I(s; \mathcal{R})) &= u(I(s; \mathcal{R}')) \\ &= (\varphi^{\mathcal{R}'} u)(I(s; \mathcal{R}')) = (\varphi^{\mathcal{R}'} u)(I(s; \mathcal{R})). \end{aligned}$$

This is also true when $s = \pm\infty$, so that we have

$$(\varphi^{\mathcal{R}} u)(I(s; \mathcal{R})) = (\varphi^{\mathcal{R}'} u)(I(s; \mathcal{R})) \tag{C.15}$$

whenever $s \in \bar{I}\backslash T$; we wish to establish (C.15) for all s. To this end, let $t \in T$, and apply Lemma 12.14 with

$$J = \bigcap \{I(s; \mathcal{R}) : s \in \bar{I} \backslash T \text{ and } s \,\mathcal{R}\, t\}$$

and $v = \mu$. We deduce that for each $\epsilon > 0$ there are s_1 and s_2 such that $s_1 \in J$, $s_2 \in \bar{I}\backslash J$ (whence in particular $s_1 \,\mathcal{R}\, s_2$), and

$$\mu(I(s_2; \mathcal{R})\backslash I(s_1; \mathcal{R})) < \epsilon.$$

Since $s_2 \in \bar{I}\backslash J$, it follows that there is an $s \in \bar{I}\backslash T$ such that $s \,\mathcal{R}\, t$ and $s_2 \notin I(s; \mathcal{R})$, i.e., $s_2 \overset{\mathcal{R}}{\leqq} s$. But from $s_1 \in J$ it follows that $s \,\mathcal{R}\, s_1$; hence

$$\mu(I(s; \mathcal{R})\backslash I(s_1; \mathcal{R})) < \epsilon. \tag{C.16}$$

Now clearly $t \in J$; hence if $t \overset{\mathcal{R}}{\leqq} s_1$, it follows that

$$\mu(I(s; \mathcal{R})\backslash I(t; \mathcal{R})) < \epsilon; \tag{C.17}$$

if $s_1 \,\mathcal{R}\, t$, then from $s_1 \in J$ it follows that $I(s_1; \mathcal{R})\backslash I(t; \mathcal{R}) \subset T$, hence $\mu(I(s_1; \mathcal{R})\backslash I(t; \mathcal{R})) = 0$, and this together with (C.16) again yields (C.17). Hence for each $\epsilon > 0$ there is an $s \in \bar{I}\backslash T$ such that $s \,\mathcal{R}\, t$ and (C.17) holds; this, together with $\varphi^{\mathcal{R}}u \ll \mu$, $\varphi^{\mathcal{R}'}u \ll \mu$, and the truth of (C.15) for all $s \in \bar{I}\backslash T$ yields

$$(\varphi^{\mathcal{R}}u)(I(t; \mathcal{R})) = (\varphi^{\mathcal{R}'}u)(I(t; \mathcal{R})).$$

Hence (C.15) holds for all s. But the $I(s; \mathcal{R})$ generate all the measurable sets, and $\varphi^{\mathcal{R}}$ and $\varphi^{\mathcal{R}'}$ are both measures. Hence $(\varphi^{\mathcal{R}}u)(S) = (\varphi^{\mathcal{R}'}u)(S)$ for all S, as was to be proved.

LEMMA C.18. *Let $u \in AC$, let $\mu \in NA^1$ be such that $u \ll \mu$, and let $\{\Theta_1, \Theta_2, \ldots\}$ be a μ-mixing sequence. Let $\mathcal{R}$ be a measurable order, and let $\tau \in NA^+$. Then there is a measure $\xi \in NA^1$ with $\mu + \tau \ll \xi$, and a ξ-mixing sequence $\{\Psi_1, \Psi_2, \ldots\}$, such that*

$$\varphi(u; \Psi_k^{-1}\mathcal{R}) = \varphi(u; \Theta_k\mathcal{R})$$

for all k.

Proof. Let $\tau = \tau^{ac} + \tau^{\perp}$ be the Lebesgue decomposition of τ w.r.t. μ, i.e., τ^{ac} and $\tau^{\perp}$ are non-negative measures such that $\tau^{ac} \ll \mu$ and $\tau^{\perp} \perp \mu$; then τ^{ac} and $\tau^{\perp}$ are necessarily non-atomic.

If $\tau^{\perp}$ vanishes identically, then $\tau \ll \mu$, so we may take $\xi = \mu$ and $\Psi_i = \Theta_i^{-1}$, and the proof is complete when we note that $\{\Theta_i^{-1}\}$ is a mixing sequence if and only if $\{\Theta_i\}$ is. Suppose, therefore, that $\tau^\perp$ does not vanish identically; w.l.o.g., assume that $\tau^\perp(I) = 1$. Define

$$\xi = \tfrac{1}{2}(\mu + \tau^\perp).$$

Let $W \subset I$ be such that $\mu(W) = 0$ and $\tau^\perp(W) = 1$.

The idea of the proof is to restrict $\{\Theta_i^{-1}\}$ to $I\backslash W$, to define an arbitrary $\tau^\perp$-mixing sequence $\{\Phi_i\}$ on W, and then to combine these into a ξ-mixing sequence $\{\Psi_k\}$ on all of I by "mixing them" together. The choice of the sequence $\{\Phi_i\}$ will not affect anything, because we are only interested in the effect of $\{\Psi_k\}$ on u, and because $u \ll \mu$, the set function u is "essentially defined" on $I\backslash W$ only. We now go on to the technical details of the proof.

Construct a $\tau^\perp$-mixing sequence $\{\Phi_1, \Phi_2, \ldots\}$ such that for all i,

$$\Phi_i(W) = W$$

and Φ_i is the identity on $I\backslash W$; the possibility of constructing such a sequence follows from Proposition 14.3 applied to the underlying space W. Next, recall again that since $\{\Theta_i\}$ is μ-mixing, so is $\{\Theta_i^{-1}\}$. Hence it is possible to construct a μ-mixing sequence $\{\Theta_1', \Theta_2', \ldots\}$ such that for all i,

$$\Theta_i'(I\backslash W) = I\backslash W,$$

such that Θ_i' is the identity on W, and such that Θ_i' differs from Θ_i^{-1} on a set of μ-measure 0 only; indeed, we may define

$$\Theta_i'(t) = \begin{cases} t & \text{when } t \in \bigcup_{n=-\infty}^{\infty} \Theta_i^n(W) \\ \Theta_i^{-1}(t) & \text{otherwise.} \end{cases}$$

For $i = 0, 1, 2, \ldots$, let Ξ_i be a ξ-measure-preserving automorphism of the player space such that

$$\Xi_i(W) = \bigcup_{j=1}^{2^i} \{J(j/2^i; \xi, \mathcal{R})\backslash J((j - \tfrac{1}{2})/2^i; \xi, \mathcal{R})\},$$

and such that there is a set U of ξ-measure 0 such that

$$s \,\mathcal{R}\, t \Leftrightarrow \Xi_i s \,\mathcal{R}\, \Xi_i t \tag{C.19}$$

whenever both s and t are in $W\backslash U$ and whenever both are in $(I\backslash W)\backslash U$; such an automorphism exists because of Corollary C.13. Clearly U may be chosen independent of i; moreover, it may be chosen so that $\Xi_i U = U$ for all i (otherwise replace U by $\cup_\alpha \Pi_\alpha U$, where Π_α ranges over all automorphisms formed by composing a finite string of Ξ_i's and their inverses).

For $k = 1, 2, \ldots$ define automorphisms Λ_k by

$$\Lambda_k(s) = \begin{cases} \Phi_k(s) & \text{when } s \in W \\ \Theta'_k(s) & \text{when } s \in I\backslash W; \end{cases}$$

for $k = 0$, set $\Lambda_0(s) = s$. Note that each of the Λ_k is ξ-measure-preserving, and define a sequence $\{\Psi_k\}$ of automorphisms by

$$\Psi_k = \Xi_k \Lambda_k.$$

Then we claim:

$$\text{(C.20)} \qquad \{\Psi_k\} \text{ is a } \xi\text{-mixing sequence.}$$

To prove this, we must show that

$$\text{(C.21)} \qquad \xi(S \cap \Psi_k T) \to \xi(S)\xi(T)$$

for all $S, T \subset I$. We first show this when S is of the form $J(h/2^i; \xi, \Lambda_i \mathcal{R})$. To this end, fix h and i, and consider the underlying spaces

$$I_1 = (I\backslash W)\backslash U$$

and

$$I_2 = W\backslash U.$$

On them, define orders $\mathcal{R}_1$ and $\mathcal{R}_2$, by restricting $\mathcal{R}$ to I_1 and I_2, respectively. Also, we will consider the underlying spaces $\Xi_k I_1$ and $\Xi_k I_2$, with the orders $\Xi_k \mathcal{R}_1$ and $\Xi_k \mathcal{R}_2$, respectively.

Now for $k \geqq i$, we have

$$\begin{aligned} S\backslash U &= J(h/2^i; \xi, \mathcal{R})\backslash U \\ &= \bigcup_{j=1}^{h2^{k-i}} \{J((j-\tfrac{1}{2})/2^k; \xi, \mathcal{R})\backslash J((j-1)/2^k; \xi, \mathcal{R})\}\backslash U \\ &\qquad \cup \bigcup_{j=1}^{h2^{k-i}} \{J(j/2^k; \xi, \mathcal{R})\backslash J((j-\tfrac{1}{2})/2^k; \xi, \mathcal{R})\}\backslash U \\ &= H_1^k \cup H_2^k, \end{aligned}$$

where, by (C.19),

$$H_1^k = S \cap \Xi_k I_1 \text{ is an initial set in } \Xi_k I_1,$$

and

$$H_2^k = S \cap \Xi_k I_2 \text{ is an initial set in } \Xi_k I_2.$$

Hence (again using (C.19)), $\Xi_k^{-1}H_1^k$ and $\Xi_k^{-1}H_2^k$ are initial sets in I_1 and I_2. respectively. On the other hand we have

$$\xi(\Xi_k^{-1}H_1^k) = \xi(H_1^k) = h/2^{i+1}$$

and

$$\xi(\Xi_k^{-1}H_2^k) = \xi(H_2^k) = h/2^{i+1}.$$

But since, of any two initial sets in the same underlying space, one must contain the other, it follows that initial sets of the same ξ-measure can differ by a set of ξ-measure 0 only. Hence

$$\Xi_k^{-1}H_1^k \cong_\xi \Xi_i^{-1}H_1^i$$

and

$$\Xi_k^{-1}H_2^k \cong_\xi \Xi_i^{-1}H_2^i$$

where $\cong_\xi$ means "equal except for a set of ξ-measure 0." Hence

$$\Xi_k^{-1}S \cong_\xi \Xi_k^{-1}(S\backslash U) = \Xi_k^{-1}H_1^k \cup \Xi_k^{-1}H_2^k \cong_\xi \Xi_i^{-1}(H_1^i \cup H_2^i).$$

To prove (C.21) for the special S under consideration, let $T = T_1 \cup T_2$, where $T_1 \subset I\backslash W$ and $T_2 \subset W$. Then since $\{\Theta_1', \Theta_2', \ldots\}$ is a μ-mixing sequence and $\Xi_i^{-1}H_2^i \subset I_2 \subset W$, we have, as $k \to \infty$,

$$\begin{aligned}
\xi(S \cap \Psi_k T_1) &= \xi\Xi_k^{-1}(S \cap \Psi_k T_1) \\
&= \xi(\Xi_k^{-1}S \cap \Theta_k' T_1) \\
&= \tfrac{1}{2}\mu(\Xi_i^{-1}(H_1^i \cup H_2^i) \cap \Theta_k' T_1) \\
&= \tfrac{1}{2}\mu(\Xi_i^{-1}H_1^i \cap \Theta_k' T_1) \\
&\to \tfrac{1}{2}\mu(\Xi_i^{-1}H_1^i)\mu(T_1) \\
&= \xi(\Xi_i^{-1}H_1^i)\mu(T_1) = \frac{h}{2^{i+1}}\mu(T_1) = \frac{h}{2^i}\frac{1}{2}\mu(T_1).
\end{aligned}$$

Similarly, using the fact that $\{\Phi_k\}$ is a $\tau^\perp$-mixing sequence, we obtain

$$\xi(S \cap \Psi_k T_2) \to \frac{h}{2^i}\frac{1}{2}\tau^\perp(T_2).$$

Hence

$$\begin{aligned}\xi(S \cap \Psi_k T) &= \xi(S \cap \Psi_k T_1) + \xi(S \cap \Psi_k T_2)\\ &\to \frac{h}{2^i}\left(\tfrac{1}{2}\mu(T_1) + \tfrac{1}{2}\tau^\perp(T_2)\right) = \frac{h}{2^i}\left(\tfrac{1}{2}\mu(T) + \tfrac{1}{2}\tau^\perp(T)\right)\\ &= \frac{h}{2^i}\,\xi(T) = \xi(J(h/2^i;\,\xi,\,\mathfrak{R}))\xi(T) = \xi(S)\xi(T),\end{aligned}$$

as was to be proved.

Now we have proved (C.21) for certain initial sets S of measure $h/2^i$, and hence an easy approximation argument yields it for certain initial sets S of arbitrary measure. Hence it follows for all initial sets S; in particular, therefore, for all initial *segments* S; and so, by standard approximation arguments, for all S. This completes the proof of (C.21), and hence also of (C.20). From the latter it follows at once that

(C.22) $\{\Psi_k^{-1}\}$ is a ξ-mixing sequence.

For a fixed k, let

$$V_k = (\Theta_k')^{-1}U \cup \bigcup_{n=-\infty}^{\infty} \Theta_k^n(W).$$

Then

(C.23) $$\mu(V_k) = 0.$$

Furthermore, when $s, t \in I\backslash V_k$, it follows that $\Theta_k'(s)$ and $\Theta_k'(t)$ are not in $\Theta_k'(V_k)$, hence not in $W \cup U$, hence in $(I\backslash W)\backslash U$. Hence

(C.24) $$\Theta_k's\ \mathfrak{R}\ \Theta_k't \Leftrightarrow \Xi_k\Theta_k's\ \mathfrak{R}\ \Xi_k\Theta_k't \Leftrightarrow \Theta_k's\ (\Xi_k^{-1}\mathfrak{R})\ \Theta_k't,$$

where the second equivalence follows from the definition of $\Xi_k^{-1}\mathfrak{R}$. Also, when $s, t \in I\backslash V_k$, we have

$$\Theta_k's = \Theta_k^{-1}s \text{ and } \Theta_k't = \Theta_k^{-1}t.$$

Hence for $s, t \in I\backslash V_k$ we deduce from (C.24) that

$$\begin{aligned} s\ (\Psi_k^{-1}\mathfrak{R})\ t &\Leftrightarrow s\ (\Theta_k'^{-1}\Xi_k^{-1}\mathfrak{R})\ t\\ &\Leftrightarrow \Theta_k's\ (\Xi_k^{-1}\mathfrak{R})\ \Theta_k't\\ &\Leftrightarrow \Theta_k's\ \mathfrak{R}\ \Theta_k't\\ &\Leftrightarrow \Theta_k^{-1}s\ \mathfrak{R}\ \Theta_k^{-1}t\\ &\Leftrightarrow s\ (\Theta_k\mathfrak{R})\ t.\end{aligned}$$

Applying Lemma C.14 with $T = V_k$ and recalling (C.23), we deduce that

$$\varphi(u; \Psi_k^{-1}\mathcal{R}) = \varphi(u, \Theta_k\mathcal{R}),$$

and the proof of Lemma C.18 is complete.

Proof of Proposition C.1. The sufficiency is immediate; indeed, we can choose μ_v to be any ν such that $v \ll \nu$ (such a ν exists because $v \in AC$). To prove the necessity, let $\mu \in NA^+$ be such that $v \ll \mu$, and let $\{\Theta_1, \Theta_2, \ldots\}$ be a μ-mixing sequence. Now apply Lemma C.18 with $u = v$ and $\tau = \mu_v$. Then $\mu_v \ll \xi$, and so by (14.2),

$$\varphi(v; \Theta_n\mathcal{R})(S) = \varphi(v; \Psi_n^{-1}\mathcal{R})(S) \to (\varphi v)(S)$$

as $n \to \infty$, for all $\mathcal{R}$ and S. This completes the proof.

References

Arrow, K. J., and F. H. Hahn (1971), *General Competitive Analysis*, San Francisco: Holden-Day.

Artstein, Z. (1971), "Values of games with denumerably many players," *Int. J. Game Theory*, *1*, pp. 27–37.

Aumann, R. J. (1961), "The core of a cooperative game without side payments," *Trans. Amer. Math. Soc.*, *98*, pp. 539–552.

——— (1964), "Markets with a continuum of traders," *Econometrica*, *32*, pp. 39–50.

——— (1965), "Integrals of set-valued functions," *J. Math. Anal. Appl.*, *12*, pp. 1–12.

——— (1966), "Existence of competitive equilibria in markets with a continuum of traders," *Econometrica*, *34*, pp. 1–17.

——— (1967a), "A survey of cooperative games without side payments," *Essays in Mathematical Economics in Honor of Oskar Morgenstern*, edited by M. Shubik, Princeton: Princeton University Press, pp. 3–27.

——— (1967b), "Random measure preserving transformations," *Proceedings of the Fifth Berkeley Symposium on Mathematical Statistics and Probability*, Vol. II, Part II, Berkeley and Los Angeles: University of California Press, pp. 321–326.

——— (1969), "Measurable utility and the measurable choice theorem," in *La Décision*, Paris: Editions du Centre National de la Recherche Scientifique, pp. 15–26.

Aumann, R. J., and M. Perles (1965), "A variational problem arising in economics," *J. Math. Anal. Appl.*, *11*, pp. 488–503.

Aumann, R. J., and L. S. Shapley (1968), *Values of Non-Atomic Games, Part I: The Axiomatic Approach*, RM-5468-PR, The Rand Corporation, Santa Monica, California; also RM-42, Research Program in Game Theory and Mathematical Economics, The Hebrew University of Jerusalem.

Auspitz, R., and R. Lieben (1889), *Untersuchungen über die Theorie des Preises*, Leipzig: Duncker & Humbolt.

Bakhtin, I. A., M. A. Krasnoselskii, and V. Ya. Stetzenko (1962), "On the continuity of linear positive operators," *Sibirskii Mat. Z.*, *3*, pp. 156–160.

Billera, L. J. (1970), "Some theorems on the core of an n-person game without side payments," *SIAM J. Appl. Math.*, *18*, pp. 567–579.

Böhm-Bawerk, E. von (1891), *Positive Theorie des Kapitals*, Jena: Stechert-Hafner (1923 reprint, New York: G. E. Stechert).

Bondareva, O. N. (1963), "Some applications of linear programming methods to the theory of cooperative games" (in Russian), *Prob. Kibernet.*, *10*, pp. 119–139.

Burger, E. (1964), "Bemerkungen zum Aumannschen Core-Theorem," *Z. Warsch.*, *3*, pp. 148–153.

Cornwall, R. (1969), "The use of prices to characterize the core of an economy," *J. Econ. Theory*, *1*, pp. 353–373.

Courant, R., and D. Hilbert (1953), *Methods of Mathematical Physics*, I, New York: Interscience.

Debreu, G. (1959), *Theory of Value*, New York: John Wiley & Sons.

——— (1962), "New concepts and techniques for equilibrium analysis," *Int. Econ. Rev.*, *3*, pp. 257–273.

——— (1964), "Continuity Properties of Paretian Utility," *Int. Econ. Rev.*, *5*, pp. 285–293.

Debreu, G., and H. Scarf (1963), "A limit theorem on the core of an economy," *Int. Econ. Rev.*, *4*, pp. 235–246.

Dunford, N., and J. T. Schwartz (1958), *Linear Operators, Part I*, New York: Interscience.

Dvoretzky, A., A. Wald, and J. Wolfowitz (1951), "Relations among certain ranges of vector measures," *Pac. J. Math.*, *1*, pp. 59–74.

Edgeworth, F. Y. (1881), *Mathematical Psychics*, London: C. Kegan Paul.

Gillies, D. B. (1959), "Solutions to general non-zero sum games," in *Contributions to the Theory of Games*, IV, Princeton: Princeton University Press, pp. 47–85.

Hahn, F. H., see Arrow, K. J.

Halmos, P. R. (1948), "The range of a vector measure," *Bull. Amer. Math. Soc.*, *54*, pp. 416–421.

——— (1950), *Measure Theory*, Princeton: Van Nostrand.

——— (1956), *Ergodic Theory*, New York: Chelsea.

Harsanyi, J. C. (1959), "A bargaining model for the cooperative n-person game," *Contributions to the Theory of Games*, IV, Princeton: Princeton University Press, pp. 325–356.

Hart, S. (1973), "Values of mixed games," *Int. J. Game Theory*, *2*, pp. 69–85.

Hilbert, D., see Courant, R.

Hildenbrand, W. (1968), "On the core of an economy with a measure space of economic agents," *Rev. Econ. Studies*, *35*, pp. 443–452.

——— (1970a), "Existence of equilibria for economies with production and a measure space of consumers," *Econometrica*, *38*, pp. 608–623.

——— (1970b), "On economies with many agents," *J. Econ. Theory*, *2*, pp. 161–188.

——— (1974), *Core and Equilibria in a Large Economy*, Princeton: Princeton University Press.

Kannai, Y. (1966), "Values of games with a continuum of players," *Israel J. Math.*, *4*, pp. 54–58.

——— (1969), "Countably additive measures in cores of games," *J. Math. Anal. Appl.*, *27*, pp. 227–240.

——— (1970), "Continuity properties of the core of a market," *Econometrica*, *38*, pp. 791–815.

Kelley, J. L. (1955), *General Topology*, Princeton: Van Nostrand.

Kohlberg, E. (1973), "On non-atomic games: Conditions for $f \circ \mu \in pNA$," *Int. J. Game Theory*, *2*, pp. 87–98.

Krasnoselskii, M. A. (1964), *Positive Solutions of Operator Equations*, Groningen: Noordhoff.

Krasnoselskii, M. A., see also Bakhtin, I. A.

Kuhn, H., and A. W. Tucker (1951), "Non-linear programming," *Proceedings of the Second Berkeley Symposium on Mathematical Statistics and Probability*, Berkeley: University of California Press, pp. 481–492.

Lieben, R., see Auspitz, R.

Lindenstrauss, J. (1966), "A short proof of Liapounoff's convexity theorem," *J. Math. Mech.*, *15*, pp. 971–972.

Lyapunov, A. (1940), "Sur les fonctions-vecteurs completement additives," *Bull. Acad. Sci. URSS Ser. Math.*, *4*, pp. 465–478.

Mackey, G. W. (1957), "Borel structures in groups and their duals," *Trans. Amer. Math. Soc.*, *85*, pp. 134–165.

Milnor, J. W., and L. S. Shapley (1961), *Values of Large Games II: Oceanic Games*, RM-2649, The Rand Corporation, Santa Monica, California.

Mitiagin, B., S. Rolewicz, and W. Zelazko (1962), "Entire functions in B_0-algebras," *Studia Mathematica*, *21*, pp. 291–306.

Morgenstern, O., see von Neumann, J.

Nikaido, H. (1956), "On the classical multilateral exchange problem," *Metroeconomica*, *8*, pp. 135–145.

Owen, G. (1968), "A note on the Shapley value," *Manag. Sci.*, *14*, pp. 731–732.

——— (1971), "Political games," *Nav. Res. Log. Q.*, *18*, pp. 345–355.

——— (1972), "Multilinear extensions of games," *Manag. Sci.*, *18*, pp. 64–79.

Parthasarathy, K. R. (1967), *Probability Measures on Metric Spaces*, New York: Academic Press.

Perles, M., see Aumann, R. J.

Robinson, A. (1966), *Non-Standard Analysis*, Amsterdam: North Holland.

Rolewicz, S., see Mitiagin, B.

Rosén, B. (1965), "Limit theorems for sampling," *Arkiv för Matematik*, *5*, pp. 383–424.

Rosenmüller, J. (1971), "On core and value," *Operations Research-Verfahren* (Methods of Operations Research), edited by Rudolf Henn, Meisenheim: Anton Hain, pp. 84–104.

Royden, H. L. (1968), *Real Analysis*, 2nd ed., New York: Macmillan (1st ed. 1963).

Saks, S. (1937), *Theory of the Integral*, New York: Hafner.

Scarf, H. (1967), "The core of an *n*-person game," *Econometrica*, *35*, pp. 50–69.

Scarf, H., see also Debreu, G.

Schmeidler, D. (1969), "Competitive equilibria in markets with a continuum of traders and incomplete preferences," *Econometrica*, *37*, pp. 578–585.

——— (1972), "Cores of exact games I", *J. Math. Anal. Appl.*, *40*, pp. 214–225.

Schwartz, J. T., see Dunford, N.

Shapley, L. S. (1953a), "A value for *n*-person games," in *Contributions to the Theory of Games*, II, Princeton: Princeton University Press, pp. 307–317.

——— (1953b), "Additive and Non-additive Set Functions," Doctoral Thesis, Princeton University.

——— (1961), *Values of Large Games III: A Corporation with Two Large Stockholders*, RM-2650-PR, The Rand Corporation, Santa Monica, California.

——— (1962a), "Simple games: An outline of the descriptive theory," *Behav. Sci.*, *7*, pp. 59–66.

——— (1962b), "Values of games with infinitely many players," in *Recent Advances in Game Theory*, papers delivered at a meeting of the Princeton University Conference in October 1961, privately printed for members of the conference, pp. 113–118.

——— (1964a), *Values of Large Market Games: Status of the Problem*, RM-3957-PR, The Rand Corporation, Santa Monica, California.

——— (1964b), *Values of Large Games VII: A General Exchange Economy with Money*, RM-4248-PR, The Rand Corporation, Santa Monica, California.

——— (1967), "On balanced sets and cores," *Nav. Res. Log. Q.*, *14*, pp. 453–460.

——— (1971), "Cores of convex games," *Int. J. Game Theory*, *1*, pp. 12–26.

——— (1973), "On balanced games without side payments," in *Mathematical Programming*, edited by T. C. Hu and S. M. Robinson, New York: Academic Press, pp. 261–290.

Shapley, L. S., and M. Shubik (1966), "Quasi-cores in a monetary economy with non-convex preferences," *Econometrica*, *34*, pp. 805–827.

——— (1969a), "On market games," *J. Econ. Theory*, *1*, pp. 9–25.

——— (1969b), "Pure competition, coalitional power, and fair division," *Int. Econ. Rev.*, *10*, pp. 337–362.

——— (1972), "The assignment game I: The core," *Int. J. Game Theory*, *1*, pp. 111–130.

Shapley, L. S., see also Aumann, R. J., and Milnor, J. W.

Shubik, M. (1959), "Edgeworth market games," in *Contributions to the*

Theory of Games, IV, Princeton: Princeton University Press, pp. 267–278.

——— (1968), "Extended Edgeworth bargaining games and competitive equilibrium," *Metroeconomica*, *20*, pp. 299–312.

Shubik, M., see also Shapley, L. S.

Sierpinski, W. (1956), *Hypothèse du Continu*, New York: Chelsea.

Stetzenko, V. Ya., see Bakhtin, I. A.

Titchmarsh, E. C. (1939) *The Theory of Functions*, 2nd ed., Oxford: Oxford University Press (1st ed., 1932).

Tucker, A. W., see Kuhn, H.

Vind, K. (1964), "Edgeworth allocations in an exchange economy with many traders," *Int. Econ. Rev.*, *5*, pp. 165–177.

Von Neumann, J. (1928), "Zur Theorie der Gesellschaftsspiele," *Math. Ann.*, *100*, pp. 295–320. English translation by Sonya Bargman entitled "On the theory of games of strategy," in *Contributions to the Theory of Games*, IV (1959), Princeton, Princeton University Press, pp. 13–42.

——— (1949), "On rings of operators, reduction theory," *Ann. of Math.*, *50*, pp. 401–485.

Von Neumann, J., and O. Morgenstern (1953), *Theory of Games and Economic Behavior*, Princeton: Princeton University Press. (1st ed., 1944.)

Wald, A., see Dvoretzky, A.

Whitney, H. (1934), "Functions differentiable on boundaries of regions," *Ann. Math.*, *35*, pp. 482–485.

Wilks, S. S. (1962), *Mathematical Statistics*, New York: John Wiley and Sons.

Wolfowitz, J., see Dvoretzky, A.

Zadeh, L. A. (1965), "Fuzzy Sets," *Information and Control*, *8*, pp. 338–353.

Zelazko, W., see Mitiagin, B.

Index of Special Spaces and Sets

Note: For other symbols see the main index.

Index

Note: An italicized page number indicates where a term or symbol is defined or a theorem stated.

Selected Rand Books

Arrow, Kenneth J., and Marvin Hoffenberg. *A Time Series Analysis of Interindustry Demands.* Amsterdam, Holland: North-Holland Publishing Company, 1959.

Baker, C. L., and F. J. Gruenberger. *The First Six Million Prime Numbers.* Madison, Wisconsin: The Microcard Foundation, 1959.

Bellman, Richard E., and Stuart E. Dreyfus. *Applied Dynamic Programming.* Princeton, New Jersey: Princeton University Press, 1962.

Bellman, Richard E., and Kenneth L. Cooke. *Differential-Difference Equations.* New York: Academic Press, Inc., 1963.

Bellman, Richard E. (ed.). *Mathematical Optimization Techniques.* Los Angeles, Calif.: University of California Press, 1963.

Bellman, Richard E. *Introduction to Matrix Analysis.* New York: McGraw-Hill Book Co., Inc., 1960.

Bellman, Richard E. *Adaptive Control Processes: A Guided Tour.* Princeton, New Jersey: Princeton University Press, 1961.

Bellman, Richard E. *Dynamic Programming.* Princeton, New Jersey: Princeton University Press, 1957.

Dantzig, George B. *Linear Programming and Extensions.* Princeton, New Jersey: Princeton University Press, 1963.

Dorfman, Robert, Paul A. Samuelson, and Robert M. Solow. *Linear Programming and Economic Analysis.* New York: McGraw-Hill Book Co., 1956.

Dresher, Melvin. *Games of Strategy; Theory and Applications.* Englewood Cliffs, New Jersey: Prentice-Hall Inc., 1961.

Dreyfus, Stuart. *Dynamic Programming and the Calculus of Variations.* New York: Academic Press Inc., 1965.

Fishman, George S. *Spectral Methods in Economics.* Cambridge, Mass.: Harvard University Press, 1969.

Ford, L. R., Jr. and D. R. Fulkerson. *Flows in Networks.* Princeton, New Jersey: Princeton University Press, 1962.

Gale, David. *The Theory of Linear Economic Models.* New York: McGraw-Hill Book Co., Inc. 1960.

Harris, Theodore E. *The Theory of Branching Processes.* Berlin, Germany: Springer-Verlag, 1963: Englewood Cliffs, N.J.: Prentice-Hall, Inc., 1964.

Hastings, Cecil, Jr. *Approximations for Digital Computers.* Princeton, New Jersey: Princeton University Press, 1955.

Jorgenson, D. W., J. J. McCall and R. Radner. *Optimal Replacement Policy.* Amsterdam, Holland: North-Holland Publishing Company; Chicago, Ill.: Rand McNally & Co., 1967.

McKinsey, J. C. C. *Introduction to the Theory of Games.* New York: McGraw-Hill Book Company, Inc., 1952.

The Rand Corporation. *A Million Random Digits With* 100,000 *Normal Deviates.* Glencoe, Ill.: The Free Press, 1955.

Tutte, William T. *Introduction to the Theory of Matroids.* New York: American Elsevier Publishing Co., 1971.

Williams, John D. *The Compleat Strategyst: Being a Primer on the Theory of Games of Strategy.* New York: McGraw-Hill Book Co., Inc. 1954.